Nanoinformatics

Isao Tanaka
Editor

Nanoinformatics

 Springer

Editor
Isao Tanaka
Kyoto University
Kyoto
Japan

ISBN 978-981-13-5661-2   ISBN 978-981-10-7617-6   (eBook)
https://doi.org/10.1007/978-981-10-7617-6

Printed on acid-free paper

This Springer imprint is published by Springer Nature
The registered company is Springer Nature Singapore Pte Ltd.
The registered company address is: 152 Beach Road, #21-01/04 Gateway East, Singapore 189721, Singapore

Preface

This book focuses on state-of-the-art ideas and tools in informatics that are currently being used in materials science, or are expected to be used in the future. Collaborative research between materials science and information science is growing actively, creating new trends in materials science and engineering. Areas utilizing "big data," generated by experiments and computations to accelerate the discovery of new materials, key factors, and design rules, have rapidly progressed. Data-intensive approaches are indispensable in advanced materials characterization.

"Material informatics" is the central paradigm in this new trend. An essential subset is "nanoinformatics," which focuses on the nanostructures of materials, such as surfaces, interfaces, dopants, and point defects. Experimental and computational techniques to characterize and gain quantitative information about nanostructures have significantly advanced, enabling nanoinformatics to play a critical role in determining material properties.

Most of this book is derived from the collaborative research projects supported by the Grant-in-Aid for Scientific Research on Innovative Areas "Nano Informatics" from the Japan Society for the Promotion of Science (JSPS). This five-year project, which was launched in 2013, aims to accelerate the exploration of frontiers in materials science and promote the integration of information and utilization of accumulated knowledge regarding nanostructures for the design and innovation of actual materials. Project researchers represent diverse disciplines, such as materials science, applied physics, solid-state chemistry, catalytic chemistry, and information science. In addition to those working in the collaborative program, three research groups actively working on data-centric materials science were invited to contribute to the book. With their participation, the subjects in the book are well balanced.

This book is composed of three parts. The first part reviews the ideas and tools of materials informatics as well as actual applications of machine-learning techniques for materials problems. Chapter 1 shows how compounds in materials datasets can be represented as descriptors and applied to machine-learning models. Chapter 2 focuses on a method to discover the potential energy surface of solid-state ionic conductors via a combination of first principles calculations and machine-learning techniques. Chapter 3 describes the machine-learning predictions of factors affecting the activity of heterogeneous metal catalysts. Chapter 4 discusses the applications of optimal experimental design algorithms for materials science. Chapters 5 and 6 are dedicated to the topological analyses of the atomic structure data of materials. One method is called persistent homology. The other uses polyhedron and polychoron codes. They have been successfully used to analyze amorphous structures.

In the second part, data-centric approaches used for nanoscale analyses of materials data are described. Chapter 7 shows topological data analyses for atom probe tomography (APT) images. Chapter 8 describes the combined efforts of scanning transmission electron microscopy (STEM) experiments, first principles calculations, and informatics approaches to analyzing the atomic structures of materials interfaces. Chapter 9 is based on nanoscale STEM spectroscopic datasets that are analyzed by machine-learning techniques.

The third part is composed of four chapters. Each chapter focuses on a specific target of nanoinformatics approaches. Chapter 10 describes high-quality epitaxial films of materials called "nanolayers" for a variety of functional applications, including thermoelectrics, batteries, memories, and superconductors. Chapter 11 focuses on the grain boundary engineering of alumina ceramics for use as protective films in the hot-section components of airplane engines, gas turbines, and heat treatment furnaces in combustion environments. Chapter 12 shows the structural relaxation of high-pressure oxide compounds, which is important for quenching high-pressure phases in ambient conditions. Chapter 13 describes the syntheses and structures of novel lithium-ion and hydride-ion conductors for use as solid-state electrolytes in electrochemical devices.

This book is an efficient overview of current progress in emerging and inter-disciplinary research areas. It will benefit experimentalists and theorists in both academic and industry sectors. All the authors and steering committee members of the collaborative program "Nano Informatics" are gratefully acknowledged. Without their devoted efforts, this book would not be possible.

Financial support for the open access publication of this book by a Grant-in-Aid for Scientific Research on Innovative Areas "Nano Informatics" (Grant No. 25106001) from the JSPS is gratefully acknowledged.

Kyoto, Japan   Isao Tanaka

Contents

Part I
Materials Informatics

Chapter 1
Descriptors for Machine Learning of Materials Data

Atsuto Seko, Atsushi Togo and Isao Tanaka

Abstract Descriptors, which are representations of compounds, play an essential role in machine learning of materials data. Although many representations of elements and structures of compounds are known, these representations are difficult to use as descriptors in their unchanged forms. This chapter shows how compounds in a dataset can be represented as descriptors and applied to machine-learning models for materials datasets.

Keywords Machine-learning interatomic potential · Lattice thermal conductivity · Recommender system · Gaussian process · Bayesian optimization

1.1 Introduction

Recent developments of data-centric approaches should accelerate the progress in materials science dramatically. Thanks to the recent advances in computational power and techniques, the results from numerous density functional theory (DFT) calculations with predictive performances have been stored as databases. A combination of such databases and an efficient machine-learning approach should realize prediction and classification models of target physical properties. Consequently, machine-learning techniques are becoming ubiquitous. They are used to explore materials and structures from a huge number of candidates and to extract meaningful information and patterns from existing data.

A key factor in controlling the performance of a machine-learning approach is how compounds are represented in a data set. Representations of compounds are called "descriptors" or "features". To perform machine-learning modeling, available descriptors must be determined according to the evaluation cost of the target property

A. Seko (✉) · I. Tanaka
Department of Materials Science and Engineering, Kyoto University, Kyoto, Japan
e-mail: seko@cms.mtl.kyoto-u.ac.jp

A. Togo
Centre for Elements Strategy Initiative for Structure Materials (ESISM),
Kyoto University, Kyoto, Japan

I. Tanaka (ed.), *Nanoinformatics*, https://doi.org/10.1007/978-981-10-7617-6_1

3

and the extent of the exploration space. Based on these considerations, we aim to select "good" descriptors. Prior or experts' knowledge, including a well-known correlation between the target property and the other properties, can be used to select good descriptors. However, the set of descriptors in many cases is examined by trial-and-error because the predictive performance (i.e., the prediction error and efficiency of the model) strongly depends on the quality and data-size of the target property.

Section 1.2 shows how to prepare descriptors of compounds. Sections 1.3 and 1.4 introduce representations of chemical elements (elemental representations) and atomic arrangements (structural representations) required to generate compound descriptors. Sections 1.5, 1.6, 1.7, and 1.8 provide applications of machine-learning models for materials datasets, including the construction of a machine-learning prediction model for the DFT cohesive energy, the construction of the machine-learning interatomic potential (MLIP) for elemental metals, materials discovery of low lattice thermal conductivity (LTC), and materials discovery based on the recommender system approach.

1.2 Compound Descriptors

Most candidate descriptors can be classified into three groups. The first is the physical properties of a compound in a library and/or their derivative quantities, which are less available. The second is the physical properties of a compound computed by DFT calculations or their derivative quantities. The third is the properties of elements and the structure of a compound and/or their derivative quantities. Combinations of different groups of descriptors can also be useful.

A set of compound descriptors should satisfy the following conditions: (i) the same-dimensional descriptors express compounds with a wide range of chemical compositions. (ii) The same-dimensional descriptors express compounds with a wide range of crystal structures. This is an important feature because crystals are generally composed of unit cells with different numbers of atoms. (iii) A set of descriptors satisfies the translational, rotational, and other invariances for all compounds included in the dataset.

Candidates for compound descriptors based on DFT calculations include volume, band gap, cohesive energy, elastic constants, dielectric constants, etc. The electronic structure and phonon properties can also be used as descriptors. Although a few first-principles databases are available, the numbers of compounds and physical properties in the databases remain limited. Nevertheless, when a set of descriptors that can well explain a target property is discovered, a robust prediction model can be derived for the target property. Examples can be found in the literature (e.g., Refs. [1–4]). Other candidates are simply a binary digit representing the presence of each element in a compound (Fig. 1.1) [5]. When training data is composed of m kinds of elements, a compound is described by an m-dimensional binary vector with elements of one or zero. As a simple extension, a binary digit can be replaced with the chemical composition. Such an application is shown in Sect. 1.7.

	H	Li	Be	B	C	N	O	F	...
LiH	1	1	0	0	0	0	0	0	...
LiF	0	1	0	0	0	0	0	1	...
BeO	0	0	1	0	0	0	1	0	...
BN	0	0	0	1	0	1	0	0	...
⋮	⋮	⋮	⋮	⋮	⋮	⋮	⋮	⋮	...

Fig. 1.1 Binary elemental descriptors representing the presence of chemical elements. The number of binary elemental descriptors corresponds to the number of element types included in the training data

Another useful strategy is to use a set of quantities derived from elemental and structural representations of a compound as descriptors. However, it is difficult to use elemental and structural representations as descriptors in their unchanged forms when the training data and search space cover a wide range of chemical compositions and crystal structures. Consequently, it is essential to consider combined forms as compound descriptors.

Here we provide compound descriptors derived from elemental and structural representations satisfying the above conditions. These descriptors can be applied not only to crystalline systems but also to molecular systems [6]. Figure 1.2 schematically illustrates the procedure to generate such descriptors for compounds. First, the compound is considered to be a collection of atoms, which are described by element types and neighbor environments that are determined by other atoms. Assuming the atoms are represented by $N_{x,\mathrm{ele}}$ elemental representations and $N_{x,\mathrm{st}}$ structural representations, each atom is described by $N_x = N_{x,\mathrm{ele}} + N_{x,\mathrm{st}}$ representations. Therefore, compound ξ is expressed by a collection of atomic representations as a matrix with $(N_a^{(\xi)}, N_x)$-dimensions, where $N_a^{(\xi)}$ is the number of atoms in the unit cell of compound ξ. The representation matrix for compound ξ, $X^{(\xi)}$, is written as

$$X^{(\xi)} = \begin{pmatrix} x_1^{(\xi,1)} & x_2^{(\xi,1)} & \cdots & x_{N_x}^{(\xi,1)} \\ x_1^{(\xi,2)} & x_2^{(\xi,2)} & \cdots & x_{N_x}^{(\xi,2)} \\ \vdots & \vdots & \ddots & \vdots \\ x_1^{(\xi,N_a^{(\xi)})} & x_2^{(\xi,N_a^{(\xi)})} & \cdots & x_{N_x}^{(\xi,N_a^{(\xi)})} \end{pmatrix}, \tag{1.1}$$

where $x_n^{(\xi,i)}$ denotes the nth representation of atom i in compound ξ.

Since the representation matrix is only a representation of the unit cell of compound ξ, a procedure to transform the representation matrix into a set of descriptors is needed to compare different compounds. One approach for this transformation is to regard the representation matrix as a distribution of data points in an N_x-dimensional space (Fig. 1.2). To compare the distributions themselves, representative quantities are subsequently introduced to characterize the distribution as descriptors, such as the mean, standard deviation (SD), skewness, kurtosis, and

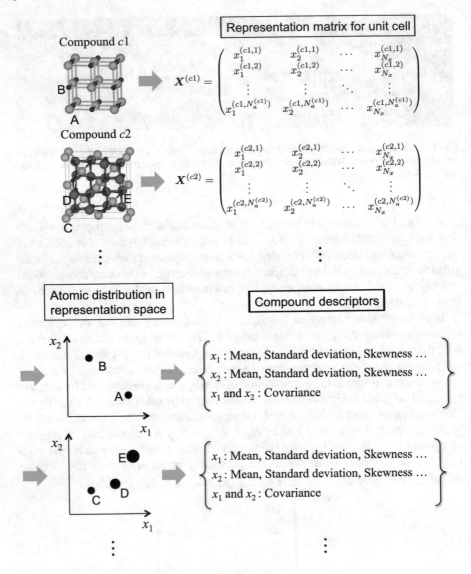

Fig. 1.2 Schematic illustration of how to generate compound descriptors

covariance. The inclusion of the covariance enables the interaction between the element type and crystal structure to be considered.

A universal or complete set of representations is ideal because it can derive good machine-learning prediction models for all physical properties. However, finding a universal set of representations is nearly impossible. On the other hand, many elemental and structural representations have been proposed for a long time, not only in the literature on the machine-learning prediction but also in the literature on the standard physics and chemistry. Using these representations, many phenomena in physics and chemistry have been explained. Therefore, it is a good way for generating descriptors to make effective use of the existing representations.

1.3 Elemental Representations

The literature contains numerous quantities that can be used as elemental representations. This chapter employs a set of elemental representations composed of the following: (1) atomic number, (2) atomic mass, (3) period and (4) group in the periodic table, (5) first ionization energy, (6) second ionization energy, (7) electron affinity, (8) Pauling electronegativity, (9) Allen electronegativity, (10) van der Waals radius, (11) covalent radius, (12) atomic radius, (13) pseudopotential radius for the s orbital, (14) pseudopotential radius for the p orbital, (15) melting point, (16) boiling point, (17) density, (18) molar volume, (19) heat of fusion, (20) heat of vaporization, (21) thermal conductivity, and (22) specific heat. These representations can be classified into the intrinsic quantities of elements (1)–(7), the heuristic quantities of elements (8)–(14), and the physical properties of elemental substances (15)–(22). Such elemental representations should capture essential information about compounds. Therefore, they should assist in building models with a high predictive performance, as shown in Sects. 1.5, 1.7 and 1.8.

1.4 Structural Representations

The literature contains many structural representations that are not intended for machine-learning applications. Examples include the simple coordination number, Voronoi polyhedron of a central atom, angular distribution function, and radial distribution function (RDF). Here, we introduce two kinds of pairwise structural representations and two kinds of angular-dependent structural representations i.e., histogram representations of the partial radial distribution function (PRDF), generalized radial distribution function (GRDF), bond-orientational order parameter (BOP) [7], and angular Fourier series (AFS) [8].

The PRDF is a well-established representation for various structures. To transform the PRDF into structural representations applicable to machine learning, a histogram representation of the PRDF is adopted with a given bin width and cutoff radius (Fig. 1.3). The number of counts for each bin is used as the structural representation.

The GPRF, which is a pairwise representation similar to the PRDF histogram representation, is expressed as

$$\mathrm{GRDF}_n^{(i)} = \sum_j f_n(r_{ij}) \tag{1.2}$$

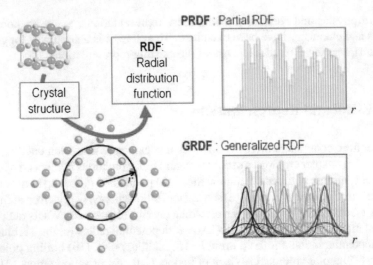

Fig. 1.3 Partial radial distribution functions (PRDFs) and generalized radial distribution functions (GRDFs)

where $f_n(r_{ij})$ denotes a pairwise function of the distance r_{ij} between atoms i and j. For example, a pairwise Gaussian-type function is expressed as

$$f_n(r) = \exp\left[-p_n(r - q_n)^2\right] f_c(r) \tag{1.3}$$

where $f_c(r)$ denotes the cutoff function. p_n and q_n are given parameters. The GRDF can be regarded as a generalization of the PRDF histogram because the PRDF histogram is obtained using rectangular functions as pairwise functions.

The BOP is also a well-known representation for local structures. The rotationally invariant BOP $Q_l^{(i)}$ for atomic neighborhoods is expressed as

$$Q_l^{(i)} = \left[\frac{4\pi}{2l+1} \sum_{m=-l}^{l} |Q_{lm}^{(i)}|^2\right]^{1/2} \tag{1.4}$$

where $Q_{lm}^{(i)}$ corresponds to the average spherical harmonics for neighbors of atom i. The third-order invariant BOP $W_l^{(i)}$ for atomic neighborhoods is expressed by

$$W_l^{(i)} = \sum_{m_1,m_2,m_3=-l}^{l} \begin{pmatrix} l & l & l \\ m_1 & m_2 & m_3 \end{pmatrix} Q_{lm_1}^{(i)} Q_{lm_2}^{(i)} Q_{lm_3}^{(i)}, \tag{1.5}$$

where the parentheses are the Wigner $3j$ symbol, satisfying $m_1 + m_2 + m_3 = 0$. A set of both $Q_l^{(i)}$ and $W_l^{(i)}$ up to a given maximum l is used as the structural representations.

The AFS is the most general among the four representations. The AFS can include both the radial and angular dependences of an atomic distribution, and is given by

$$\text{AFS}_{n,l}^{(i)} = \sum_{j,k} f_n(r_{ij}) f_n(r_{ik}) \cos(l\theta_{ijk}) \tag{1.6}$$

where θ_{ijk} denotes the bond angle between three atoms.

1.5 Machine Learning of DFT Cohesive Energy

The performances of the descriptors derived from elemental and structural representations have been examined by developing kernel ridge regression (KRR) prediction models for the DFT cohesive energy [6]. The dataset is composed of the cohesive energy for 18093 binary and ternary compounds computed by DFT calculations. First, descriptor sets derived only from elemental representations, which are expected to be more dominant than structural representations in the prediction of the cohesive energy, are adopted. Since the elemental representations are incomplete for some of the elements in the dataset, only elemental representations, which are complete for all elements, are considered. The root-mean-square error (RMSE) is estimated for the test data. The test data is comprised of 10% of the randomly selected data. This random selection of the test data is repeated 20 times, and the average RMSE is regarded as the prediction error.

The simplest option is to use only the mean of each elemental representation as a descriptor. The prediction error, in this case, is 0.249 eV/atom. Figure 1.4a compares the cohesive energy calculated by DFT calculations to that by the KRR model, where only the test data in one of the 20 trials are shown. Numerous data points deviate from the diagonal line, which represents equal DFT and KRR energies. When considering the means, SDs, and covariances of the elemental representations, the prediction model has a slightly smaller prediction error of 0.231 eV/atom. Additionally, skewness and kurtosis are not important descriptors for the prediction.

Next, descriptors related to structural representations are introduced. They can be computed from the crystal structure optimized by the DFT calculations or the initial prototype structures. The former is only useful for machine-learning predictions when a target observation is expensive. Since the optimized structure calculation requires the same computational cost as the cohesive energy calculation, the benefit of machine learning is lost when using the optimized structure. The structural representations are computed from the optimized crystal structure only to examine the limitation of the procedure and representations introduced here. KRR models are constructed using many descriptor sets, which are composed of elemental and structural representations. The cutoff radius is set to 6 Å for the PRDF, GRDF, and AFS, while the cutoff radius is set to 1.2 times the nearest neighbor distance for the BOP. This nearest neighbor definition is common for the BOP.

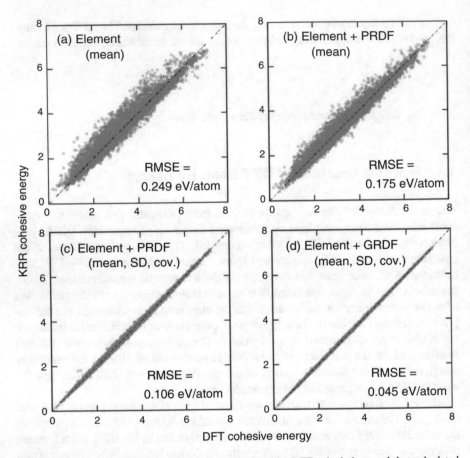

Fig. 1.4 Comparison of the cohesive energy calculated by DFT calculations and that calculated by the KRR prediction model. Only one test dataset is shown. Descriptor sets are composed of **a** the mean of the elemental representation, **b** the means of the elemental and PRDF representations, **c** the means, SDs, and covariances of the elemental and PRDF representations and **d** the means, SDs, and covariances of the elemental and 20 trigonometric GRDF representations. Mean of the PRDF corresponds to the RDF. Structure representations are computed from the optimized structure for each compound

Figure 1.4 compares the DFT and KRR cohesive energies, where the KRR models are constructed by (b) a set of the means of the elemental and PRDF histogram representations and (c) a set of the means, standard deviations, and covariances of the elemental and PRDF histogram representations. When considering the means of the elemental and PRDF representations, the lowest prediction error is as large as 0.166 eV/atom. This means that simply employing the PRDF histogram does not yield a good model for the cohesive energy. However, including the covariances of the elemental and PRDF histogram representations produces a much better prediction model and the prediction error significantly decreases to 0.106 eV/atom.

Considering only the means of the GRDFs, prediction models are obtained with errors of 0.149–0.172 eV/atom. These errors are similar to those of prediction models considering the means of the PRDFs. Similar to in the case of the PRDF, the prediction model improves upon considering the SDs and covariances of the elemental and structural representations. The best model shows a prediction error of 0.045 eV/atom, which is about half that of the best PRDF model. This is also approximately equal to the "chemical accuracy" of 43 meV/atom (1 kcal/mol).

Figure 1.4d compares the DFT and KRR cohesive energies, where a set of the means, SDs, and covariances of the elemental and trigonometric GRDF representations is adopted. Most of the data are located near the diagonal line. We also obtain the best prediction model with a prediction error of 0.041 eV/atom by considering the means, SDs, and covariances of the elemental, 20 trigonometric GRDF, and 20 BOP representations. Therefore, the present method should be useful to search for compounds with diverse chemical properties and applications from a wide range of chemical and structural spaces without performing exhaustive DFT calculations.

1.6   Construction of MLIP for Elemental Metals

A wide variety of conventional interatomic potentials (IPs) have been developed based on prior knowledge of chemical bonds in some systems of interest. Examples include Lennard-Jones, embedded atom method (EAM), modified EAM (MEAM), and Tersoff potentials. However, the accuracy and transferability of conventional IPs are often lacking due to the simplicity of their potential forms. On the other hand, the MLIP based on a large dataset obtained by DFT calculations is beneficial to improve the accuracy and transferability. In the MLIP framework, the atomic energy is modeled by descriptors corresponding to structural representations, as shown in Sect. 1.4. Once the MLIP is established, it has a similar computational cost as conventional IPs. MLIPs have been applied to a wide range of materials, regardless of chemical bonding nature of the materials. Recently, frameworks applicable to periodic systems have been proposed [9–11].

The Lasso regression has been used to derive a sparse representation for the IP. In this section, we demonstrate the applicability of the Lasso regression to derive the IPs of 12 elemental metals (Na, Mg, Ag, Al, Au, Ca, Cu, Ga, In, K, Li, and Zn) [11, 12]. The features of linear modeling of the atomic energy and descriptors using the Lasso regression include the following. (1) The accuracy and computational cost of the energy calculation can be controlled in a transparent manner. (2) A well-optimized sparse representation for the IP, which can accelerate and increase the accuracy of atomistic simulations while decreasing the computational costs, is obtained. (3) Information on the forces acting on atoms and stress tensors can be included in the training data in a straightforward manner. (4) Regression coefficients are generally determined quickly using the standard least-squares technique.

The total energy of a structure can be regarded as the sum of the constituent atomic energies. In the framework of MLIPs with only pairwise descriptors, the atomic energy of atom i is formulated as

$$E^{(i)} = F\left(b_1^{(i)}, b_2^{(i)}, \ldots, b_{n_{max}}^{(i)}\right), \tag{1.7}$$

where $b_n^{(i)}$ denotes a pairwise descriptor. Numerous pairwise descriptors are generally used to formulate the MLIP. We use the GRDF expressed by Eq. (1.2) as the descriptors. For the pairwise function f_n, we introduce Gaussian, cosine, Bessel, Neumann, modified Morlet wavelet, Slater-type orbital, and Gaussian-type orbital functions. Although artificial neural network and Gaussian process black-box models have been used as functions F, we use a polynomial function to construct the MLIPs for the 12 elemental metals. In the approximation considering only the power of $b_n^{(i)}$, the atomic energy is expressed as

$$E^{(i)} = w_0 + \sum_n w_n b_n^{(i)} + \sum_n w_{n,n} b_n^{(i)} b_n^{(i)} + \cdots, \tag{1.8}$$

where w_0, w_n, and $w_{n,n}$ denote the regression coefficients. Practically, the formulation is truncated by the maximum value of power, p_{max}.

The vector w composed of all the regression coefficients can be estimated by a regression, which is a machine-learning method to estimate the relationship between the predictor and observation variables using a training dataset. For the training data, the energy, forces acting on atoms, and stress tensor computed by DFT calculations can be used as the observations in the regression process since they all are expressed by linear equations with the same regression coefficients [12]. A simple procedure to estimate the regression coefficients employs a linear ridge regression [13]. This is a shrinkage method where the number of regression coefficients is reduced by imposing a penalty. The ridge coefficients minimize the penalized residual sum of squares and are expressed as

$$L(w) = ||Xw - y||_2^2 + \lambda ||w||_2^2, \tag{1.9}$$

where X and y denote the predictor matrix and observation vector, respectively, which correspond to the training data. λ, which is called the regularization parameter, controls the magnitude of the penalty. This is referred to as L2 regularization. The regression coefficients can easily be estimated while avoiding the well-known multicollinearity problem that occurs in the ordinary least-squares method.

Although the linear ridge regression is useful to obtain an IP from a given descriptor set, a set of descriptors relevant to the system of interest is generally unknown. Moreover, an MLIP with a small number of descriptors is desirable to decrease the computational cost in atomistic simulations. Therefore, a combination of the Lasso regression [13, 14] and a preparation involving a considerable number of descriptors is used. The Lasso regression provides a solution to the linear regression as well as a sparse representation with a small number of nonzero regression coefficients. The solution is obtained by minimizing the function that includes the L1 norm of regression coefficients and is expressed as

$$L(w) = ||Xw - y||_2^2 + \lambda ||w||_1. \tag{1.10}$$

Simply adjusting the values of λ for a given training dataset controls the accuracy of the solution.

To begin with, training and test datasets are generated from DFT calculations. The test dataset is used to examine the predictive power for structures that are not included in the training dataset. For each elemental metal, 2700 and 300 configurations are generated for the training and test datasets, respectively. The datasets include structures made by isotropic expansions, random expansions, random distortions, and random displacements of ideal face-centered-cubic (fcc), body-centered-cubic (bcc), hexagonal-closed-packed (hcp), simple-cubic (sc), ω and β-tin structures, in which the atomic positions and lattice constants are fully optimized. These configurations are made using supercells constructed by the $2 \times 2 \times 2$, $3 \times 3 \times 3$, $3 \times 3 \times 3$, $4 \times 4 \times 4$, $3 \times 3 \times 3$ and $2 \times 2 \times 2$ expansions of the conventional unit cells for fcc, bcc, hcp, sc, ω, and β-tin structures, which are composed of 32, 54, 54, 64, 81, and 32 atoms, respectively.

For a total of 3000 configurations for each elemental metal, DFT calculations have been performed using the plane-wave basis projector augmented wave (PAW) method [15] within the Perdew–Burke–Ernzerhof exchange-correlation functional [16] as implemented in the VASP code [17–19]. The cutoff energy is set to 400 eV. The total energies converge to less than 10^{-3} meV/supercell. The atomic positions and lattice constants are optimized for the ideal structures until the residual forces are less than 10^{-3} eV/Å.

For each MLIP, the RMSE is calculated between the energies for the test data predicted by the DFT calculations and those predicted using the MLIP. This can be regarded as the prediction error of the MLIP. Table 1.1 shows the RMSEs of linear ridge MLIPs with 240 terms for Na and Mg, where the RMSE converges as the number of terms increases. The MLIPs with only pairwise interactions have low

Table 1.1 RMSEs for the test data of linear ridge MLIPs using 240 terms (Unit: meV/atom)

Function type for f_n and p_{max}	Na	Mg
Cosine ($p_{max} = 1$)	7.3	11.8
Cosine ($p_{max} = 2$)	1.6	2.6
Cosine ($p_{max} = 3$)	1.4	1.6
Cosine, Gaussian ($p_{max} = 3$)	1.4	1.1
Cosine, Bessel ($p_{max} = 3$)	1.4	1.3
Cosine, Gaussian, Bessel ($p_{max} = 3$)	1.4	0.9

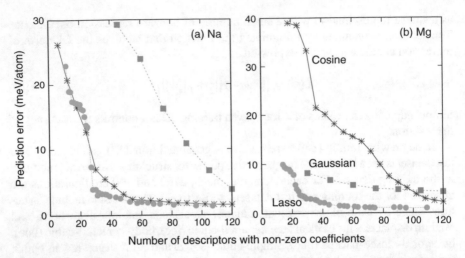

Fig. 1.5 RMSEs for the test data of the linear ridge MLIP using cosine-type and Gaussian-type descriptors with $p_{max} = 3$, $R_c = 7.0$ Å and $\lambda = 0.001$ for **a** Na and **b** Mg. RMSEs of the Lasso MLIPs are also shown

predictive powers for both Na and Mg. Increasing pmax improves the predictive power of the MLIPs substantially. Using cosine-type functions with $p_{max} = 3$ and cutoff radius $R_c = 7.0$ Å, the RMSEs are 1.4 and 1.6 meV/atom for Na and Mg, respectively. By increasing the cutoff radius to $R_c = 9.0$ Å, the RMSE reaches a very small value of 0.4 meV/atom for Na, but the RMSE remains almost unchanged for Mg. The RMSE for Na is not improved, even after considering all combinations of the Gaussian, cosine, Bessel, and Neumann descriptor sets. In contrast, the combination of Gaussian, cosine, and Bessel descriptor sets provides the best prediction for Mg with an RMSE of 0.9 meV/atom.

The Lasso MLIPs have been constructed using the same dataset. Candidate terms for the Lasso MLIPs are composed of numerous Gaussian, cosine, Bessel, Neumann, polynomial, and GTO descriptors. Sparse representations are then extracted from a set of candidate terms by the Lasso regression. Figure 1.5 shows the RMSEs of the Lasso MLIPs for Na and Mg, respectively. The RMSEs of the Lasso MLIP decrease faster than those of the linear ridge MLIPs constructed from a single-type of descriptors. In other words, the Lasso MLIP requires fewer terms than the linear ridge MLIP. For Na, a sparse representation with an RMSE of 1.3 meV/atom is obtained using only 107 terms. This is almost the same accuracy as the linear ridge MLIP with 240 terms based on the cosine descriptors. It is apparent that the Lasso MLIP is more advantageous for Mg than for Na. The obtained sparse representation with 95 terms for Mg has an RMSE of 0.9 meV/atom. This is almost half the terms for the linear ridge MLIP based on the cosine descriptors, which requires 240 terms.

Figure 1.6a shows the dependence of the RMSE for the energy and stress tensor of the Lasso MLIP on the number of nonzero regression coefficients for the other ten elemental metals. The number of selected terms tends to increase as the

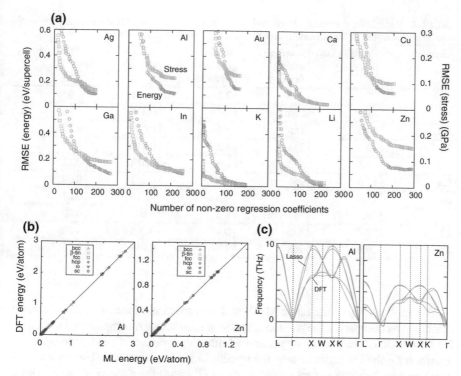

Fig. 1.6 **a** Dependence of RMSEs for the energy and stress tensor of the Lasso MLIP on the number of nonzero regression coefficients for ten elemental metals. Orange open circles and blue open squares show RMSEs for the energy and stress tensor, respectively. **b** Comparison of the energies predicted by the Lasso MLIP and DFT for Al and Zn measured from the energy of the most stable structure. **c** Phonon dispersion relationships for FCC-Al and FCC-Zn. Blue solid and orange broken lines show the phonon dispersion curves obtained by the Lasso MLIP and DFT, respectively. Negative values indicate imaginary modes

regularization parameter λ decreases. The RMSEs for the energy and stress tensor tend to decrease. Although multiple MLIPs with the same number of terms are sometimes obtained from different values of λ, only the MLIP with the lowest criterion score with the same number of terms is shown in Fig. 1.6a. Table 1.2 shows the RMSEs for the energy, force, and stress tensor of the optimal Lasso MLIP. The MLIPs are obtained with the RMSE for the energy in the range of 0.3–3.5 meV/atom for the ten elemental metals using only 165–288 terms. The RMSEs for the force and stress are within 0.03 eV/Å and 0.15 GPa, respectively.

Figure 1.6b compares the energies of the test data predicted by the Lasso MLIP and DFT for Al and Zn. Both the largest and second largest RMSEs for the energy are shown. Regardless of the crystal structure, the DFT and Lasso MLIP energies are similar. In addition, the RMSE is clearly independent of the energy despite the wide range of structures included in both the training and test data.

Table 1.2 RMSEs for the energy, force, and stress tensor of the Lasso MLIPs showing the minimum criterion score. Optimal cutoff radius for each element is also shown

Element	Cutoff radius (Å)	Number of basis functions	RMSE (energy) (meV/atom)	RMSE (force) (eV/Å)	RMSE (stress) (GPa)
Ag	7.5	190	2.2	0.011	0.07
Al	8.0	210	3.5	0.020	0.12
Au	6.0	165	2.4	0.030	0.15
Ca	9.5	234	1.2	0.010	0.03
Cu	7.5	202	2.6	0.018	0.12
Ga	10.0	266	2.2	0.017	0.09
In	10.0	253	2.3	0.019	0.07
K	10.0	197	0.3	0.001	0.00
Li	8.5	222	0.4	0.005	0.02
Zn	10.0	288	2.9	0.016	0.15

The applicability of the Lasso MLIP to the calculation of the force has been also examined by comparing the phonon dispersion relationships computed by the Lasso MLIP and DFT. The phonon dispersion relationships are calculated by the supercell approach for the fcc structure with the equilibrium lattice constant. The phonon calculations use the phonopy code [20]. Figure 1.6c shows the phonon dispersion relationships of the fcc structure for elemental Al and Zn computed by both the Lasso MLIP and DFT. The phonon dispersion relationships calculated by the Lasso MLIP agree well with those calculated by DFT. This demonstrates that the Lasso MLIP is sufficiently accurate to perform atomistic simulations with an accuracy similar to DFT calculations.

It is important to use an extended approximation for the atomic energy in transition metals [21, 22]. The extended approximation also improves the predictive power for the above elemental metals. The MLIPs are constructed by a second-order polynomial approximation with the AFSs described by Eq. (1.6) and their cross terms. For elemental Ti, the optimized angular-dependent MLIP is obtained with a prediction error of 0.5 meV/atom (35245 terms), which is much smaller than that of the Lasso MLIP with only the power of pairwise descriptors of 17.0 meV/atom. This finding demonstrates that it is very important to consider angular-dependent descriptors when expressing interatomic interactions of elemental Ti. The angular-dependent MLIP can predict the physical properties much more accurately than existing IPs.

1.7   Discovery of Low Lattice Thermal Conductivity Materials

Thermoelectric generators are essential to utilize waste heat. The thermoelectric figure of merit should be increased to improve the conversion efficiency. Since the figure of merit is inversely proportional to the thermal conductivity, many works have strived to reduce the thermal conductivity, especially the LTC. To evaluate LTCs with an accuracy comparable to the experimental data, a method that greatly exceeds ordinary DFT calculations is required. Since multiple interactions among phonons, or anharmonic lattice dynamics, must be treated, the computational cost is many orders of magnitudes higher than ordinary DFT calculations of primitive cells. Such expensive calculations are feasible only for a few simple compounds. High-throughput screening of a large DFT database of the LTC is an unrealistic approach unless the exploration space is narrowly confined.

Recently, Togo et al. reported a method to systematically obtain the theoretical LTC through first-principles anharmonic lattice dynamics calculations [23]. Figure 1.7a shows the results of first-principles LTCs for 101 compounds as functions of the crystalline volume per atom, V. PbSe with the rocksalt structure shows the lowest LTC, 0.9 W/mK (at 300 K). Its trend is similar to that in a recent report on low LTC for lead- and tin-chalcogenides.

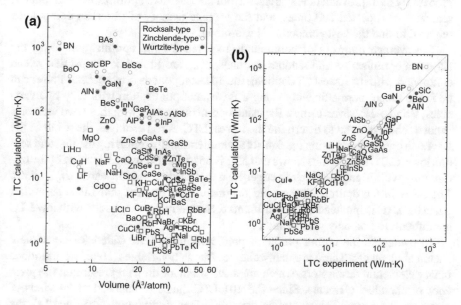

Fig. 1.7   a LTC calculated from the first-principles calculations for 101 compounds along with volume, V. **b** Experimental LTC data are shown for comparison when the experimental LTCs are available

Figure 1.7b compares the computed results with the available experimental data. The satisfactory agreement between the experimental and computed results demonstrates the usefulness of the first-principles LTC data for further studies. A phenomenological relationship has been proposed where $\log \kappa_L$ is proportional to $\log V$ [24]. Although a qualitative correlation is observed between our LTC and V, it is difficult to predict the LTC quantitatively or discover new compounds with low LTCs only from the phenomenological relationship. It should be noted that the dependence on V differs remarkably between rocksalt-type and zincblende- or wurtzite-type compounds. However, zincblende- and wurtzite-type compounds show a similar LTC for the same chemical composition. The 101 first-principles LTC data has been used to create a model to predict the LTCs of compounds within a library [5]. First, a Gaussian process (GP)-based Bayesian optimization [25] is adopted using two physical quantities as descriptors: V and density, ρ. These quantities are available in most experimental or computational crystal structure databases. Although a phenomenological relationship is proposed between $\log \kappa_L$ and V, the correlation between them is low. Moreover, the correlation between $\log \kappa_L$ and ρ is even worse.

We start from an observed data set of five compounds that are randomly chosen from the dataset. The Bayesian optimization searches for the compound with a maximum probability of improvement [26] among the remaining data. That is, the compound with the highest Z-score derived from GP is searched. The compound is included into the observed dataset. Then another compound with the maximum probability of improvement is searched. Both the Bayesian optimization and random searches are repeated 200 times, and the average number of observed compounds required to find the best compound is examined.

The average numbers of compounds required for the optimization using the Bayesian optimization and random searches, N_{ave}, are 11 and 55, respectively. The compound with the lowest LTC among the 101 compounds (i.e., rocksalt PbSe) can be found much more efficiently using a Bayesian optimization with only two variables, V and ρ. However, using a Bayesian optimization only with these two variables is not a robust method to determine the lowest LTC. As an example, the result of the Bayesian optimization using the dataset after intentionally removing the first and second lowest LTC compounds shows that N_{ave} is 65 to find LiI using Bayesian optimization only with V and ρ, which is larger than that of the random search ($N_{ave} = 50$). The delay in the optimization should originate from the fact that LiI is an outlier when the LTC is modeled only with V and ρ. Such outlier compounds with low LTC are difficult to find only with V and ρ.

To overcome the outlier problem, predictors have been added for constituent chemical elements. There are many choices for such variables. Here, we introduce binary elemental descriptors, which are a set of binary digits representing the presence of chemical elements. Since the 101 LTC data is composed of 34 kinds of elements, there are 34 elemental descriptors. When finding both PbSe and LiI, the compound with the lowest LTC is found with $N_{ave} = 19$. The use of binary elemental descriptors improves the robustness of the efficient search.

Better correlations with LTC can be found for parameters obtained from the phonon density of states. Figure 1.8 shows the relationships between the LTC and the

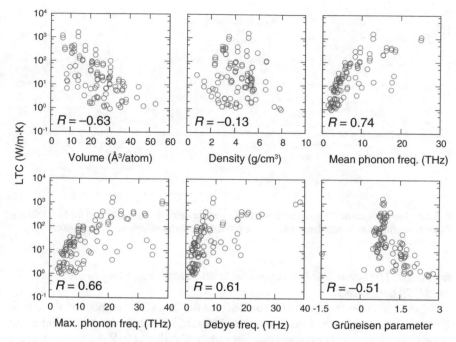

Fig. 1.8 Relationship between $\log \kappa_L$ and the physical properties derived from the first-principles electronic structure and phonon calculations. Correlation coefficient, R, is shown in each panel

physical properties. Other than volume and density, the following quantities are obtained by our phonon calculations: mean phonon frequency, maximum phonon frequency, Debye frequency, and Grüneisen parameter. The Debye frequency is determined by fitting the phonon density of states for a range between 0 and 1/4 of the maximum phonon frequency to a quadratic function. The thermodynamic Grüneisen parameter is obtained from the mode-Grüneisen parameters calculated with a quasi-harmonic approximation and mode-heat capacities. The correlation coefficients R between $\log \kappa_L$ and these physical properties are shown in the corresponding panels. The present study does not use such phonon parameters as descriptors because a data library for such phonon parameters for a wide range of compounds is unavailable. Hereafter, we show results only with the descriptor set composed of 34 binary elemental descriptors on top of V and ρ.

A GP prediction model has been used to screen for low-LTC compounds in a large library of compounds. In the biomedical community, a screening based on a prediction model is called a "virtual screening" [27]. For the virtual screening, all 54779 compounds in the Materials Project Database (MPD) library [28], which is composed mostly of crystal structure data available in ICSD [29], are adopted. Most of these compounds have been synthesized experimentally at least once. On the basis of the GP prediction model made by V, ρ, and the 34 binary elemental descriptors

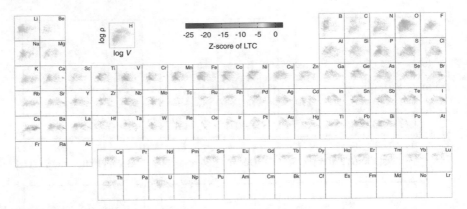

Fig. 1.9 Dependence of the Z-score on the constituent elements for compounds in the MPD library. Color along the volume and density for each element denote the magnitude of the Z-score

for the 101 LTC data, low-LTC compounds are ranked according to the Z-score of the 54779 compounds.

Figure 1.9 shows the distribution of Z-scores for the 54779 compounds along with V and ρ. The magnitude of the Z-score is plotted in the panels corresponding to the constituent elements. The compounds are widely distributed in $V - \rho$ space. Thus, it is difficult to identify compounds without performing a Bayesian optimization with elemental descriptors. The widely distributed Z-scores for light elements such as Li, N, O, and F imply that the presence of such light elements has a negligible effect on lowering the LTC. When such light elements form a compound with heavy elements, the compound tends to show a high Z-score. It is also noteworthy that many compounds composed of light elements such as Be and B tend to show a high LTC. Pb, Cs, I, Br, and Cl exhibit special features. Many compounds composed of these elements exhibit high Z-scores. Most compounds showing a positive Z-score are a combination of these five elements. On the other hand, elements in the periodic table neighboring these five elements do not show analogous trends. For example, compounds of Tl and Bi, which neighbor Pb, rarely exhibit high Z-scores. This may sound odd since Bi_2Te_3 is a famous thermoelectric compound, and some compounds containing Tl have a low LTC. This may be ascribed to our selection of the training dataset, which is composed only of AB compounds with 34 elements and three kinds of simple crystal structures. In other words, the training dataset is somehow "biased". Currently, this bias is unavoidable because first-principles LTC calculations are still too expensive to obtain a sufficiently unbiased training dataset with a large enough number of data points to cover the diversity of the chemical compositions and crystal structures. Nevertheless, the usefulness of biased training dataset to find low-LTC materials will be verified in the future. Due to the biased training dataset, all low-LTC materials in the library may not be discovered. However, some of them can be discovered. A ranking of LTCs from the Z-score does not necessarily correspond to the true first-principles ranking. Therefore, a verification process

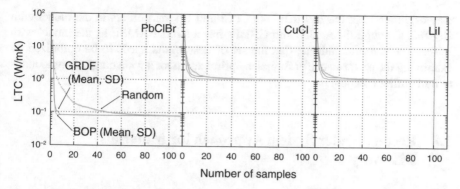

Fig. 1.10 Behavior of the Bayesian optimization for the LTC data to find PbClBr, CuCl, and LiI

for candidates of low-LTC compounds after the virtual screening is one of the most important steps in "discovering" low-LTC compounds. First-principles LTCs have been evaluated for the top eight compounds after the virtual screening. All of them are considered to form ordered structures. However, the LTC calculation is unsuccessful for Pb_2RbBr_5 due to the presence of imaginary phonon modes within the supercell used in the present study. All of the top five compounds, $PbRbI_3$, $PbIBr$, $PbRb_4Br_6$, $PbICl$, and $PbClBr$, show a LTC of <0.2 W/mK (at 300 K), which is much lower than that of the rocksalt PbSe, [i.e., 0.9 W/mK (at 300 K)]. This confirms the powerfulness of the present GP prediction model to efficiently discover low-LTC compounds. The present method should be useful to search for materials in diverse applications where the chemistry of materials must be optimized.

Finally, the performance of Bayesian optimization has been examined using the compound descriptors derived from elemental and structural representations for the LTC dataset containing the compounds identified by the virtual screening. GP models are constructed using (1) the means and SDs of the elemental representations and GRDFs and (2) the means and SDs of elemental representations and BOPs. Figure 1.10 shows the behavior of the lowest LTC during Bayesian optimization relative to a random search. The optimization aims to find PbClBr with the lowest LTC. For the GP model with the BOP, the average number of samples required for the optimization, N_{ave}, is 5.0, which is ten times smaller than that of the random search, $N_{ave} = 50$. Hence, the Bayesian optimization more efficiently discovers PbClBr than the random search.

To evaluate the ability to find a wide variety of low-LTC compounds, two datasets have been prepared after intentionally removing some low-LTC compounds. In these datasets, CuCl and LiI, which respectively show the 11th-lowest and 12th-lowest LTCs, are solutions of the optimizations. For the GP model with BOPs, the average number of observations required to find CuCl and LiI is $N_{ave} = 15.1$ and 9.1, respectively. These numbers are much smaller than those of the random search. On the other hand, for the GP model with GRDFs, the average number of observations required to find CuCl and LiI is $N_{ave} = 40.5$ and 48.6, respectively. The delayed optimization may originate from the fact that both CuCl and LiI are outliers in the model with GRDFs, although the model with GRDFs has a similar RMSE as the model with BOPs. These results indicate that the set of descriptors needs to be optimized by examining the performance of Bayesian optimization for a wide range of compounds to find outlier compounds.

1.8 Recommender System Approach for Materials Discovery

Many atomic structures of inorganic crystals have been collected. Of the few available databases for inorganic crystal structures, the ICSD [29] contains approximately 10^5 inorganic crystals, excluding duplicates and incompletes. Although this is a rich heritage of human intellectual activities, it covers a very small portion of possible inorganic crystals. Considering 82 nonradioactive chemical elements, the number of simple chemical compositions up to ternary compounds $A_a B_b C_c$ with integers satisfying $\max(a, b, c) \leq 15$ is approximately 10^8, but increases to approximately 10^{10} for quaternary compounds $A_a B_b C_c D_d$. Although many of these chemical compositions do not form stable crystals, the huge difference between the number of compounds in ICSD and the possible number of compounds implies that many unknown compounds remain. Conventional experiments alone cannot fill this gap. Often, first-principles calculations are used as an alternative approach. However, systematic first-principles calculations without a priori knowledge of the crystal structures are very expensive.

Machine learning is a different approach to consider all chemical combinations. A powerful machine-learning strategy is mandatory to discover new inorganic compounds efficiently. Herein we adopt a recommender system approach to estimate the relevance of the chemical compositions where stable crystals can be formed [i.e., chemically relevant compositions (CRCs)] [30, 31]. The compositional similarity is defined using the procedure shown in Sect. 1.2. A composition is described by a set of 165 descriptors composed of the means, SDs, and covariances of the established elemental representations. The probability for CRCs is subsequently estimated on the basis of a machine-learning two-class classification using the compositional similarity. This approach significantly accelerates the discovery of currently unknown CRCs that are not present in the training database.

References

1. K. Fujimura, A. Seko, Y. Koyama, A. Kuwabara, I. Kishida, K. Shitara, C.A.J. Fisher, H. Moriwake, I. Tanaka, Adv. Energy Mater. **3**, 980 (2013)
2. A. Seko, T. Maekawa, K. Tsuda, I. Tanaka, Phys. Rev. B **89**, 054303 (2014)
3. J. Lee, A. Seko, K. Shitara, K. Nakayama, I. Tanaka, Phys. Rev. B **93**, 115104 (2016)

4. K. Toyoura, D. Hirano, A. Seko, M. Shiga, A. Kuwabara, M. Karasuyama, K. Shitara, I. Takeuchi, Phys. Rev. B **93**, 054112 (2016)
5. A. Seko, A. Togo, H. Hayashi, K. Tsuda, L. Chaput, I. Tanaka, Phys. Rev. Lett. **115**, 205901 (2015)
6. A. Seko, H. Hayashi, K. Nakayama, A. Takahashi, I. Tanaka, Phys. Rev. B **95**, 144110 (2017)
7. P.J. Steinhardt, D.R. Nelson, M. Ronchetti, Phys. Rev. B **28**, 784 (1983)
8. A.P. Bartók, R. Kondor, G. Csányi, Phys. Rev. B **87**, 184115 (2013)
9. J. Behler, M. Parrinello, Phys. Rev. Lett. **98**, 146401 (2007)
10. A.P. Bartók, M.C. Payne, R. Kondor, G. Csányi, Phys. Rev. Lett. **104**, 136403 (2010)
11. A. Seko, A. Takahashi, I. Tanaka, Phys. Rev. B **90**, 024101 (2014)
12. A. Seko, A. Takahashi, I. Tanaka, Phys. Rev. B **92**, 054113 (2015)
13. T. Hastie, R. Tibshirani, J. Friedman, *The Elements of Statistical Learning*, 2nd edn. (Springer, New York, 2009)
14. R. Tibshirani, J. R. Stat. Soc. B **58**, 267 (1996)
15. P.E. Blöchl, Phys. Rev. B **50**, 17953 (1994)
16. J.P. Perdew, K. Burke, M. Ernzerhof, Phys. Rev. Lett. **77**, 3865 (1996)
17. G. Kresse, J. Hafner, Phys. Rev. B **47**, 558 (1993)
18. G. Kresse, J. Furthmüller, Phys. Rev. B **54**, 11169 (1996)
19. G. Kresse, D. Joubert, Phys. Rev. B **59**, 1758 (1999)
20. A. Togo, I. Tanaka, Scr. Mater. **108**, 1 (2015)
21. A. Takahashi, A. Seko, I. Tanaka, Phys. Rev. Mater. **1**, 063801 (2017)
22. A. Takahashi, A. Seko, I. Tanaka (2017), arXiv:1710.05677
23. A. Togo, L. Chaput, I. Tanaka, Phys. Rev. B **91**, 094306 (2015)
24. G.A. Slack, *Solid State Physics*, vol. 34 (Academic Press, New York, 1979), pp. 1–71
25. C.E. Rasmussen, C.K.I. Williams, *Gaussian Processes for Machine Learning* (MIT Press, Cambridge, 2006)
26. D. Jones, J. Global Optim. **21**, 345 (2001)
27. D.B. Kitchen, H. Decornez, J.R. Furr, J. Bajorath, Nat. Rev. Drug Discov. **3**, 935 (2004)
28. A. Jain, S.P. Ong, G. Hautier, W. Chen, W.D. Richards, S. Dacek, S. Cholia, D. Gunter, D. Skinner, G. Ceder et al., APL Mater. **1**, 011002 (2013)
29. G. Bergerhoff, I.D. Brown, *Crystallographic Databases*, edited by F.H. Allen et al. (International Union of Crystallography, Chester, 1987)
30. A. Seko, H. Hayashi, H. Kashima, I. Tanaka (2017), arXiv:1710.00659
31. A. Seko, H. Hayashi, I. Tanaka (2017), arXiv:1711.06387

Chapter 2
Potential Energy Surface Mapping of Charge Carriers in Ionic Conductors Based on a Gaussian Process Model

Kazuaki Toyoura and Ichiro Takeuchi

Abstract The potential energy surface (PES) of a charge carrier in a host crystal is an important concept to fundamentally understand ionic conduction. Such PES evaluations, especially by density functional theory (DFT) calculations, generally require vast computational costs. This chapter introduces a novel selective sampling procedure to preferentially evaluate the partial PES characterizing ionic conduction. This procedure is based on a machine learning method called the Gaussian process (GP), which reduces computational costs for PES evaluations. During the sampling procedure, a statistical model of the PES is constructed and sequentially updated to identify the *region of interest* characterizing ionic conduction in configuration space. Its efficacy is demonstrated using a model case of proton conduction in a well-known proton-conducting oxide, barium zirconate ($BaZrO_3$) with the cubic perovskite structure. The proposed procedure efficiently evaluates the partial PES in the region of interest that characterizes proton conduction in the host crystal lattice of $BaZrO_3$.

Keywords Gaussian process · Bayesian optimization · Ionic conduction
Potential energy surface

2.1 Introduction

Atomic transport phenomena in solids such as atomic diffusion and ionic conduction are generally governed by thermally activated processes. Based on transition state theory (TST) [1–3], the mean frequency of an elementary process (ν) with

K. Toyoura (✉)
Department of Materials Science and Engineering, Kyoto University,
Yoshida, Sakyo, Kyoto 606-8501, Japan
e-mail: toyoura.kazuaki.5r@kyoto-u.ac.jp

I. Takeuchi
Department of Computer Science, Nagoya Institute of Technology,
Gokiso, Showa, Nagoya 466-8555, Japan

© The Author(s) 2018
I. Tanaka (ed.), *Nanoinformatics*, https://doi.org/10.1007/978-981-10-7617-6_2

a single saddle point state, a so-called an atomic or ionic jump, is approximated by $\nu = \nu_0 \exp(-\Delta E^{mig}/k_B T)$, where ν_0 is the vibrational prefactor, k_B is the Boltzmann constant, T is the temperature, and ΔE^{mig} is the potential barrier, i.e., the change in the potential energy (PE) from the initial state to the saddle point state. ν_0 is typically a constant value in the range of 10^{12}–10^{13} s^{-1} associated with a lattice vibration [3–8]. Consequently, ΔE^{mig} mainly determines the rate of an atomic jump in a solid.

In general, atomic transfer is composed of several types of atomic jumps, which form a complicated three-dimensional (3D) network in the crystal lattice. Therefore, it is necessary to grasp the entire potential energy surface (PES) of a mobile atom or ion. However, a theoretical PES evaluation, e.g., based on density functional theory (DFT), generally requires huge computational costs, particularly in the case of a host crystal with a low crystallographic symmetry. The nudged elastic band (NEB) method [9, 10] is a well-established technique to avoid evaluating the entire PES, in which only the minimum energy paths (MEPs) are focused on in the PES. Because of its efficiency and versatility, the NEB method is used conventionally to clarify the atomic-scale-picture and the kinetics of atomic transfer in crystals.

However, the NEB method has some practical limitations. First, the initial and final states of all elementary paths in a crystal must be specified. That is, all local energy minima in the crystal and all conceivable elementary paths between adjacent local energy minima must be known in advance. As the crystallographic symmetry of the host crystal decreases, the number of local energy minima and conceivable elementary paths rapidly increase. Consequently, satisfying the requirements in the NEB method is very difficult without a priori information on the entire PES. In cases without a priori information, physical and chemical knowledge (e.g., ionic radii, chemical bonding states, electrostatic interaction, and interstitial and bottle-neck sizes) are generally used. However, a key elementary path determining the rate of atomic diffusion or ionic conduction is sometimes missed in such an arbitrary manner. In addition, the NEB method requires huge computational costs for low-symmetry crystals, even if only the MEPs in the PES are evaluated. For example, in our recent study on proton conduction in tin pyrophosphate (SnP_2O_7) with space group of $P2_{1/C}$, we evaluated 143 possible elementary paths connecting 15 local energy minima by the NEB method [11]. An alternative method that is both robust and efficient is desirable to analyze complicated atomic transfers consisting of many elementary paths in a low-symmetry crystal.

This chapter introduces a novel selective sampling procedure for PES mapping based on a machine learning technique [12]. This sampling procedure preferentially evaluates a partial PES in the region of interest characterizing ionic conduction. The region of interest is defined in two ways: (1) a *low-PE region* forming long-range migration pathways throughout the crystal lattice in the PES and (2) a *low-force norm region* (*low-FN region*), which includes all the local minima and saddle points in the PES. It should be noted that other mathematically definable and efficient choices could be considered as the region of interest. See the synthetic 2D PES and FN surface (FNS) for the definitions of the region of interest (Fig. 2.1).

(a) Potential energy surface **(b)** Force norm surface

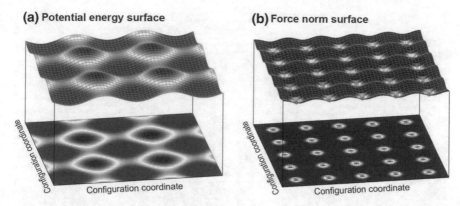

Fig. 2.1 Synthetic two-dimensional (**a**) PES and (**b**) FNS of a charge carrier in a host crystal lattice. Region of interest is defined as the low-PE region in the PES and the low-FN region in the FNS

The proposed sampling procedure has three key features. (1) A statistical model of the PES or FNS is developed as a *Gaussian process* (GP) [13, 14]. The statistical model is iteratively updated by repeatedly (i) sampling at a point where the predicted PE or FN is low and (ii) incorporating the newly calculated PE or FN value at the sampled point. (2) The statistical PES or FNS model is used to identify the subset of grid points at which the PEs or FNs are relatively low. Here a selection criterion is introduced for this advanced purpose, because GP applications have generally targeted the single global minimum or maximum point (not a subset). (3) The procedure allows us to estimate how many points in the region of interest remain unsampled, i.e., lets us know when sampling should be terminated.

These features are possible by exploiting an advantage of the GP that it provides not only the predicted PE or FN value but also the uncertainty at each grid point. Figure 2.2 illustrates selective sampling sequences using a one-dimensional synthetic PES where nine grid points in the low-PE region should be selectively sampled from all (50) points as an example. Roughly speaking, the grid point most likely to be located in the low-PE region is sampled at each step based on the predicted PEs (red solid curve) and the uncertainties (pale blue area). In the early steps, the predicted PEs are uncertain with large discrepancies from the true PES (black solid curve), resulting in selecting grid points with large uncertainties. As the sampling proceeds, the predicted PE curve gradually approaches the true one and the uncertainty decreases. Eventually, the grid points in the low-PE region are selectively sampled in the latter steps.

The uncertainty in the GP model is useful also to determine when to terminate sampling. The termination criterion should be determined based on the existence probability of unsampled low-PE points, for which the information on the uncertainty is indispensable. As a model case, herein the efficacy of the proposed procedure is demonstrated using proton conduction in a proton-conducting oxide, barium zirconate ($BaZrO_3$) [15–18].

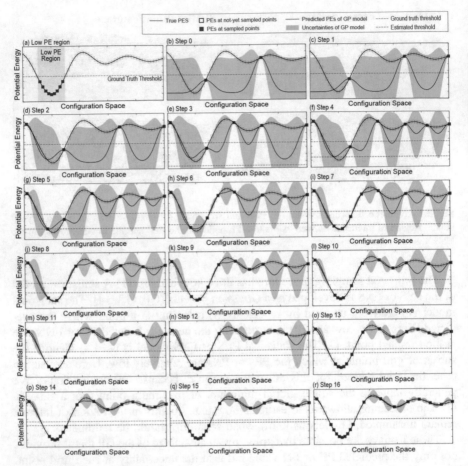

Fig. 2.2 Schematic illustration of the proposed selective sampling procedure in a one-dimensional configuration space with synthetic data [12]. In each plot, the x- and y-axes represent the configuration space and the PEs, respectively. Red area in plot (**a**) represents the low-PE region. In this example, the goal is to efficiently identify and evaluate the PEs at the nine points in the low-PE region. Plot (**b**) indicates the initialization step, where two points (filled red squares) are randomly selected and their PEs are evaluated. Remaining 16 plots [plots (**c**) to plot (**r**)] indicate steps 1–16 of the procedure

2.2   Problem Setup

2.2.1   Entire Proton PES in BaZrO₃

The entire PES of a proton in $BaZrO_3$ evaluated using DFT calculations with structural optimization is initially shown for the problem setup of the demonstration study. Figure 2.3 shows the crystal structure of $BaZrO_3$ [space group: $Pm\bar{3}m$ (221)] and its asymmetric unit satisfying $0 \leq x, y, z \leq 0.5, y \leq x$, and $z \leq y$. x, y, and z

Fig. 2.3 Crystal structure and the asymmetric unit of BaZrO₃ with the cubic perovskite structure

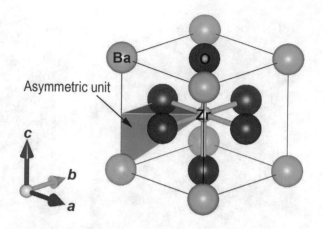

denote the 3D fractional coordinates of a proton introduced into the host lattice. Ba, Zr, and O ions occupy the $1a$, $1b$, and $3c$ sites, respectively, using the origin setting shown in Fig. 2.3. A $40 \times 40 \times 40$ grid is introduced in the unit cell (the grid interval is nearly equal to 0.1 Å), which contains 64,000 grid points in total. Due to the high crystallographic symmetry of BaZrO₃, the asymmetric unit has only 1771 grid points. Among these points, three coincide with Ba, Zr, or O ion. Removing these three points reduces the remaining grid points to 1768.

The DFT calculations for the PES (and FNS) evaluation of a proton in BaZrO₃ are based on the projector augmented wave (PAW) method as implemented in the VASP code [19–22]. The generalized gradient approximation (GGA) parameterized by Perdew, Burke, and Ernzerhof is used for the exchange-correlation term [23]. The $5s$, $5p$, and $6s$ orbitals for Ba, $4s$, $4p$, $5s$, and $4d$ for Zr, $2s$ and $2p$ for O, and $1s$ for H are treated as valence states. The supercell consisting of $3 \times 3 \times 3$ unit cells (135 atoms) is used with a $2 \times 2 \times 2$ mesh for the k-point sampling. Only the atomic positions in a limited region corresponding to the $2 \times 2 \times 2$ unit cells around the introduced proton are optimized with fixing all other atoms and the proton. The atomic positions are optimized until the residual forces converge to less than 0.02 eV/Å.

Figure 2.4a shows the calculated proton PES in the low-PE region below 0.3 eV. The blue regions around the O ions are the most stable proton sites and are located ∼1 Å from the O ions. The OH distance is almost equivalent to that in water, indicating that OH bond formation stabilizes the protons in BaZrO₃. There are four equivalent proton sites per O ion, which are connected by the low-PE points around the O ions. The rotational path around the O ions consists of four equivalent quarter-rotational paths, where the calculated potential barrier is 0.18 eV. On the other hand, the hopping path connecting adjacent rotational orbits is located at the periphery of the edges of the ZrO₆ octahedra. The calculated potential barrier of the hopping path is 0.25 eV, which is higher than that of the rotational path. The two kinds of paths form a three-dimensional proton-conducting network throughout the crystal lattice. Consequently, protons exhibit a long-range

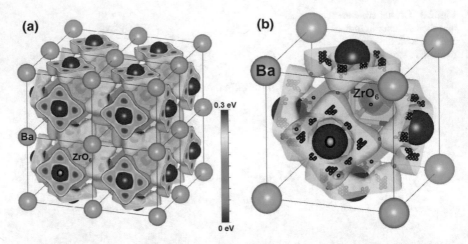

Fig. 2.4 (**a**) Calculated proton PES in the low PE region below 0.3 eV in reference to the most stable point [12]. (**b**) Grid points at which the force norm acting on a proton is less than 0.2 eV/Å

migration via repeated rotation and hopping, where the hopping path is the rate-determining path in proton conduction.

2.2.2   Problem Statement

Figure 2.4(a) indicates that the partial PES of a proton in the low-PE region below 0.3 eV is necessary and sufficient to estimate the proton diffusivity and conductivity in the crystal lattice of $BaZrO_3$. In the low-PE region, there are 353 grid points to be evaluated by DFT calculations, corresponding to the lowest 20% of the grid points. Therefore, the first task is to selectively sample all the low-PE grid points as efficiently as possible. Hereafter this is referred to task 1. Task 2 is based on the force norm (FN) acting on a proton at each grid point. The FN is calculated along with the PE by the DFT calculations. In this task, the region of interest is defined as grid points with an FN below a threshold (i.e., 0.2 eV/Å in the present study), denoted by blue spheres in Fig. 2.4(b). There are only 15 grid points in the low-FN region in the asymmetric unit. The region of interest in task 2 is much smaller than that in task 1, hopefully leading to more efficient sampling.

Prior to the detailed description of the proposed procedure in Sect. 2.3, this problem is generalized and mathematically formulated using the identification of the low-PE region as an example. There are N grid points, $i = 1, \ldots, N$, in the asymmetric unit of the host crystal lattice. The PE of a proton at grid point i is denoted by E_i. Using the parameter $0 < \alpha < 1$, the low-PE region is defined as the set of αN points where the PEs are lower than those at other $(1-\alpha)N$ points. The goal is to identify all αN grid points in the low-PE region as efficiently as possible.

For simplicity, α is assumed to be prespecified. However, it can be adaptively determined, as demonstrated in Sect. 2.4.3.

When θ_α represents the PE threshold of the low-PE region, the subsets of P_α and N_α are defined as

$$P_\alpha := \{i \in \{1, \ldots, N\} | E_i < \theta_\alpha\} \tag{2.1}$$

$$N_\alpha := \{i \in \{1, \ldots, N\} | E_i \geq \theta_\alpha\}. \tag{2.2}$$

The task is formally stated as the problem of identifying all points in P_α. Using statistical terminology, the points in P_α and N_α are called *"positive"* and *"negative"* points, respectively. Note that P_α, N_α, and θ_α are *unknown* unless the PEs at all grid points are actually computed. During the sampling process, these quantities are estimated based on the PEs at points sampled in the earlier steps. Our estimates of positive and negative sets are denoted as $\hat{P}_\alpha$ and $\hat{N}_\alpha$, respectively. The former indicates the set of points at which the PEs have been sampled and computed in earlier steps. The latter represents the set of points at which the PEs have yet to be computed. The proposed selective sampling procedure can be interpreted as the process of sequentially updating these two sets of points. Specifically, we begin at $\hat{P}_\alpha = \varnothing$ and $\hat{N}_\alpha = \{1, \ldots, N\}$. The two sets are updated as

$$\hat{P}_\alpha \leftarrow \hat{P}_\alpha \cup \{i'\}, \tag{2.3}$$

$$\hat{N}_\alpha \leftarrow \hat{N}_\alpha \backslash \{i'\}, \tag{2.4}$$

where i' is the sampled point in the step. When the termination criterion is satisfied, $\hat{P}_\alpha$ has a high probability of containing all points in P_α. The estimated θ_α is also defined as $\hat{\theta}_\alpha$. Section 2.3.3 shows how to estimate θ_α from the prespecified α. Note that the θ_α estimation is unnecessary in task 2 because the FN threshold is directly specified by the FN value.

2.3 GP-Based Selective Sampling Procedure

Here the proposed sampling procedure based on the GP is described using the PES-based task (task 1) as an example. Specifically, the key features are explained in the following subsections: the GP-based PE statistical model (Sect. 2.3.1), the selection criterion of the next grid point (Sect. 2.3.2), the estimation of the PE threshold (Sect. 2.3.3), and the criterion for sampling termination (Sect. 2.3.4). Note that the threshold estimation (Sect. 2.3.3) is irrelevant to task 2 for the low-FN identification.

2.3.1 Gaussian Process Models

We adopt a GP model [13, 14] as the statistical model of the PES. Using a GP model, the potential energy E_i is represented as

$$E_i \sim N(\mu_i, \sigma_i^2), \quad i = 1, \ldots, N, \tag{2.5}$$

where $N(\mu_i, \sigma_i^2)$ denotes the normal distribution with mean μ_i and variance σ_i^2. A GP model is a type of regression model. Consider a d-dimensional vector of descriptors for each point, where the vector is denoted as $\chi_i \in \mathbb{R}^d$ for $i = 1, \ldots, N$. The mean and variance of the PE at the ith point, which are given in Eqs. (2.8) and (2.9), respectively, are represented as functions of χ_i. The GP model employs the so-called kernel function $k: \mathbb{R}^d \times \mathbb{R}^d \rightarrow \mathbb{R}$. For two different points indexed by i and j, $k(\chi_i, \chi_j)$ is roughly interpreted as the similarity between these two points. One of the most commonly used kernel functions is the RBF kernel, which is given by

$$k(\chi, \chi') = \sigma_f^2 \exp(-\|\chi - \chi'\|/2l^2), \tag{2.6}$$

where $\sigma_f, l > 0$ are tuning parameters, and $\|\cdot\|$ represents the L_2 norm. Furthermore, for n points indexed by $1, \ldots, n$, let $\mathbf{K} \in \mathbb{R}^{nn}$ be the so-called kernel matrix defined as

$$\mathbf{K} := \begin{bmatrix} k(\chi_1, \chi_1) & \cdots & k(\chi_1, \chi_n) \\ \vdots & \ddots & \vdots \\ k(\chi_n, \chi_1) & \cdots & k(\chi_n, \chi_n) \end{bmatrix}. \tag{2.7}$$

For any point in the configuration space whose descriptor vector is represented as $\chi \in \mathbb{R}^d$, the GP model provides the predictive distribution of its PE in the form of a normal distribution $N[\mu(\chi), \sigma^2(\chi)]$. Here, the mean function $\mu: \mathbb{R}^d \rightarrow \mathbb{R}$ is given as

$$\mu(\chi) := \kappa(\chi)^{\mathrm{T}} \mathbf{K}^{-1} \mathbf{E}, \tag{2.8}$$

where $\kappa(\chi) := [k(\chi, \chi_1), \ldots, k(\chi, \chi_n)]^{\mathrm{T}}$ and $\mathbf{E} := [E_1, \ldots, E_n]$, while the variance function $\sigma^2: \mathbb{R}^d \rightarrow \mathbb{R}$ is given as

$$\sigma^2(\chi) := k(\chi, \chi) - \kappa(\chi)^{\mathrm{T}} \mathbf{K}^{-1} \kappa(\chi). \tag{2.9}$$

At each step, the GP model of PES is fitted based on $\{(\chi_i, E_i)\}_{i \in \hat{P}_a}$, which is the set of points whose PEs have already been computed by DFT calculations in earlier steps.

2.3.2 Selection Criterion

Given a GP model in the form of Eq. (2.5) for each point, the subsequent task is to select the next point at which the PE is most likely to be lower than the estimated threshold $\hat{\theta}_\alpha$. (The following subsection discusses how to estimate the threshold.) For this task, some techniques developed in the context of Bayesian optimization [24, 25], which are used to minimize or maximize an unknown function, can be borrowed. There are two main options that can be adapted for our task in the Bayesian optimization literature. The first is to select the point at which the probability that the PE is lower than $\hat{\theta}_\alpha$ is maximized. This is called the "probability of improvement", which is formulated as

$$i' := \arg\max_{i \in \tilde{N}_\alpha} \Phi[\hat{\theta}_\alpha; \mu(\chi_i), \sigma^2(\chi_i)], \tag{2.10}$$

where $\Phi(\cdot; \mu, \sigma^2)$ is the cumulative distribution function of $N(\mu, \sigma^2)$. The second option is the "expected improvement". Similarly, it is formulated as

$$i' := \arg\min_{i \in \tilde{N}_\alpha} \int_{-\infty}^{\hat{\theta}_\alpha} E\phi[E; \mu(\chi_i), \sigma^2(\chi_i)]dE, \tag{2.11}$$

where $\phi(\cdot; \mu, \sigma^2)$ is the probability density function of $N(\mu, \sigma^2)$. This study employs the second option, although the performance difference between Eqs. (2.10) and (2.11) is negligible in our experience.

2.3.3 PE Threshold

PE threshold θ_α should be estimated because it is unknown prior to evaluating the entire PES. The contingency table (Table 2.1) is here considered to obtain an estimate $\hat{\theta}_\alpha$ of the threshold θ_α. TP, FP, FN, and TN denote the true positive, false positive, false negative, and true negative, respectively. The notation # indicates the event number. Note that the FN is not the "force norm" acting on a proton in this context. The numbers for these four events can be rephrased as:

- #TP: The number of sampled points in the low-PE region.
- #FP: The number of sampled points in the high-PE region.

Table 2.1 Contingency table defining TP, FP, FN, and TN

	P_α	N_α
$\hat{P}_\alpha$	$\#TP(\hat{\theta}_\alpha)$	$\#FP(\hat{\theta}_\alpha)$
$\hat{N}_\alpha$	$\#FN(\hat{\theta}_\alpha)$	$\#TN(\hat{\theta}_\alpha)$

- #FN: The number of not-yet-sampled points in the low-PE region.
- #TN: The number of not-yet-sampled points in the high-PE region.

These four numbers depend on the estimated PE threshold $\hat{\theta}_\alpha$. Recalling the equation of $P_\alpha/(P_\alpha + N_\alpha) = \alpha$, the following relationship should be maintained

$$[\#TP(\hat{\theta}_\alpha) + \#FN(\hat{\theta}_\alpha)]/N = \alpha. \tag{2.12}$$

Because E_i for $i \in \hat{P}_\alpha$ is already evaluated, we simply obtain

$$\#TP(\hat{\theta}_\alpha) = \sum_{i \in \hat{P}_\alpha} I(E_i < \hat{\theta}_\alpha), \tag{2.13}$$

where $I(\cdot)$ is the indicator function defined by $I(z) = 1$ if z is true and $I(z) = 0$ if z is false. On the other hand, $\#FN(\hat{\theta}_\alpha)$ must be estimated based on the statistical model Eq. (2.6) because E_i is unknown for $i \in \hat{N}_\alpha$

$$\#FN(\hat{\theta}_\alpha) \approx \widehat{FN}(\hat{\theta}_\alpha) := \sum_{i \in N_\alpha} \Phi[\hat{\theta}_\alpha; \mu(\boldsymbol{\chi}_i), \sigma^2(\boldsymbol{\chi}_i)]. \tag{2.14}$$

The estimate of the threshold $\hat{\theta}_\alpha$ is determined for each step so that it satisfies Eq. (2.12) where the quantities on the left-hand side are given by Eqs. (2.13) and (2.14).

2.3.4 Termination Criterion

When sampling is terminated, $\hat{P}_\alpha$ should ideally contain all the points in P_α, i.e., $\hat{P}_\alpha \supseteq P_\alpha$. As easily noted from the contingency table, this requirement can be rewritten as $\#FN(\hat{\theta}_\alpha) = 0$. This indicates that the estimated false negative rate (FNR) defined as

$$\widehat{FNR} := \frac{\#\widehat{FN}(\hat{\theta}_\alpha)}{\#TP(\hat{\theta}_\alpha) + \#\widehat{FN}(\hat{\theta}_\alpha)}, \tag{2.15}$$

can assess the badness of the sampled points. $\widehat{FNR}$ in Eq. (2.15) can be interpreted as the proportion of points where the PEs have yet to be evaluated. At each step, $\#TP(\hat{\theta}_\alpha)$ is computed by Eq. (2.13) and $\#FN(\hat{\theta}_\alpha)$ is estimated by Eq. (2.14). Then, the sampling is terminated if $\widehat{FNR}$ is close to zero (e.g., $<10^{-6}$).

2.4   Results of Selective Sampling

2.4.1   Low-PE Region Identification

The performances of several sampling procedures for $\alpha = 0.2$ are compared in the low-PE region identification problem. Specifically, the following six sampling methods are assessed: (1) GP1(xyz), (2) GP2(xyz + 1st NNs), (3) GP3(xyz + prePES), (4) random, (5) prePES, and (6) ideal. The first three are the proposed GP-based selective sampling methods with different descriptors. In GP1, the 3D coordinates (x_i, y_i, and z_i) in the host crystal lattice are used as the descriptors of the ith point (denoted as xyz). In GP2, the first nearest neighbor (1st NN) distances to the Ba, Zr, and O atoms from each point are used as additional descriptors (denoted as 1st NNs). In GP3, a preliminary PES (denoted as prePES) is used as an additional descriptor. The preliminary PES means a rough but quick approximation of the PES obtained using less accurate but more efficient computational methods. For prePES, the PE values at all N points obtained by single-point DFT calculations are used. Random indicates naive random sampling, where a point is selected randomly at each step. prePES denotes a selective sampling method based only on the preliminary PES. Specifically, points are sequentially selected in ascending order of the preliminary PEs obtained by single-point DFT calculations. Finally, ideal indicates the ideal sampling method, which can only be realized when the actual PEs at all the points are known in advance.

In GP1 to GP3, two points are randomly selected to initialize the GP model. The average and the standard deviation over ten runs with different random seeds are discussed. The tuning parameters of the GP models are set to $\sigma_f = l = 0.5$. According to our preliminary experiments, the performances are insensitive to the tuning parameter choices.

Figure 2.5 compares the efficiencies of the six sampling methods. The number of points successfully sampled from the low-PE region (#TP) is plotted as a function of the number of PE computations based on DFT (#TP + #FP). The results of the three different GP-based sampling methods (GP1 to GP3) indicate the importance of choosing the descriptors. Using the 3D coordinates (GP1) as the descriptors is only slightly better than using the random method. On average, GP1 requires 1539.6 ± 31.2 DFT computations until all the points in the low-PE region are identified. GP2 has an improved performance, suggesting that additional appropriate descriptors are generally advantageous. GP2 requires 1269.4 ± 100.3 DFT computations to identify all the low-PE grid points. GP3 has a markedly enhanced performance. GP3 requires only 394.1 ± 5.2 DFT computations, indicating that the preliminary PES is a very helpful descriptor. On the other hand, prePES has a much poorer performance and requires 1479 DFT computations. Thus, the preliminary PES alone is insufficient to effectively identify the low-PE region. The importance of the preliminary PES is discussed in more detail below.

Figure 2.6 demonstrates the differences between the sampling sequences of the GP1, GP3, prePES, and ideal methods. GP1 erroneously selects many points in the

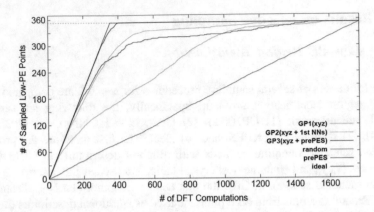

Fig. 2.5 Efficiencies of the six sampling methods for $\alpha = 0.20$ [12]. Number of grid points successfully sampled from the low-PE region (#TP) is plotted versus the number of PE evaluations by DFT (#TP + #FP)

high-PE region. In contrast, GP3 preferentially selects points in the low-PE region, and only a small number of points are mistakenly selected from the high-PE region. Although the prePES method preferentially selects points in the low-PE region, it fails to identify all of them. Surprisingly, the sampling sequence of GP3 is almost identical to that in the ideal sampling, despite the unknown low-PE region beforehand. This indicates that the GP model in GP3 successfully estimates the PES in the low-PE region.

Figure 2.5 indicates that the preliminary PES obtained by single-point DFT calculations is highly valuable as a descriptor when it is used along with three-dimensional coordinates (x, y, z) in GP modeling. However, using the preliminary PES alone cannot identify the low-PE region in the prePES sampling. The results are only slightly better than random. In the earlier steps, the sampling curve of prePES almost overlaps with the ideal sampling curve, but it gradually deviates as the sampling proceeds. Eventually, 1479 steps are necessary to find all points in the low-PE region using prePES. This is 4.2-fold decline compared to the ideal sampling case (353 points). The inefficiency of prePES is attributed to the relationship between the DFT calculations with and without structural optimization.

Figure 2.7 shows the rank correlation between the actual and preliminary PEs, where the points with low PEs are located below the horizontal dotted line. The prePES sampling method selects points in ascending order of the preliminary PEs, meaning that the points are selected from left to right in Fig. 2.7(a). Therefore, most of the N grid points (all points located in the shaded region) must be sampled to select all the points in the low-PE region. On the other hand, in GP3 with xyz and prePES as descriptors, the average number of sampling steps required to identify all the points in the low-PE region is only 394.1, which is only a 1.1-fold increase compared to the ideal sampling method.

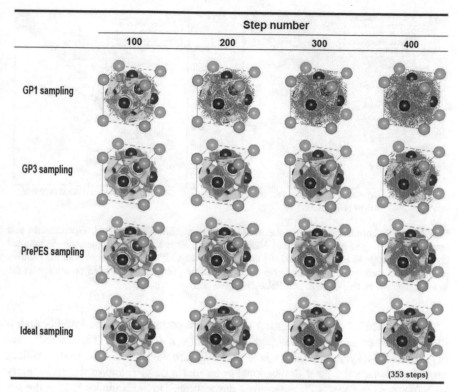

Fig. 2.6 Selected grid points (gray dots) at 100, 200, 300, and 400 steps by the different sampling methods in the model crystal lattice of $BaZrO_3$ for $\alpha = 0.20$ [12]. Yellow surface in each plot is the isosurface corresponding to the PE threshold at $\alpha = 0.20$

2.4.2 Low-FN Region Identification

The previous subsection demonstrates several types of sampling methods, which use different descriptors to identify the low-PE region. GP3, which employs descriptors of xyz and prePES, exhibits the best performance and is comparable to ideal sampling. However, the region of interest (i.e., the low-PE region) comprises 20% of the configuration space. Thus, the computational cost can be reduced by 80% at most.

To further reduce computational costs, it is necessary to redefine a smaller region of interest. The mean frequency of atomic or ionic jumps in a solid is determined mainly by the change in PE from the initial point to the saddle point. As both of these points can be mathematically defined as points with a zero gradient in the PES, the region of interest can be redefined as the region where the force norm (FN) acting on a proton is small. In this model case, the FN threshold is set to 0.2 eV/Å, which leads to 15 grid points in the low-FN region (See Fig. 2.4b).

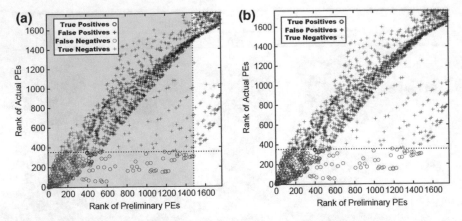

Fig. 2.7 Rank correlation between the actual and the preliminary PEs [12]. Open circles and crosses show the grid points in P_α and N_α, respectively. Blue and red symbols indicate sampled points at 400 steps in (**a**) prePES and (**b**) GP4, respectively. GP4 method samples all the positive points at 400 steps with a small number of False ositive points (i.e., sampled points not in the low-PE region). In (**b**), there are no False Negative points

The efficiencies of four sampling methods are compared for the low-FN region identification problem: (1) GP4(xyz), (2) GP5(xyz + preFNS), (3) preFNS, and (4) ideal. GP4(xyz) and GP5(xyz + preFNS) are GP-based selective sampling procedures where the three-dimensional coordinates (x, y, z) and/or the preliminary FNS (denoted as "preFNS") are used as descriptors. The preliminary FNS is the FN values at all N points computed by single-point DFT calculations, which should have a higher contribution to the sampling performance. The preFNS method indicates a selective sampling where the grid points are sequentially selected in the ascending order of the preliminary FNs. The average and the standard deviation over ten runs with different random seeds are discussed for GP4(xyz) and GP5 (xyz + preFNS). The tuning parameters of the GP models σ_f and l are optimized for each method.

Figure 2.8 compares the performances of several sampling methods. The GP-based sampling (GP4 and GP5) can selectively sample the grid points in the low-FN region requiring 199.7 ± 68.6 and 116.0 ± 30.6 DFT computations to identify all the low-FN grid points, respectively. Both methods show higher efficiencies than that of PES-based GP3(xyz + prePES). These enhanced performances are due to the smaller region of interest defined on the basis of the FNS. Analogous to the preliminary PES, the preliminary FNS evaluated by single-point DFT calculations is a valuable descriptor, which improves the sampling performance in GP-based sampling. However, the naive sampling based on the preFNS shows a much worse performance as it requires 955 DFT computations.

Figure 2.9 shows the rank correlation between the actual and preliminary FNs. The open red circles denote the 15 grid points in the low-FN region. In the preFNS sampling, the points are selected from left to right in the figure. Consequently,

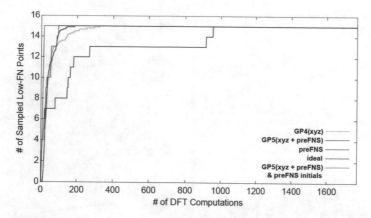

Fig. 2.8 Efficiencies of the four FN-based samplings: GP4(xyz), GP5(xyz + preFNS), preFNS, and ideal. Number of grid points successfully sampled from the low-FN region (#TP) is plotted versus the number of FN computations by DFT (#TP + #FP). Green line is the result in GP5 using the 16 lowest FN points in preFNS as the initial grid points

all 955 points located in the shaded region must be sampled to select all positive points. The difference in the rank depends on whether structural optimization is performed, implying that the local structural relaxation around a proton in oxides is important.

Although using the low-FN region as the region of interest improves the sampling performance, the performance still deviates from that of ideal sampling. Figure 2.10 shows the step numbers where each of the low-FN grid points (Nos. 1–15)

Fig. 2.9 Rank correlation between the actual and the preliminary FNs. Red open circles and crosses show grid points in P_α and N_α, respectively. Blue and black crosses denote False Positive and True Negative points at step 955 in the preFNS sampling method, respectively

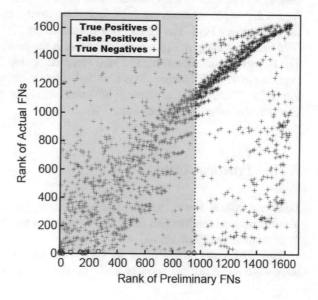

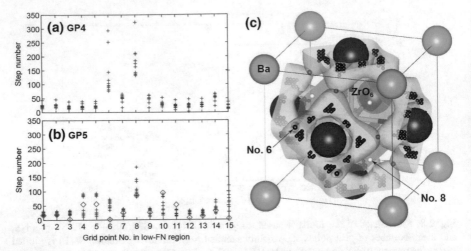

Fig. 2.10 Step numbers where each of the low-FN grid points is sampled in ten runs of (**a**) GP4 and (**b**) GP5 (red crosses). Green open diamonds in (**b**) are the results in GP5 using the 16 lowest FN points in preFNS (bottom 1%) as the initial grid points. (**c**) Two most difficult grid points to sample among the low-FN grid points (No. 6: black spheres, No. 8: white spheres)

are sampled in ten runs of GP4 and GP5. Two grid points (Nos. 6 and 8) are relatively difficult to sample, degrading the sampling performance. This is probably because these points are isolated from the other low-FN points (Fig. 2.10c). Consequently, the FNS statistical model cannot predict that these two points are likely to be in the low-FN region.

To overcome this difficulty, information on the preliminary FNS is exploited not only as a descriptor in the FNS statistical model. Specifically, the initial grid points for the GP-based methods are not selected randomly, but in the ascending order of the preliminary FNs. The green open diamonds in Fig. 2.10b show the results by GP5 sampling using the 16 lowest FN points in the preFNS as the initial grid points. The two grid points (Nos. 6 and 8) are sampled at step 2 and step 86, respectively, resulting in 95 DFT computations to sample all low-FN points (See the green line in Fig. 2.8). Thus, fully exploiting information about the preliminary FNS can improve the sampling performance.

2.4.3 Practical Issues

Here two critical issues, which limit practicality, are discussed in the case of the low-PE region identification (task 1): (1) when to terminate sampling and (2) how to determine the PE threshold α.

The first issue is common in GP-based sampling methods. One practical advantage of statistical models such as the GP model is that the number of

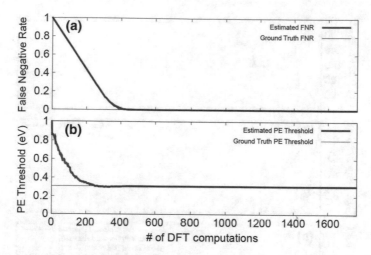

Fig. 2.11 Profiles of the estimated (**a**) FNRs and (**b**) PE thresholds for GP3 sampling [12]

remaining points to be sampled can be estimated by estimating the FNR. Figure 2.11 shows the profiles of the estimated FNR and threshold as functions of the number of DFT computations in GP3. These plots indicate that the estimated FNR almost coincides with the ground truth line. Additionally, the estimated threshold converges to the true value as the sampling proceeds. These results suggest that the estimated FNR should be a useful termination criterion.

Another practical issue is how to choose an appropriate α, which depends on the focused system. In the case of proton conduction in an oxide, the low-PE region should be defined such that a proton-conducting network exists throughout the crystal lattice within the region. According to the actual PEs, the low-PE regions are isolated when $\alpha < 0.15$, but they are abruptly connected when $\alpha = 0.20$. This means that a proper α value should be around 0.20 in the present study. If such an appropriate α value is initially unknown, the α value can be set in a stepwise manner. To demonstrate this approach, the performance of GP3 is investigated as α is increased from 0.05 to 0.20 in a stepwise manner (The results are shown in Fig. 2.12). In this scenario, α is increased by 0.05 when the estimated FNR becomes smaller than 10^{-6}.

Figure 2.12(b) indicates that the convergence of the estimated FNR is slightly slower than the ground truth FNR in the first step with $\alpha = 0.05$. This is why more than 250 DFT computations are required to ensure that all points in $P_{0.05}$ are successfully sampled. On the other hand, when $\alpha = 0.10$, 0.15, or 0.20, the convergences of the FNRs are almost as fast as the ground truth FNRs. It should be noted that the true positive points abruptly increase when the α value is switched, indicating that the positive points for higher α are sampled in earlier steps. Although this stepwise strategy is less efficient than directly specifying $\alpha = 0.20$, it is much more efficient than the prePES and random sampling methods.

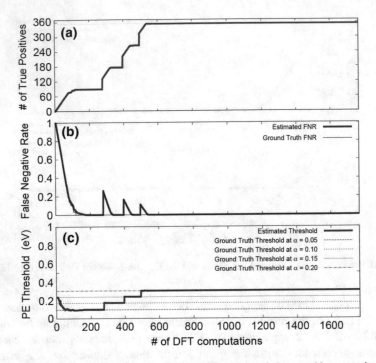

Fig. 2.12 (**a**) Efficiency of GP3(xyz + prePES) sampling when α is increased in a stepwise manner from 0.05 to 0.20 in 0.05 increments [12]. Number of grid points successfully sampled from the low-PE region (#TP) is plotted versus the number of DFT computations (#TP + #FP). (**b**), (**c**) Profiles of the estimated FNRs and PE thresholds versus the number of DFT computations [12]

2.5 Conclusions

In this chapter, a machine learning-based selective sampling procedure for PES evaluation is introduced and applied to proton conduction in $BaZrO_3$ to demonstrate its efficacy. The region of interest governing the ionic conduction is defined in the two ways: (1) a low-PE region and (2) a low-FN region.

For the low-PE region, the performance of the selective sampling based on the GP model greatly depends on the descriptors. Employing the preliminary PES (prePES) is significantly effective, which is evaluated by single-point DFT computations in a smaller supercell. The GP3(xyz + prePES) sampling requires 394 DFT computations to sample all the low-PE grid points (353 points) in a grid with 1768 points for the asymmetric unit of $BaZrO_3$ crystal. This is a 78% reduction in the computational costs. However, the defined region of interest, i.e., the low-PE region, comprises 20% of the configuration space. Consequently, the reducible computational cost is limited to 80%.

The region of interest should, therefore, be redefined as it becomes smaller in the configuration space. For the low-FN region, the region of interest contains only 15 grid points, whose volume is less than 1% of the configuration space. Among the several sampling methods to identify the low-FN region, GP5(xyz + preFNS) shows the best performance. It requires only 116 DFT computations to identify all grid points in the low-FN region. Furthermore, the computational cost can be further reduced to 95 DFT computations using the 16 lowest FN grid points in the preFNS as the initial points. This means that exploiting the information on the preFNS can reduce the computational cost by 95%.

Thus, preliminary information (i.e., prePES and preFNS) significantly contributes to the sampling performance. Therefore, a machine learning-based approach hybridized with a low-cost PES and/or FNS evaluation should be a solid methodology for preferential PES evaluation in the region of interest. In addition, using the FNR, which is defined in Eq. (2.15), solves two critical issues, which are when to terminate sampling and how to determine an appropriate α value (equivalent to the PE threshold).

Acknowledgements We recognize Mr. Daisuke Hirano and Mr. Makoto Otsubo for their contributions and Dr. Atsuto Seko for the insightful comments and suggestions. This work is financially supported by JSPS KAKENHI (Grant Nos. 25106002 and 26106513).

References

1. S. Glasstone, K.J. Laidler, H. Eyring, *The Theory of Rate Processes* (McGraw-Hill, New York, 1941)
2. G.H. Vineyard, J. Phys. Chem. Solids **3**, 121 (1957)|
3. K. Toyoura, Y. Koyama, A. Kuwabara, F. Oba, I. Tanaka, Phys. Rev. B **78**, 214303 (2008)
4. T. Vegge, Phys. Rev. B **70**, 035412 (2004)
5. A. Van der Ven, G. Ceder, Phys. Rev. Lett. **94**, 045901 (2005)
6. C.O. Hwang, J. Chem. Phys. **125**, 226101 (2006)
7. L.T. Kong, L.J. Lewis, Phys. Rev. B **74**, 073412 (2006)
8. K. Toyoura, Y. Koyama, A. Kuwabara, I. Tanaka, J. Phys. Chem. C **114**, 2375 (2010)
9. G. Henkelman, B.P. Uberuaga, H. Jonsson, J. Chem. Phys. **113**, 9978 (2000)
10. G. Henkelman, B.P. Uberuaga, H. Jonsson, J. Chem. Phys. **113**, 9901 (2000)
11. K. Toyoura, J. Terasaka, A. Nakamura, K. Matsunaga, J. Phys. Chem. C **121**, 1578 (2017)
12. K. Toyoura, D. Hirano, A. Seko, M. Shiga, A. Kuwabara, M. Karasuyama, K. Shitara, I. Takeuchi, Phys. Rev. B **93**, 054112 (2016)
13. C.E. Rasmussen, C.K.I. Williams, *Gaussian Processes for Machine Learning* (The MIT Press, Cambridge, 2006)
14. M.L. Stein, *Interpolation of Spatial Data: Some Theory for Kriging* (Springer Science & Business Media, New York, 2012)
15. H. Iwahara, T. Yajima, T. Hibino, K. Ozaki, H. Suzuki, Solid State Ionics **61**, 65 (1993)
16. W. Münch, K.-D. Kreuer, G. Seifert, J. Maier, Solid State Ionics **136–137**, 183 (2000)
17. M.S. Islam, J. Mater. Chem. **10**, 1027 (2000)
18. M.E. Björketun, P.G. Sundell, G. Wahnström, D. Engberg, Solid State Ionics **176**, 3035 (2005)
19. P.E. Blöchl, Phys. Rev. B **50**, 17953 (1994)
20. G. Kresse, J. Hafner, Phys. Rev. B **48**, 13115 (1993)
21. G. Kresse, J. Furthmüller, Comput. Mater. Sci. **6**, 15 (1996)
22. G. Kresse, D. Joubert, Phys. Rev. B **59**, 1758 (1999)

23. J.P. Perdew, K. Burke, M. Ernzerhof, Phys. Rev. Lett. **77**, 3865 (1996)
24. J. Mockus, J. Global Optim. **4**, 347 (1994)
25. E. Brochu, V.M. Cora, N. De Freitas (2010), arXiv:1012.2599

Chapter 3
Machine Learning Predictions of Factors Affecting the Activity of Heterogeneous Metal Catalysts

Ichigaku Takigawa, Ken-ichi Shimizu, Koji Tsuda
and Satoru Takakusagi

Abstract The ultimate goal in heterogeneous catalytic science is to accurately predict trends in catalytic activity based on the electronic and geometric structures of active metal surfaces. Such predictions would allow the rational design of materials having specific catalytic functions without extensive trial-and-error experiments. The d-band center values of metals are well known to be an important parameter affecting the catalytic activity of these materials, and activity trends in metal surface catalyzed reactions can be explained based on the linear Brønsted–Evans–Polanyi relationship and the Hammer–Nørskov d-band model. The present work demonstrates the possibility of employing state-of-the-art machine learning methods to predict the d-band centers of metals and bimetals while using negligible CPU time compared to the more common first-principles approach.

Keywords Heterogeneous catalysis · d-band center · Machine learning

I. Takigawa
Graduate School of Information Science and Technology, Hokkaido University,
Sapporo 060-0814, Japan
e-mail: takigawa@ist.hokudai.ac.jp

K. Shimizu · S. Takakusagi (✉)
Institute for Catalysis, Hokkaido University, Sapporo 001-0021, Japan
e-mail: takakusa@cat.hokudai.ac.jp

K. Shimizu
e-mail: kshimizu@cat.hokudai.ac.jp

K. Tsuda
Graduate School of Frontier Sciences, University of Tokyo, Kashiwa 277-8561, Japan
e-mail: tsuda@k.u-tokyo.ac.jp

© The Author(s) 2018
I. Tanaka (ed.), *Nanoinformatics*, https://doi.org/10.1007/978-981-10-7617-6_3

3.1   Introduction

Heterogeneous catalysis plays a key role in the industrial production of various chemicals. Over 80% of catalytic processes use heterogeneous catalysts to achieve high conversion and/or selectivity through lowering the activation barriers leading to the desired products [1, 2]. The majority of these materials consist of active transition metal or alloy nanoparticles dispersed on oxide supports, such as Al_2O_3, SiO_2, MgO, TiO_2, and CeO_2.

Heterogeneous catalysis is a surface phenomenon that involves a sequence of elementary steps, including adsorption, surface diffusion, chemical rearrangement of the adsorbed intermediates (the actual reaction), and desorption of the products, as shown in Fig. 3.1. Thus, detailed experimental and theoretical characterizations of the surface electronic/geometric structures during these steps are indispensable in order to understand the reaction mechanisms and to be able to enhance the activity and selectivity of the catalysts. However, even though surface characterization techniques have improved dramatically in recent years, new catalytic materials are still primarily developed through trial-and-error experiments. This is because catalytic reactions are actually more complicated than the process illustrated in Fig. 3.1 due to the complexity of catalyst surface structures and the effects of a large number of parameters (such as temperature, pressure, metal particle size/shape, and metal–support interactions). Unfortunately, the empirical development of catalytic materials is typically time-consuming and expensive with no guarantee of success.

For these reasons, the theory-based rational prediction of activity trends in catalysis is one of the ultimate goals in catalytic science. Such predictions would allow the design of surfaces with specific catalytic properties without extensive experimentation. To this end, it is important to elucidate the factors that control activity, also known as descriptors. To date, the bond energies derived from bulk oxide properties or the adsorption energies of reactants have been used as descriptors to predict the activity of metal/metal oxide surfaces [1]. Activity–descriptor plots typically exhibit a so-called volcano shape due to several effects. First, the strong

Fig. 3.1 Potential energy diagram of a heterogeneous catalytic reaction (A + B → P) with gaseous reactants (A, B), product (P), and solid metal catalyst

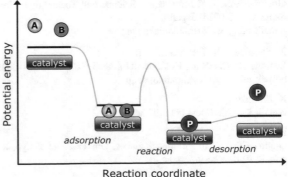

binding of an intermediate can result in surface poisoning, whereas weak binding leads to low coverage of the surface; in both cases, the catalytic rates are less than optimal. Consequently, moderate interactions produce the highest reactivity (representing Sabatier's principle). In addition, a linear relationship between activation energy and adsorbate–surface interaction energy, known as the Brønsted–Evans–Polanyi relation, has been demonstrated by several groups based on theoretical calculations [3–8]. The above effects allow a semiquantitative understanding of the activity trends in heterogeneous catalytic systems by simply considering the bond energies derived from bulk oxide properties and/or the adsorption energies of a reactant to a first approximation.

Recently, a simple but powerful approach based on machine learning (ML) techniques combined with density functional theory (DFT) calculations has attracted much attention as a novel tool for the rapid screening of metal catalyst reactivity. This method makes it possible to predict various catalyst properties typically calculated using DFT, such as reactant gas adsorption energies on various metal or alloy surfaces. This is done by constructing an appropriate regression model and using explanatory variables (often termed descriptors) that correlate with intrinsic properties of the constituent metals and/or reactant gases. Once the regression model is successfully constructed, it permits the rapid identification of the optimal catalyst for a target reaction by interpolation without calculating results for all the other candidates. Ras and Rothenberg et al. presented a simple and efficient model based on genetic algorithm variable selection and Partial Least Squares (PLS) regression for predicting the adsorption of molecules (heats of adsorption) on metal surfaces [9]. Their model used six descriptors for each metal (number of d-electrons, surface energy, first ionization potential, as well as atomic radius, volume, and mass) and three for each adsorptive species (HOMO–LUMO energy gap, molecular volume, and mass). This method was found to accurately predict the chemisorption of a range of adsorptive compounds (H_2, HO, N_2, CO, NO, O_2, H_2O, CO_2, NH_3, and CH_4) on a variety of metals (Fe, Co, Ni, Cu, Mo, Ru, Rh, Pd, Ag, W, Ir, Pt, and Au) as calculated using DFT or reported in the literature. This group also acquired experimental adsorption data for CO, CO_2, CH_4, H_2, N_2, and O_2 on Ni, Pt, and Rh supported on TiO_2, and confirmed that their model, using the same descriptors, generated results in good agreement with the data. Ma and Xin et al. systematically calculated CO adsorption energies on 250–300 {100}-terminated multimetallic alloy surfaces and presented an ML-augmented chemisorption model for CO_2 electroreduction catalyst screening [10, 11]. They demonstrated that artificial neural networks are able to reproduce the complex, nonlinear interactions of CO adsorbed on multimetallic alloy surfaces with an error of approximately 0.1 eV. The associated results identified multimetallic alloys that show promise with regard to improving the efficiency and selectivity during the electrochemical reduction of CO_2 to C_2 species. Okamoto developed a method based on a combination of DFT calculations and data mining to find the optimum composition for PtRu alloys to minimize the CO adsorption energy, since CO poisoning of the alloy catalysts tends to deactivate the catalytic function in proton exchange membrane fuel cells (PEMFCs) [12]. He first calculated the CO adsorption energies on 44 PtRu(111) bimetallic slabs having various compositions and subsequently employed multiple regression analysis for the data mining. This work determined that the resulting model accurately predicted CO adsorption energies on PtRu surfaces. This regression model also identified the optimum composition associated with a minimum CO adsorption energy, which was later confirmed by DFT calculations using the same alloy composition.

The above examples demonstrate that ML techniques can effectively predict the interaction energy between a specific adsorbate and a given metal surface for a particular reaction, and can sometimes assist in finding the optimal catalytic material. However, the interaction energy may not always be used as a universal descriptor for predicting activity trends in different catalytic reactions by various transition metal catalysts. For this reason, the present work focused on the so-called d-band center, which is one of the most important activity-controlling factors and can be used to explain activity trends in various types of catalytic reactions. In this study, we employed state-of-the-art ML techniques to predict the DFT-calculated d-band centers for metals and bimetals.

3.2 The d-Band Center: A Widely Accepted Indicator Explaining Activity Trends in Metal Catalysts

Nørskov et al. performed a series of systematic DFT calculations and proposed the semiempirical concept of the d-band model [3–5]. The model assumes that the d-electrons of transition metals play the most important role in chemisorption. This approach involves linear scaling between the energy of the d-band center (ε_d) relative to the Fermi level (E_F) and the adsorption energy for a given adsorbate. The higher the d-states are in energy relative to the Fermi level, the emptier the anti-bonding states and the larger the adsorption energy of an adsorbed species on a surface. A calorimetric study by Lu et al. [13] subsequently provided experimental evidence to support the d-band model. This work showed moderate linear correlations between the experimental heats of adsorption of CO, H_2, O_2, and C_2H_4 on various metal surfaces and the positions of the d-band centers as calculated by Hammer and Nørskov [3]. The d-band model also predicts that adsorbate binding energies should correlate with one another [5]. Since the transition-state structures on different metals tend to be rather similar, the activation energy for an elementary reaction should exhibit a linear relationship with the energy change for the elementary reaction. Thus, the kinetic parameter for a catalytic reaction involving a metal can be written as $\varepsilon_d - E_F$, equivalent to the position of the d-band center relative to E_F. Recent experimental studies have demonstrated the validity of the d-band model when describing trends in catalytic activity [14–18]. As an example, Furukawa et al. found a relationship between the d-band centers of Ni and Ni_3M (M = Ge, Nb, Sn, Ta, or Ti) intermetallics and their activation energies with regard to the H_2–D_2 equilibration [16].

To confirm whether the activity trends in multistep catalytic reactions can be understood in terms of the d-band model in combination with linear energy relations, Tamura et al. studied correlations between the reaction rates of dehydrogenation and hydrogenation reactions and the associated $\varepsilon_d - E_F$ values [17]. The activities per surface metal atom, or turnover frequency (TOF), for various metal-loaded SiO_2 samples with similar particle size ranges (8.9–11.7 nm) were plotted against the d-band center values (Fig. 3.2). In the cases of the dehydrogenation of 2-propanol adspecies (2-PrOH$_{ad}$) on the surface (Fig. 3.2a), the hydrogenation of PhNO$_{2ad}$ (Fig. 3.2b), the OH/OD exchange of surface SiOH groups under D_2 (Fig. 3.2c), and the liquid phase hydrogenation of PhNO$_2$ by M/SiO$_2$ (Fig. 3.2d), the activities generally show volcano-type variations with the d-band center values, except for the Pd catalyst in Fig. 3.2a. A common trend is

Fig. 3.2 The activities of metal catalysts for various reactions versus d-band center values

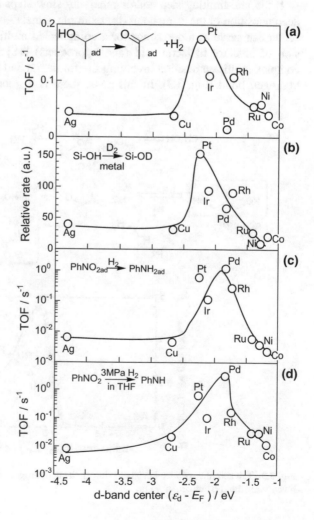

evident in which, as the d-band center moves further from E_F, the metal–hydrogen (M–H) and metal–oxygen (M–O) bond energies become weaker [4]. Each of the reactions in Fig. 3.2 includes the formation and dissociation of M–H bonds as a common elementary step. Dehydrogenation and hydrogenation reactions include the formation and dissociation of M–O bonds. Hence, these results suggest that moderate M–H and/or M–O bond strengths favor the reactions. The observation of similar volcano-type trends for different reactions demonstrates that the d-band center can serve as a general activity–descriptor for the catalytic systems shown in Fig. 3.2a. This outcome can possibly be explained by considering that the strong binding of surface intermediates via M–H and M–O bonds leads to surface poisoning, whereas weak binding limits the availability of the intermediates. In both cases, the catalytic rates are less than optimal. Consequently, Pt-group metal catalysts with moderate bond strengths give the highest activities.

If the rate-limiting step and/or relatively slow steps involve the formation and decomposition of the same bond (for example, a metal–hydrogen bond), the d-band center can serve as a descriptor for a complicated multistep reaction. The conversion of glycerol to lactic acid (and byproducts) [18] is a typical example of a complex multistep reaction involving the formation and decomposition of a metal–hydrogen bond (Fig. 3.3). In this case, there is a good volcano-type correlation

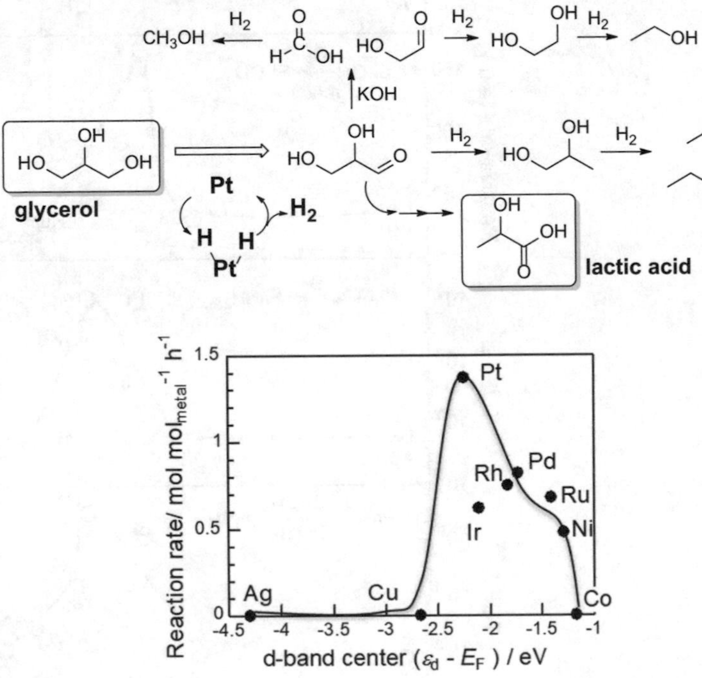

Fig. 3.3 The activities of metal catalysts during the dehydrogenation of glycerol to lactic acid as a function of the d-band center value

between the d-band center value and the catalytic activity. Pt, having an intermediate d-band center level, shows higher catalytic activity than the other metals, because the interaction between surface intermediates and the metal surface is moderately strong, which tends to favor metal-catalyzed dehydrogenation.

The above examples demonstrate that the d-band model (in combination with linear energy relations) can be used to understand the activity trends in transition metal-catalyzed multistep reactions. We can conclude that this model is an important concept with regard to assessing or predicting reactivity trends in the heterogeneous catalysis of transition metals during multistep organic reactions. Thus, as a first approximation, the reactivity of a metal catalyst for an organic reaction can be described by a single parameter: the d-band center.

3.3   Prediction of the d-Band Center Values for Mono- and Bimetallic Systems by Machine Learning

3.3.1   Data-Driven Prediction of d-Band Center Values by Machine Learning Methods

Herein, we present our most recent results regarding the ML-based predictions of d-band centers for metallic and bimetallic compounds [19]. Using DFT calculations, Nørskov's group [3, 20] determined the d-band centers for 11 different metals (Fe, Co, Ni, Cu, Ru, Rh, Pd, Ag, Ir, Pt, and Au) and for the associated 110 bimetallic pairs having two different structures (surface impurities and overlayers) on clean metal surfaces) [21]. In this case, the d-band centers were independently calculated using first principles for each metal or bimetal under typical conditions. In contrast, our own study involved a quantitative investigation of a fully data-driven approach based on ML that infers the d-band center of a metal or a bimetal from those of other metals and bimetals. As an example, it would be of significant interest to know whether or not the d-band center of the Cu–Co pair can be somehow inferred from those of Cu, Au, Cu–Fe, Ni–Ru, Pd–Co, and Rh–Pd from the materials informatics perspective. Our result shows sufficient predictability of d-band centers by ML methods using a small set of readily available properties of metals as descriptors. Given the rapid increase in data in recent years, this outcome would suggest that ML methods may possibly substitute for or complement first-principles calculations.

3.3.2   Datasets and Descriptors

To assess the accuracy of ML predictions, we employed $\varepsilon_d - E_F$ data for 11 metals (Fe, Co, Ni, Cu, Ru, Rh, Pd, Ag, Ir, Pt, and Au) and all the associated pairwise

Table 3.1 The "impurities" dataset: DFT-calculated d-band centers (eV) of metals (bold) and 1% guest metals (M_g) doped into the surfaces of host metals (M_h) as reported by Nørskov's group [3, 20]. Reproduced from Ref. [19] with permission from the Royal Society of Chemistry

M_g / M_h	Fe	Co	Ni	Cu	Ru	Rh	Pd	Ag	Ir	Pt	Au
Fe	**-0.92**	-0.87	-1.12	-1.05	-1.21	-1.46	-2.16	-1.75	-1.28	-2.01	-2.34
Co	-1.16	**-1.17**	-1.45	-1.33	-1.41	-1.75	-2.54	-2.08	-1.53	-2.36	-2.73
Ni	-1.20	-1.10	**-1.29**	-1.10	-1.43	-1.60	-2.26	-1.82	-1.43	-2.09	-2.42
Cu	-2.11	-2.07	-2.40	**-2.67**	-2.09	-2.35	-3.31	-3.37	-2.09	-3.00	-3.76
Ru	-1.20	-1.15	-1.40	-1.29	**-1.41**	-1.58	-2.23	-1.68	-1.39	-2.03	-2.25
Rh	-1.49	-1.39	-1.57	-1.29	-1.69	**-1.73**	-2.27	-1.66	-1.56	-2.08	-2.22
Pd	-1.46	-1.29	-1.33	-0.89	-1.59	-1.47	**-1.83**	-1.24	-1.30	-1.64	-1.66
Ag	-3.58	-3.46	-3.63	-3.83	-3.46	-3.44	-4.16	**-4.30**	-3.16	-3.80	-4.45
Ir	-1.90	-1.84	-2.06	-1.90	-2.02	-2.26	-2.84	-2.24	**-2.11**	-2.67	-2.85
Pt	-1.92	-1.77	-1.85	-1.53	-2.11	-2.02	-2.42	-1.81	-1.87	**-2.25**	-2.30
Au	-2.93	-2.79	-2.93	-3.01	-2.86	-2.81	-3.39	-3.35	-2.58	-3.10	**-3.56**

bimetallic alloys (110 pairs of a host metal, M_h, and a guest metal, M_g). These values were obtained from a DFT study by Refs. [3, 20] for two different structures: those having surface impurities (Table 3.1) and those with overlayers (Table 3.2). In the original datasets, the d-band centers for bimetals are given as shifts relative to the clean metal values and so have been converted to values relative to the Fermi level. In Table 3.1, the surfaces considered are the most closely packed, and 1% guest metals are doped into the topmost surfaces of the host metals. In Table 3.2, the overlayer structures are pseudomorphic and guest metal monolayers are formed on the surface of the host metals. The histograms of d-band centers for each

Table 3.2 The "overlayers" dataset: DFT-calculated d-band centers (eV) of metals (bold) and the guest metal (M_g) monolayers on the surfaces of host metals (M_h) as reported by Nørskov's group [3, 20]. Reproduced from Ref. [19] with permission from the Royal Society of Chemistry

M_g / M_h	Fe	Co	Ni	Cu	Ru	Rh	Pd	Ag	Ir	Pt	Au
Fe	**-0.92**	-0.78	-0.96	-0.97	-1.65	-1.64	-2.24	-2.17	-1.87	-2.40	-3.11
Co	-1.18	**-1.17**	-1.37	-1.23	-1.87	-2.12	-2.82	-2.53	-2.26	-3.06	-3.56
Ni	-0.33	-1.18	**-1.29**	-1.17	-1.92	-2.03	-2.61	-2.43	-2.15	-2.82	-3.39
Cu	-2.42	-2.29	-2.49	**-2.67**	-2.89	-2.94	-3.71	-3.88	-2.99	-3.82	-4.63
Ru	-1.11	-1.04	-1.12	-1.11	**-1.41**	-1.53	-1.88	-1.81	-1.54	-2.02	-2.27
Rh	-1.42	-1.32	-1.39	-1.51	-1.70	**-1.73**	-2.12	-1.81	-1.70	-2.18	-2.30
Pd	-1.47	-1.29	-1.29	-1.03	-1.94	-1.58	**-1.83**	-1.68	-1.52	-1.79	-1.97
Ag	-3.75	-3.56	-3.62	-3.68	-3.80	-3.63	-4.03	**-4.30**	-3.50	-3.93	-4.51
Ir	-1.78	-1.71	-1.78	-1.55	-2.12	-2.14	-2.53	-2.20	**-2.11**	-2.60	-2.70
Pt	-1.90	-1.72	-1.71	-1.47	-2.13	-2.01	-2.23	-2.06	-1.96	**-2.25**	-2.33
Au	-3.03	-2.82	-2.85	-2.86	-3.09	-2.89	-3.21	-3.44	-2.77	-3.13	**-3.56**

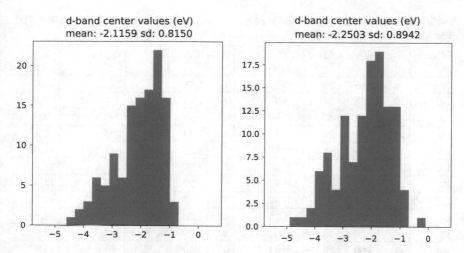

Fig. 3.4 Histogram of d-band centers for the "impurities" (left) and "overlayers" (right) datasets

structure are provided in Fig. 3.4. Although the two structures are physically very different, the Pearson's correlation coefficient between Tables 3.1 and 3.2 is 0.948 ($p < 0.001$) and thus the d-band centers exhibit significant correlation. Therefore, in order to differentiate these structure-specific values, any data-driven prediction requires a highly adaptive mechanism that can capture the subtle differences.

Regarding the choice of descriptors for metals, we pretested several candidates and chose nine physical properties (Table 3.3) that are readily available from the periodic table and a standard reference source [22]. From a practical point of view, it is important to choose readily accessible but characteristic values as descriptors in order to effectively bypass time-consuming DFT calculations while maintaining sufficient prediction accuracy. Each metal can thus be represented as a nine-dimensional vector of the descriptor values. Accordingly, an 18-dimensional concatenated vector of M_h and M_g values was used for predictions of the d-band centers of bimetals. In the case of monometallic surfaces, we employed an 18-dimensional vector by concatenating two vectors for the same metal. We also searched for smaller subsets of descriptors yielding simpler models among the 18 descriptors by assessing the relevance or redundancy of each descriptor. Table 3.4 shows the correlation matrix between descriptors and demonstrates highly correlated descriptor variables. This result prompted us to investigate variable selection with the aim of identifying a smaller nonredundant subset of the 18 descriptors. Table 3.5 indicates the correlation coefficients between each descriptor and the d-band center values. It can be seen that no single descriptor exhibits direct correlation with the d-band centers.

Table 3.3 Input features (descriptors) used for the prediction of d-band centers from Ref. [22]. Reproduced from Ref. [19] with permission from the Royal Society of Chemistry

Metal	G	R/Å	AN	AM/ g mol^{-1}	P	EN	IE/ eV	$\Delta_{fus}H$/ J g^{-1}	ρ/ g cm^{-3}
Fe	8	2.66	26	55.85	4	1.83	7.90	247.3	7.87
Co	9	2.62	27	58.93	4	1.88	7.88	272.5	8.86
Ni	10	2.60	28	58.69	4	1.91	7.64	290.3	8.90
Cu	11	2.67	29	63.55	4	1.90	7.73	203.5	8.96
Ru	8	2.79	44	101.07	5	2.20	7.36	381.8	12.10
Rh	9	2.81	45	102.91	5	2.28	7.46	258.4	12.40
Pd	10	2.87	46	106.42	5	2.20	8.34	157.3	12.00
Ag	11	3.01	47	107.87	5	1.93	7.58	104.6	10.50
Ir	9	2.84	77	192.22	6	2.20	8.97	213.9	22.50
Pt	10	2.90	78	195.08	6	2.20	8.96	113.6	21.50
Au	11	3.00	79	196.97	6	2.40	9.23	64.6	19.30

Group (G)
Bulk Wigner–Seitz radius (R) in Å
Atomic number (AN)
Atomic mass (AM) in g mol^{-1}
Period (P)
Electronegativity (EN)
Ionization energy (IE) in eV
Enthalpy of fusion ($\Delta_{fus}H$) in J g^{-1}
Density at 25 °C (ρ) in g cm^{-3}

Table 3.4 Correlation matrix of the nine descriptors for the 11 metals in Table 3.3. Reproduced from Ref. [19] with permission from the Royal Society of Chemistry

	G	R/Å	AN	AM	P	EN	IE	$\Delta_{fus}H$	ρ
G	1.00								
R/Å	0.43	1.00							
AN	0.24	0.77	1.00						
AM	0.24	0.75	1.00	1.00					
P	0.18	0.82	0.98	0.98	1.00				
EN	0.07	0.66	0.78	0.77	0.84	1.00			
IE	0.28	0.48	0.82	0.84	0.74	0.53	1.00		
$\Delta_{fus}H$	−0.75	−0.72	−0.58	−0.58	−0.53	−0.28	−0.66	1.00	
ρ	0.13	0.61	0.97	0.98	0.94	0.73	0.84	−0.46	1.00

3.3.3 Monte Carlo Cross-Validation for Assessing the Prediction Accuracies of ML Models

Our primary intent was to assess the data-driven prediction of the d-band center of a given metal (or bimetal) from the d-band centers of other metals and bimetals. To

Table 3.5 Correlation coefficients between each of the 18 descriptors and the d-band centers. Reproduced from Ref. [19] with permission from the Royal Society of Chemistry

(For host metal)

	G	R/Å	AN	AM	P	EN	IE	$\Delta_{fus}H$	ρ
Impurities	−0.63	−0.53	−0.29	−0.29	−0.26	−0.02	−0.15	0.56	−0.17
Overlayers	−0.63	−0.34	−0.11	−0.11	−0.06	0.13	−0.06	0.49	0.00

(For guest metal)

	G	R/Å	AN	AM	P	EN	IE	$\Delta_{fus}H$	ρ
Impurities	−0.24	−0.35	−0.26	−0.26	−0.27	−0.28	−0.24	0.33	−0.20
Overlayers	−0.22	−0.47	−0.41	−0.40	−0.42	−0.39	−0.31	0.36	−0.33

do so, we first randomly separated 121 targets (11 metals and 110 bimetals) into two disjoint sets: a "test set" of size n and a "training set" of size 121-n. The subsequent challenge was to evaluate the accuracy with which the d-band centers of the test set could be predicted using those of the training set. As a first step, an ML model was constructed using the training set. Following this, the model was employed to predict the d-band centers of the test set and the root-mean-square errors (RMSEs) between the predicted and true values (the ground truth) were calculated for the purposes of predictability evaluation. A single-shot random trial of this procedure could provide estimates of RMSE values, whereas those estimates vary depending on the split between the training and test sets (with a certain level of variance). For quantitative evaluations, we reduced this estimation variance by repeating the single-shot trials over 100 random test/training splits (that is, for 100 random leave-n-out trials) and used the mean of 100 RMSE estimates to assess the prediction accuracy of the ML model. The test set in each trial was never used to build the corresponding ML model in that trial, and hence simulated yet unseen targets to be predicted. Another benefit of this approach was that it was also possible to control the size, n, of the test set, and so to determine the training set size required for accurate predictions. It should be noted that this general approach is well established in statistics and is referred to as Monte Carlo cross-validation [23] or as leave-n-out [24], random permutation cross-validation (shuffle and split) [25], or random subsampling cross-validation [26]. This method was determined to be a better match for our scenario than more typical choices such as k-fold cross-validation or bootstrapping.

3.3.4   Machine Learning Methods and Hyperparameter Selection

We initially selected 11 ML regression models: five linear models (linear regression (OLS), PLS regression (PLS), L1-penalized linear regression (LASSO), L2-penalized linear regression (RIDGE), and robust linear regression (RANSAC))

Table 3.6 List of the ML regression methods and tuning parameters (the hyperparameters to be tuned)

Abbreviation	Tuning parameters [tested range]
OLS, RANSAC	(No tuning parameters) (min_samples = 50 for RANSAC)
PLS	n_components $\in$ [1, 2, ..., # of vars]
LASSO, RIDGE	alphas $\in$ [1.0, 10^{-1}, 10^{-2}, 10^{-3}, 10^{-4}, 10^{-5}]
GPR	theta0 $\in$ [1.0, 10^{-1}, 10^{-2}, 10^{-3}, 10^{-4}, 10^{-5}]
KRR	alpha, gamma $\in$ [1.0, 10^{-1}, 10^{-2}, 10^{-3}, 10^{-4}, 10^{-5}]
SVR	C $\in$ [1.0, 10, 10^{2}, 10^{3}, 10^{4}] gamma $\in$ [1.0, 10^{-1}, 10^{-2}, 10^{-3}, 10^{-4}, 10^{-5}]
RFR, ET	max_depth $\in$ [4, 6, 8, 10] n_estimators $\in$ [100, 250, 300]
GBR	learning_rate $\in$ [1.0, 10^{-1}, 10^{-2}, 10^{-3}, 10^{-4}, 10^{-5}] max_depth $\in$ [4, 6, 8, 10] n_estimators $\in$ [100, 250, 300]

and six nonlinear models (Gaussian-process regression (GPR), kernel ridge regression (KRR), support vector regression (SVR), random forest regression (RFR), extra-trees regression (ET), and gradient boosting regression (GBR)). Each of these approaches is based on a popular, easy-to-use, off-the-shelf ML package: scikit-learn (http://scikit-learn.org) [24]. The models selected herein are those most commonly used in the ML field, and details concerning the individual methods can be found in standard ML references [27, 28].

In practice, some models include tuning parameters called *hyperparameters* in addition to the target parameters to be estimated from the training set, and these hyperparameters must be determined prior to training. In such cases, the appropriate setting of these parameters is the key to successful predictions. These hyperparameters were determined by assessing a reasonable range of candidate values in an exhaustive manner (via a grid search), as shown in Table 3.6, and choosing the best parameters by threefold cross-validation with the training set. It should be noted that this selection process was performed for each training/test split independently, and the test data in each split were never used to select a hyperparameter.

3.3.5   Screening and Evaluation of Predictive ML Methods

We evaluated the prediction performance of 11 ML models using Monte Carlo cross-validation with 100 random leave-25%-out splits, in conjunction with internal threefold cross-validation with the training set to ensure selection of the optimum model (Table 3.6). That is, the following random trials were each performed 100 times. Assuming that 25% of Tables 3.1 or 3.2 has not yet been obtained, the ML method statistically infers those values using the other available 75% of the values. Following this, the RMSE of the difference between the predicted values and the ground truth is calculated for each trial, and these RMSEs are then averaged to

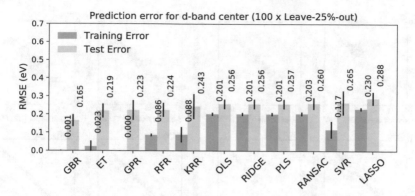

Fig. 3.5 Prediction performances of the 11 ML models for the "impurities" dataset

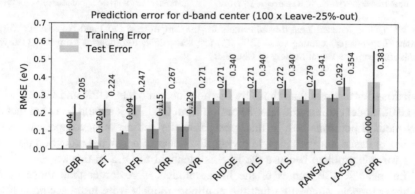

Fig. 3.6 Prediction performances of the 11 ML models for the "overlayers" dataset

obtain the mean RMSE and its standard deviation. The results of d-band center prediction evaluations are shown in Figs. 3.5 and 3.6 for the surface impurity and surface overlayer trials, respectively. Among the various methods examined, the GBR approach exhibited the best prediction performance. This was not unexpected, since GBR [29] is widely used and has performed well in top-level data prediction contests such as the Kaggle competition in recent years [30, 31]. Technically, it is an ensemble model composed of boosted regression trees, which often give accurate and stable predictions.

Figure 3.7 illustrates the predictive performances of four typical ML methods (OLS, PLS, GPR, and GBR) in a single-shot cross-validation with a 75% training set (●) and a 25% test set (○). These trials used all 18 descriptors: nine for the host and nine for the guest metal. The x-axis in these plots represents the DFT-calculated d-band center values (the ground truth values), while the y-axis gives the predictions from the ML methods. Deviations from the x = y line indicate prediction errors. Clearly, the predictions by linear models (OLS and PLS) exhibit larger

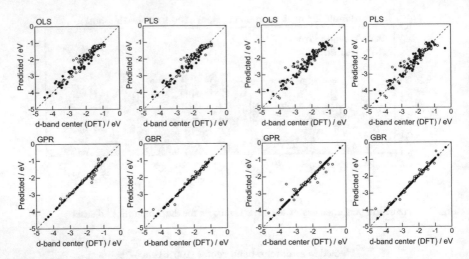

Fig. 3.7 DFT-calculated local d-band centers for the "impurities" (left) and "overlayers" (right) datasets. Legend: (●) training set = 75%, (○) test set = 25%. Reproduced from Ref. [19] with permission from the Royal Society of Chemistry

deviations for test sets than those by the nonlinear models (GPR and GBR), while the GBR model exhibits the least deviation from the line. It should be noted that the PLS method performs best at the hyperparameter setting of n_components = the number of descriptor variables, implying that linear dimensional reduction does not work for this problem because the PLS is identical to OLS in this setting.

The mean RMSE values of the linear models were larger than those of the nonlinear models, suggesting that the nonlinear models were more accurate. From these results, we concluded that GBR would be the best choice for the prediction of the d-band centers. It is known that the GBR model is more flexible than the linear regression models and exhibits greater stability than the GPR model, which is more sensitive to the hyperparameter settings. It should be noted that the linear models show lower standard deviations than the nonlinear models, which are more flexible but also more dependent on hyperparameter settings. Thus, the GPR approach worked well in the case of the "impurities" dataset but gave poor results with the "overlayers" values (due to overfitting only of the given training set), even though GPR achieved zero training errors in both cases. These results suggest that the incorrect choice of hyperparameters for GPR can significantly affect the performance of this method, while controlling the method simply by cross-validation could be difficult in real-world scenarios. Conversely, we observed that the tree-ensemble-based methods such as GBR, ET, and RFR were not greatly affected by the hyperparameter choices, so it is relatively simple to understand what is controlled by each hyperparameter. Therefore, the use of highly adaptive methods such as GPR, SVR, and KRR requires careful tuning and could be difficult in practice given that the d-band centers of the "impurities" and "overlayers" datasets are highly correlated (with a correlation coefficient of 0.948), as discussed in

Sect. 3.3.2. The ensemble approaches used in the GBR, ET, and RFR methods could mitigate possible large deviances in predictions by stabilizing the prediction variances.

3.3.6   The Importance of Descriptors to GBR Predictions

We subsequently investigated the relevance or redundancy of each of the 18 descriptors for the host and guest metals in Table 3.3 that had been used in the GBR model. GBR is based on an ML ensemble technique referred to as "boosting". This technique adaptively combines large numbers of relatively simple regression-tree models that recursively partition the data using a single selected descriptor. Thus, it provides a feature importance score for each descriptor: a weighted average of the number of times (or the extent of contribution to the entire prediction) the descriptor is selected for partitioning. This score can be used to assess the relative importance

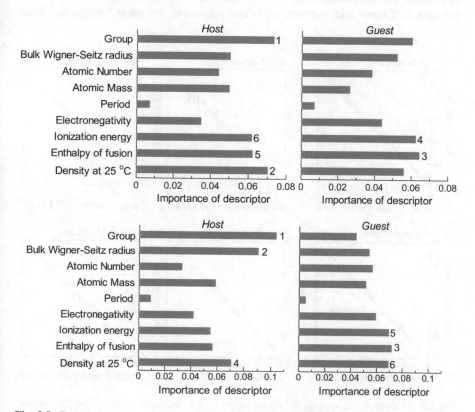

Fig. 3.8 Feature importance scores of the descriptors for the GBR prediction of the d-band centers using the "impurities" (upper) and "overlayers" (lower) datasets. Reproduced from Ref. [19] with permission from the Royal Society of Chemistry

of that descriptor with respect to the predictability of the d-band center values. Note that these values are only employed with GBR, and the statistical importance of descriptors varies with the ML method used.

Figure 3.8 shows the feature importance scores of all 18 descriptors with regard to predicting the "impurities" and "overlayers" datasets. The six most important descriptors are highlighted, with a rank next to their bars. In the case of the "impurities," these were (1) the group in the periodic table in which the host metal is found, (2) the density of the host metal at 25 °C, (3) the guest metal enthalpy of fusion, (4) the guest metal ionization energy, (5) the host metal enthalpy of fusion, and (6) the host metal ionization energy. In contrast, the most important factors for the "overlayers" dataset were (1) the host metal group in the periodic table, (2) the bulk Wigner–Seitz radius of the host metal, (3) the guest metal enthalpy of fusion, (4) the host metal density at 25 °C, (5) the guest metal ionization energy, and (6) the guest metal density at 25 °C.

To evaluate the effect of the number of descriptors on the predictive performance of GBR, the prediction results with 18 (that is, all), the top six and the top four descriptors are compared in Fig. 3.9. For quantitative evaluations, we also repeated the tests 100 times with random splits and calculated the mean RMSEs for these

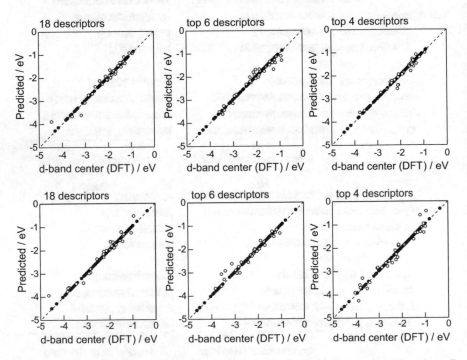

Fig. 3.9 DFT-calculated local d-band center values for the "impurities" (upper) and "overlayers" (lower) datasets correlated with the values predicted by GBR with 18 (all), the top six and the top four descriptors. Legend: (●) training set = 75%, (○) test set = 25%. Reproduced from Ref. [19] with permission from the Royal Society of Chemistry

predictions. The resultant values for the "impurities" dataset were 0.17 ± 0.04, 0.18 ± 0.04, and, 0.16 ± 0.04 eV for 18, 6 and, 4 descriptors, respectively. For the "overlayers" dataset, the respective values were 0.19 ± 0.04, 0.19 ± 0.04, and 0.23 ± 0.05 eV. These data demonstrate that the ML prediction performance remained moderately good even when employing only four descriptors. Furthermore, it was found that the prediction accuracy obtained with six descriptors was superior to that with four descriptors when the test set proportion was increased above 25%. Based on these results, we used the GBR model with the top six descriptors for the subsequent analysis.

3.3.7 Model Estimations Using Different Test/Training Splits

Finally, we attempted to determine the size of training set required for ML to achieve sufficient prediction performance. Figure 3.10 shows the predictive performances using GBR with the top six descriptors when employing various test/training set ratios (25%/75%, 50%/50%, and 75%/25%). That is, we withheld 25%,

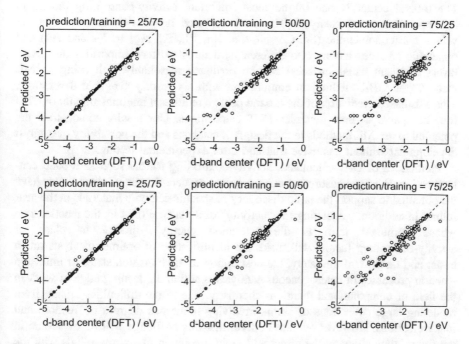

Fig. 3.10 DFT-calculated local d-band center values for the "impurities" dataset (upper) and the "overlayers" dataset (lower) and the values predicted by GBR with the top six descriptors. Legend: (●) training set = 75%, (○) test set = 25%. Reproduced from Ref. [19] with permission from the Royal Society of Chemistry

50%, or 75% of the "impurities" and "overlayers" datasets as the test sets, and predicted these using GBR based on the remaining values. In Table 3.1 ("impurities") and Table 3.2 ("overlayers"), we have 121 values in total, and 25%/75% corresponds to sets with size 30/91, 50%/50% to 61/60, and 25%/75% to 90/31. For quantitative evaluation, we also calculated the mean RMSEs for the 100 random splits for each setting. The resultant values for the "impurities" dataset were 0.18 ± 0.04, 0.23 ± 0.05, and 0.38 ± 0.07 eV for the 25%/75%, 50%/50%, and 75%/25% tests. In the case of the "overlayers" dataset, these values were 0.19 ± 0.04, 0.27 ± 0.05, and 0.41 ± 0.08 eV. These results quantitatively exhibit a general trend of ML such that a greater quantity of data generates better results and also show that d-band center values can be predicted with a moderate level of accuracy (RMSE = 0.38 ± 0.07 eV for the "impurities" set, RMSE = 0.41 0.08 eV for "overlayers"), even when only 25% of the data are available and 75% are missing. This result provides a useful guideline for the trade-off between the predictive performance and data availability.

3.4 Conclusion and Future Prospects

The d-band center is one of the most important activity-controlling factors in heterogeneous metal catalysts. The work reported herein demonstrates that the values for monometallic (Fe, Co, Ni, Cu, Ru, Rh, Pd, Ag, Ir, Pt, and Au) and bimetallic surfaces having two different structures (surface impurities and overlayers on clean metal surfaces) can be predicted reasonably well using an ML method (the GBR method) in conjunction with six readily available descriptors. This ML-based prediction of the d-band centers requires a minimal amount of CPU time compared to first-principles DFT calculations. Our results demonstrate the potential to use ML methods in the design of catalysts and the possibility of catalyst development without extensive trial-and-error experimental testing.

Predictions of DFT-calculated, activity-controlling factors such as d-band centers, and reactant gas adsorption energy values by data-driven ML techniques have the potential to support the rapid discovery of specific catalytic materials in the near future. In addition, calculating the "activity" of a material (that is, the reaction rate whose dominant term is the activation energy) directly in place of activity-controlling factors will make the identification of optimal catalysts much easier and faster. Unfortunately, the description of the transition states of multistep reaction processes in heterogeneous catalysis is still at the leading edge of work in the field of computational quantum chemistry, due to the difficulties arising from modeling large collections of atoms involving numerous degrees of freedom and many electrons. However, we are hopeful that this challenge will be overcome in the future, thus allowing the direct and rapid prediction of activity trends with the aid of ML techniques.

Acknowledgements This work was supported by a Grant-in-Aid for Scientific Research on Innovative Areas "Nano Informatics" (Grant No. 25106010) from the Japan Society for the Promotion of Science (JSPS).

References

1. G. Ertl, H. Knözinger, F. Schüth, J. Weitkamp, *Handbook of Heterogeneous Catalysis*, vol. 1, 2nd edn. (Wiley–VCH, Weinheim, 2008)
2. G.A. Somorjai, Y. Li, *Introduction to Surface Chemistry and Catalysis*, 2nd edn. (Wiley, 2010)
3. B. Hammer, J.K. Nørskov, Adv. Catal. **45**, 71 (2000)
4. V. Pallassana, M. Neurock, L.B. Hansen, B. Hammer, J.K. Nørskov, Phys. Rev. B **60**, 6146 (1999)
5. T. Bligaard, J.K. Nørskov, S. Dahl, J. Matthiesen, C.H. Christensen, J. Sehested, J. Catal. **224**, 206 (2004)
6. J.K. Nørskov, T. Bligaard, J. Rossmeisl, C.H. Christensen, Nat. Chem. **1**, 37 (2009)
7. Z.W. Seh, J. Kibsgaard, C.F. Dickens, I. Chorkendorff, J.K. Nørskov, T.F. Jaramillo, Science **355**, 146 (2017)
8. F. Calle-Vallejo, D. Loffreda, M.T.M. Koper, P. Sautet, Nat. Chem. **7**, 403 (2015)
9. E.-J. Ras, M.J. Louwerse, M.C. Mittelmeijer-Hazelegerb, G. Rothenberg, Phys. Chem. Chem. Phys. **15**, 4436 (2013)
10. X. Ma, Z. Li, L.E.K. Achenie, H. Xin, J. Phys. Chem. C **6**, 3528 (2015)
11. Z. Li, X. Ma, H. Xin, Catal. Today **280**, 232 (2017)
12. Y. Okamoto, Chem. Phys. Lett. **395**, 279 (2004)
13. C. Lu, I.C. Lee, R.I. Masel, A. Wieckowski, C. Rice, J. Phys. Chem. A **106**, 3084 (2002)
14. E. Toyoda, R. Jinnouchi, T. Hatanaka, Y. Morimoto, K. Mitsuhara, A. Visikovskiy, Y. Kido, J. Phys. Chem. C **115**, 21236 (2011)
15. T. Anniyev, S. Kaya, S. Rajasekaran, H. Ogasawara, D. Nordlund, A. Nilsson, Angew. Chem. Int. Ed. **51**, 7724 (2012)
16. S. Furukawa, K. Ehara, K. Ozawa, T. Komatsu, Phys. Chem. Chem. Phys. **16**, 19828 (2014)
17. M. Tamura, K. Kon, A. Satsuma, K.I. Shimizu, ACS Catal. **2**, 1904 (2012)
18. S.M.A.H. Siddiki, A.S. Touchy, K. Kon, T. Toyao, K. Shimizu, ChemCatChem **2017**(9), 2816 (2017)
19. I. Takigawa, K. Shimizu, K. Tsuda, S. Takakusagi, RSC Adv. **6**, 52587 (2016)
20. A. Ruban, B. Hammer, P. Stoltze, H.L. Skriver, J.K. Nørskov, J. Mol. Catal. A **115**, 421 (1997)
21. A. Vojvodic, J.K. Nørskov, F. Abild-Pedersen, Top. Catal. **57**, 25 (2014)
22. *CRC Handbook of Chemistry and Physics*, 83rd edn., edited by D.R. Lide (CRC Press, London, 2002)
23. R.R. Picard, R.D. Cook, J. Am. Stat. Assoc. **79**(387), 575 (1984)
24. J. Shao, J. Am. Stat. Assoc. **88**(422), 486 (1993)
25. F. Pedregosa, G. Varoquaux, A. Gramfort, V. Michel, B. Thirion, O. Grisel, M. Blondel, P. Prettenhofer, R. Weiss, V. Dubourg, J. Vanderplas, A. Passos, D. Cournapeau, M. Brucher, M. Perrot, E. Duchesnay, J. Mach. Learn. Res. **12**, 2825 (2011)
26. N. Japkowicz, M. Shah, *Evaluating Learning Algorithms: A Classification Perspective* (Cambridge University Press, 2011)
27. K. Murphy, *Machine Learning: A Probabilistic Perspective* (The MIT Press, 2012)
28. T. Hastie, R. Tibshirani, J. Friedman, *The Elements of Statistical Learning: Data Mining, Inference, and Prediction* (Springer, 2013)

29. J. Friedman, Ann. Statist. **29**(5), 1189 (2001)
30. T. Chen, T. He, JMLR. Workshop Conf. Proc. **42**, 69 (2015)
31. T. Chen, C. Guestrin (2016), arXiv:1603.02754

Chapter 4
Machine Learning-Based Experimental Design in Materials Science

Thaer M. Dieb and Koji Tsuda

Abstract In materials design and discovery processes, optimal experimental design (OED) algorithms are getting more popular. OED is often modeled as an optimization of a black-box function. In this chapter, we introduce two machine learning-based approaches for OED: Bayesian optimization (BO) and Monte Carlo tree search (MCTS). BO is based on a relatively complex machine learning model and has been proven effective in a number of materials design problems. MCTS is a simpler and more efficient approach that showed significant success in the computer Go game. We discuss existing OED applications in materials science and discuss future directions.

Keywords Materials design · Optimal experiment design · Machine learning

4.1 Introduction

Materials design and discovery is a fundamental issue in materials science and engineering. The design of composite material structure, that achieves certain quality metrics, is often the problem of selecting the optimal solution from a search space [1, 2]. Traditionally, this process depends on personal experience and expensive trial-and-error experiments. To accelerate this process, several optimal experimental design (OED) algorithms have been proposed aiming to reduce the number of required experiments [3–8]. Figure 4.1 illustrates the materials design process by an optimal experimental design approach. Given a space of candidates S, OED aims to

T. M. Dieb · K. Tsuda (✉)
National Institute for Materials Science, Tsukuba, Japan
e-mail: tsuda@k.u-tokyo.ac.jp

T. M. Dieb
e-mail: MOUSTAFADIEB.Thaer@nims.go.jp

T. M. Dieb · K. Tsuda
Graduate School of Frontier Sciences, The University of Tokyo, Kashiwa, Japan

K. Tsuda
Center for Advanced Intelligence Project, RIKEN, Tokyo, Japan

Fig. 4.1 Optimal
experimental design (OED)
algorithm process. For a
predetermined number of
iterations, OED algorithm
selects a candidate set from
the candidate space for
experimentation. The
experimental outcomes are
then exploited for a better
selection in the next iteration

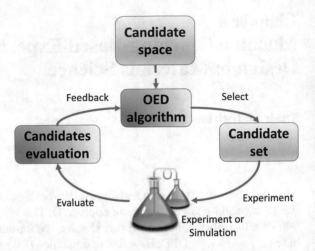

find the best candidate that optimizes a black-box function $f(s)$, whose evaluation
is possible only by an experiment. Starting from a random set of candidate solutions, an OED algorithm iteratively selects a set of candidate solutions for experiments. Experimental results are fed back to the OED algorithm to make further deci-
sions. In many cases, experiments are replaced by simulators such as first-principle
calculation.

In this chapter, we review the applications of two OED algorithms in the materials
science domain. The first is Bayesian optimization (BO) [9], which has been proven
effective in many materials design and discovery studies [1, 2, 6, 7, 10–13]. In BO
methods, a machine learning model is employed to reconstruct the black-box function $f(s)$. In addition, the uncertainty of prediction is also taken into consideration
in candidate selection. The second is Monte Carlo tree search (MCTS) that showed
exceptional performance in computer Go [14]. MCTS explores a tree-shaped search
space and is more efficient than BO in most cases. In a recent study [8], MCTS was
applied to a Si-Ge alloy design problem and shown to be applicable to large-scale
design problems.

This chapter is organized into four sections. Section 4.2 discusses the Bayesian
optimization method and its applications in materials design and discovery, while
Sect. 4.3 is dedicated to Monte Carlo tree search. Section 4.4 concludes this chapter
with a brief look at other available OED approaches.

4.2 Bayesian Optimization

In machine learning communities, Bayesian Optimization (BO), aka kriging, has
become a very popular tool for optimization problems recently [15–17]. BO is a
sequential design strategy to optimize an expensive black-box function $f(s)$. Deriva-
tives of f are not required. The difference between Bayesian optimization and earlier

models that used regression [18] is that, BO methods not only consider the predicted merit of candidates, but also quantify uncertainty as the predictive variance. Based on this variance, BO can determine where to query $f(s)$ next to achieve maximum performance. In this section, we will briefly describe a basic BO method, then review several applications in the domain of materials design and discovery.

4.2.1 Method

Assume that each candidate is represented using a set of N descriptors. The candidate set is then described as a set of points $S = \{s_1, ...s_m\}$ in an N-dimensional space. We are looking for the best point $s_{opt} \in S$ that maximizes a target black-box function $f(s)$. It is very common, particularly in materials science and engineering domain, that the cost of querying $f(s)$ is very high. It is necessary to find the optimal solution s_{opt} with as few queries as possible.

Bayesian optimization methods maintain a probabilistic model of $f(s)$, most commonly Gaussian process (GP) [19] (Fig. 4.2). Initially, a number of candidates are randomly selected and $f(s)$ is obtained for each of them. GP is trained using these data and the user obtains a nonlinear regression function and its predictive variance. In BO, an aquisition function quantifies how promising a candidate is, and depends both on the regression function and predictive variance. There are three typical choices: maximum probability of improvement, maximum expected improvement, and Thompson sampling [9]. The aquisition function is applied to all remaining candidates and the one with the largest value is selected for next experimentation.

The importance of uncertainty evaluation was investigated by Balachandran et al. [2]. They aimed to find the optimal design of M_2AX family of compounds, where the interest is focused on elastic properties [bulk (B), shear (G), and Young's (E) modulus]. Balachandran et al. compared BO with the selection with predicted values of support vector machines and showed that using uncertainty lead to better performance.

Fig. 4.2 Illustration of Bayesian optimization (BO). Gaussian process provides a regression function (red curve) and its variance (blue curves). Candidate points are shown as red triangles. The next candidate is selected based on an aquisition function

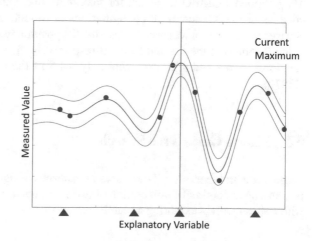

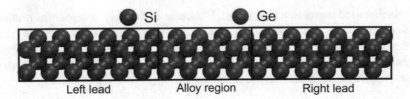

Left lead   Alloy region   Right lead

Fig. 4.3 Si-Ge interfacial structure between two Si leads. In this case, the interface region is made up of 16 atoms

4.2.2 COMBO: Bayesian Optimization Package

With the increasing popularity of applications of Bayesian optimization to materials design problems, there was a need to develop an efficient tool to support this process. We implemented an open source package for Bayesian optimization in python (COMBO: COMmon Bayesian Optimization library, https://github.com/tsudalab/combo) [11]. Thompson sampling, random feature maps and one-rank Cholesky update made it particularly suitable to handle large training datasets. It was shown that COMBO is more efficient than a GP implementation in scikit-learn (http://scikit-learn.org). To make it usable by non-experts, COMBO is parameter-free and can easily be used in various materials design problems. COMBO was first applied to optimize crystalline interface structures [10], where the aim is to find the best translation parameters with lowest grain boundary energy. It is reported that more than 50 times speedup was observed in comparison to random design.

4.2.3 Designing Phonon Transport Nanostructures

In a recent paper, Ju et al. [7] studied thermal conductivity in Si-Ge nanostructures. They applied COMBO to search for maximum and minimum interfacial thermal conductance (ITC) across all configurations of Silicon and Germanium (Fig. 4.3). Binary representation was used to describe the position of each atom in the structure: 1 and 0 represent the Ge and Si atom respectively. It is reported that the optimal solution was reached after exploring only 3.4% of the total number of candidates (12870).

4.3 Monte Carlo Tree Search

Large-scale problems are not rare cases in materials design and discovery. For example, finding the optimal configuration of two elements in a materials crystal structure with x sites involves exploring a search space with the size 2^x. When $x = 10$, the size of the space is 1024. The space size increases exponentially with the number of sites x (for $x = 20$, the size becomes 1048576). Since BO applies an aquisition function to all candidates, the computational time becomes inhibitive for large x.

The significant success of Monte Carlo tree search (MCTS) [20] in computer Go game [14] inspired researchers to develop similar approaches in different research areas including other type of games [21–24]. MCTS is a guided-random best-first search method that models the search space as a gradually expanded tree. Additionally, MCTS does not involve costly matrix operation like GP, making it very scalable for large-scale search spaces. We recently applied MCTS to atom assignment problems in Fig. 4.3 and showed that MCTS is more efficient in BO in large-scale problems [8].

4.3.1 Method

Assume a material structure s with p positions. Each position has to be assigned by an atom from set A. We are looking for the best assignment of length p from the set of all possible assignments. The evaluation of a structure is given by a black-box function $f(s)$ corresponding to either an experiment or simulation.

MCTS uses a tree data structure to represent the search space (Fig. 4.4). A node at level n of the tree corresponds to the assignment of $a \in A$ into n-th position. The maximum depth of the tree is p. A solution is defined by a path from the root to a leaf node at level p. MCTS constructs only a top part of the search tree and it is expanded gradually to promising areas. At a node at depth $n < p$, only a part of the solution is obtained. To obtain a full solution, MCTS uses a technique called *rollout*, i.e., completing the solution by random assignment of atoms in the remaining positions. After a full solution is made, $f(s)$ is evaluated and recorded as the immediate merit of the node that the rollout started.

At the beginning, only the root node exists. The search continues until a pre-requested number of iterations are finished. In each iteration, MCTS has four steps (Fig. 4.4): selection, expansion, simulation, and backpropagation. The pseudo-code of MCTS is shown as Algorithm 1. In the selection step, MCTS starts from the root and traverses down following the path of the most promising child. Children of the node are scored with different methods. The most common one is the Upper Confidence Bound (UCB) score [20],

$$ucb_i = \frac{z_i}{v_i} + C\sqrt{\frac{2\ln v_{parent}}{v_i}}, \tag{4.1}$$

where z_i is the accumulated merit of the node, i.e., the sum of immediate merits of the all downstream nodes, v_i is the visit count of the node, v_{parent} is the visit count of the parent node, and C is the constant to balance exploration and exploitation. In the expansion step, one or more child (depending on the implementation) are created

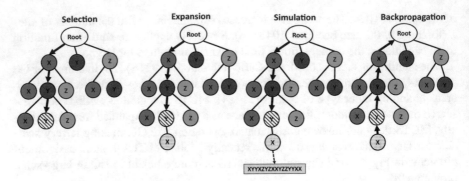

Fig. 4.4 Monte Carlo tree search (MCTS) for a three atom assignment problem. Atoms are to be assigned to a set of available positions. The search space is modeled as a decision tree where each node denotes a possible assignment. MCTS repeats four steps in each iteration: In the selection step, a promising leaf node is chosen by following the child with the best score. The expansion step adds a number of children nodes to the selected one. In simulation, a full solution is created by random rollout for each expanded node. The backpropagation step updates nodes' information along the path back to the root for a better selection in the next iteration

under the selected node. For each expanded child, a full solution is obtained through rollout, then evaluated using $f(s)$ and recorded in the simulation step. Finally, in the backpropagation step, the node information z_i, v_i is updated to be used for better selection in the next iteration.

4.3.2 MDTS: A Python Package for MCTS

We developed a python package of the MCTS algorithm that solves atom assignment problems [8]. The package named MDTS (Materials Design using Tree Search) is available at https://github.com/tsudalab/MDTS. MDTS is a parameter-free tool that automatically sets the only hyperparameter of MCTS algorithm (C) to obtain the best performance based on the target application. Following a similar idea to [25], MDTS controls C adaptively at each node as follows:

$$C = \frac{\sqrt{2}J}{4}(f_{max} - f_{min}), \tag{4.2}$$

where J is a meta-parameter initially set to one and increased whenever the algorithm encounters a so-called *dead-end* leaf to allow more exploration. f_{max} and f_{min} are the maximum and minimum immediate merits in downstream nodes.

To investigate the efficiency of MDTS, we compared the application of MDTS and an efficient Bayesian optimization package [11] to design optimal Silicon-Germanium (Si-Ge) alloy interfacial structures (Si:Ge = 1:1) in order to achieve both minimum and maximum thermal conductance [7]. The total computation time was

Start
    make root node *root*   ▷ Each node has 2 values, *z*: accumulated merit, *v*: visit count
    solutions_set ← ∅
    while within number of iterations **do**
        n ← SELECTION(*root*)
        if *n* is not a maximum depth leaf **then**
            children ← EXPANSION(*n*)
            for all *child* ∈ *children* **do**
                solution ← SIMULATION(*child*)
                e ← evaluate *solution* using experiment or computation
                BACKPROPAGATION(*child*, *e*)
                solutions_set ← [*solutions_set*, *solution*]
            end for
        end if
    end while
    return argmax(*solutions_set*)
Finish

    function SELECTION(*node*)
        if *node* has no children **then**
            return *node*
        else
            $bst\_child \leftarrow argmax(\frac{node.z}{node.v} + C\sqrt{\frac{2ln(parent.v)}{node.v}})$ ▷ *parent* is the parent of *node*
            return SELECTION(*bst_child*)
        end if
    end function

    function EXPANSION(*node*)
        for all possible children **do**
            make node *child*
            add *child* to children of the *node*
        end for
        return all children of the *node*
    end function

    function SIMULATION(*node*)
        structure ← the path from the root to *node*
        if *node* is not a maximum depth leaf **then**
            structure ← complete the solution randomly   ▷ random rollout
        end if
        return *structure*
    end function

    function BACKPROPAGATION(*node*, *e*)
        node.z ← *node.z* + *e*
        node.v ← *node.v* + 1
        if *parent* is not None **then**   ▷ *parent* is the parent of *node*
            return BACKPROPAGATION(*parent*, *e*)
        end if
    end function

Algorithm 1: Monte Carlo tree search

divided into design time and simulation time. The former is the time needed by the OED algorithm to select the next candidates, and the later is the time needed to query the target function $f(s)$, i.e., time to compute the thermal conductance for the candidate solution in this particular application. When the number of positions is smaller than 24, Bayesian optimization showed better efficiency due to its sophisticated machine learning algorithm. However, for larger problems, the design time of BO gets prohibitively long and MDTS was better in finding the best solution quickly.

4.3.3 Discussion

Use of the rollout is the basis of MCTS. It enables systematic space exploration without needing to generate the whole search space. In MDTS, the rollout is random, but it can possibly be improved using machine learning. For example, Yee et al. proposed a new MCTS algorithm with machine learning in continuous action spaces [26], where the UCB score is modified using kernel regression. It should be possible to apply this approach to materials science as well.

It is important to consider the balance between design time and simulation time. MCTS methods are most useful when the simulation time is short. The long design time of a more inefficient machine learning-based approach can appear less problematic when the simulation time is longer [8].

4.4 Concluding Remarks

Optimal experimental design (OED) methods are gaining more importance recently in the field of materials science and engineering due to popular need to reduce the cost of materials design and discovery. In this chapter, we presented two OED methods and their applications in materials design. Bayesian optimization (BO) is a well-established method with several successful applications; however, it struggles with large-scale problems. A new approach using Monte Carlo tree search (MCTS) has emerged with competitive search efficiency and superior scalability. In the future, a hybrid approach combining machine learning and MCTS may achieve even better design efficiency.

Other available OED methods include evolutionary algorithms such as genetic algorithms [27, 28]. Such methods are scalable, but they have many parameters to tune (such as crossover and mutation rates). With limited data available a priori, as in most cases in materials design and discovery, tuning parameters may be difficult. Other sequential learning (SL) methodologies have been proposed. For example Ling et al. have implemented a new OED approach based on random forests with uncertainty estimates [29]. The proposed framework is scalable to high-dimensional parameter spaces. Wang et al. proposed a nested-batch-mode sequential learning method that suggests experiments in batches [30]. In order to increase the efficiency of BO, some researchers proposed a new surrogate model which combines independent Gaussian Processes with a linear model that encodes a tree-based dependency structure, which can transfer information between overlapping decision sequences [31]. In their approach, Jenatton et al. designed a specialized a two-step acquisition function that explores the search space more effectively.

Acknowledgements This work was supported by a Grant-in-Aid for Scientific Research on Innovative Areas 'Nano Informatics' (Grant No. 25106005) from the Japan Society for the Promotion of Science (JSPS).

References

1. A. Seko, A. Togo, H. Hayashi, K. Tsuda, L. Chaput, I. Tanaka, Phys. Rev. Lett. **115**, 205901 (2015)
2. P.V. Balachandran, D. Xue, J. Theiler, J. Hogden, T. Lookman, Sci. Rep. **6**, 19660 (2016)
3. D. Reker, S.G. Drug, Discov. Today **20**, 458 (2015)
4. A.R. Oganov, C.W. Glass, J. Chem. Phys. **124**, 244704 (2006)
5. M. Ahmadi, M. Vogt, P. Iyer, J. Bajorath, H. Frhlich, J. Chem. Inf. Model. **53**, 553 (2013)
6. A. Seko, T. Maekawa, K. Tsuda, T. I. Phys. Rev. B **89**, 054303 (2014)
7. S. Ju, T. Shiga, L. Feng, Z. Hou, K. Tsuda, J. Shiomi, Phys. Rev. X **7**, 021024 (2017)
8. T.M. Dieb, S. Ju, K. Yoshizoe, Z. Hou, J. Shiomi, K. Tsuda, Sci. Tech. Adv. Mater. **18**, 498 (2017)
9. J. Snoek, H. Larochelle, R. Adams, *Advances in Neural Information Processing Systems*, pp. 2951–2959, 2012
10. S. Kiyohara, H. Oda, K. Tsuda, T. Mizoguchi, Jpn. J. Appl. Phys. **55**, 045502 (2016)
11. T. Ueno, T. Rhone, Z. Hou, T. Mizoguchi, K. Tsuda, Mater. Discov. **4**, 18 (2016)
12. R. Aggarwal, M.J. Demkowicz, Y.M. Marzouk, Modelling Simul. Mater. Sci. Eng. **23**, 015009 (2015)
13. T. Lookman, F. Alexander, K. Rajan (eds.), *Information Science for Materials Discovery and Design, Springer Series in Materials Science*, vol. 225 (Springer International Publishing, Switzerland, 2016)
14. D. Silver, A. Huang, C. Maddison, A. Guez, L. Sifre, G. van den Driessche, J. Schrittwieser, I. Antonoglou, V. Panneershelvam, E.A. Lanctot, M. Nature **529**, 484 (2016)
15. D.R. Jones, M. Schonlau, W.J. Welch, J. Glob. Optim. **13**, 455 (1998)
16. S. Streltsov, P. Vakili, J. Glob. Optim. **14**, 283 (1999)
17. M.J. Sasena, Flexibility and efficiency enhancement for constrained global design optimization with kriging approximations. Ph.D. thesis, University of Michigan, 2002
18. D. Coulinga, R. Bernotb, K.M. Dochertyb, J.K. Dixona, E.J. Maginn, Green Chem. **8**, 82 (2006)
19. C.E. Rasmussen, C.K.I. Williams (eds.), *Gaussian Processes for Machine Learning* (MIT Press, 2006)
20. C. Browne, E. Powley, D. Whitehouse, S. Lucas, P. Cowling, P. Rohlfshagen et al., IEEE Trans. Comput. Intell. AI Games **4**(1), 1 (2012)
21. B. Arneson, R.B. Hayward, P. Henderson, I.E.E.E. Trans, Comput. Intell. AI Games **2**, 251 (2010)
22. J. Mehat, T. Cazenave, I.E.E.E. Trans, Comput. Intell. AI Games **2**, 271 (2010)
23. A. Rimmel, F. Teytaud, T. Cazenave, Appl. Evol. Comput. 501–510 (2011)
24. C. Mansley, A. Weinstein, M.L. Littman, Int. Conf. Automat. Plan. Sched 335–338 (2011)
25. L. Kocsis, C. Szepesvári, *Machine Learning: ECML 2006* (Springer, Berlin, 2006), pp. 282–293

26. T. Yee, V. Lisy, M. Bowling, in *International Joint Conference on Artificial Intelligence*, pp. 690–696, 2016
27. Patra, T.K., Meenakshisundaram, V., Hung, J., Simmons, D, Comb, A.C.S. Sci. *19*(2), 96 (2017). https://doi.org/10.1021/acscombsci.6b00136
28. W. Paszkowicz, K.D. Harris, R.L. Johnston, Comput. Mater. Sci. **45**(1), ix (2009). https://doi.org/10.1016/j.commatsci.2008.07.008
29. J. Ling, M. Hutchinson, E. Antono, S. Paradiso, B. Meredig, Integr. Mater. Manuf. Innov. (2017). https://doi.org/10.1007/s40192-017-0098-z
30. Y. Wang, K.G. Reyes, K.A. Brown, C.A. Mirkin, W.B. Powell, SIAM J. Sci. Comput. **37**, B361 (2015)
31. R. Jenatton, C. Archambeau, J. Gonzalez, M. Seeger, in *International Conference on Machine Learning*, pp. 1655–1664, 2017

Chapter 5
Persistent Homology and Materials Informatics

Mickaël Buchet, Yasuaki Hiraoka and Ippei Obayashi

Abstract This paper provides an introduction to persistent homology and a survey of its applications to materials science. Mathematical prerequisites are limited to elementary linear algebra. Important concepts in topological data analysis such as persistent homology and persistence diagram are explained in a self-contained manner with several examples. These tools are applied to glass structural analysis, crystallization of granular systems, and craze formation of polymers.

Keywords Persistent homology · Materials informatics · Topological data analysis

5.1 Introduction

Because of the rapid growth of computers, internet, and experimental measurement devices, huge amounts of data are now available and they induce drastic changes in scientific activities. Namely, data-driven science has recently emerged and this new trend also applies to materials science, leading to a new concept of materials informatics. The basic strategy is to try to capture meaningful information embedded in the database using machine learning. The readers can discover results at the frontiers of materials informatics from some papers in this book.

A key to the success of materials informatics is to select compact descriptors of data to appropriately study materials properties. Available data is large and compli-

M. Buchet · Y. Hiraoka (✉) · I. Obayashi
Advanced Institute for Materials Research (WPI-AIMR), Tohoku University, 2 Chome-1-1 Katahira, Aoba Ward, Sendai 980-8577, Japan
e-mail: hiraoka@tohoku.ac.jp

Y. Hiraoka
Center for Materials research by Information Integration (CMI2), Research and Services Division of Materials Data and Integrated System (MaDIS), National Institute for Materials Science (NIMS), 1 Chome-2-1 Sengen, Ibaraki Prefecture, Tsukuba 305-0047, Japan

Y. Hiraoka
Center for Advanced Intelligence Project, RIKEN, Tokyo 103-0027, Japan

I. Tanaka (ed.), *Nanoinformatics*, https://doi.org/10.1007/978-981-10-7617-6_5

cated. Therefore, good descriptors are required for efficient applications of machine learning, expanding the possibilities beyond conventional descriptors.

This story applies not only to materials science, but also to various communities in science and technology. Topological data analysis (TDA) has emerged in this century [1] and shed a new light on data science. A distinguishing property of TDA is that it provides tools for capturing the *shape of data* in a multi-scale way. They capture topological and geometric features embedded in data and enable the study of relationships of those detected features in different scales. Nowadays, topological data analysis is applied to a wide variety of scientific and industrial areas (e.g., materials science, life science, neuroscience, and social networks).

A particularly important tool in TDA is persistent homology and persistence diagrams. Briefly speaking, these tools describe topological features characterized by holes in data (components, rings, cavities, etc.). Practically, the input to persistent homology is usually given as a finite point set in a Euclidean space or digital images of any dimension. In materials science, atomic (or particle) configurations obtained by molecular dynamics simulations as well as digital images observed by experiments can be studied by these tools. The persistence diagram is a two-dimensional histogram compactly expressing the output of persistent homology. Based on this visualization, we can easily study higher dimensional topological features in a multi-scale way.

The purpose of this paper is to provide a self-contained introduction to persistent homology and survey several applications to materials science [2–5]. We only assume knowledge of elementary linear algebra and show several examples to help the readers' understanding. We hope that this paper will be useful for materials scientists to get used to persistent homology.[1]

5.2 Mathematical Background

First, we review the mathematical background behind topological data analysis. Our goal is to provide both a rigorous mathematical development and easily understandable intuition. The aim of topological data analysis is to provide an understanding of the structure of data. For that, we first need to define what we are looking for and then describe how to extract this information.

5.2.1 Homology

The structure we study is called homology. While homology is not as descriptive as the maybe more classical concept of homotopy, it does present the undeniable

---

[1] The readers can obtain further information of materials TDA project organized by our group from the website http://www.wpi-aimr.tohoku.ac.jp/hiraoka_labo/index.html.

advantage of being computable. For the sake of simplicity, we will only introduce
the concept of simplicial homology.

We will endeavor to present the concept from the algebraic side while maintaining
a geometric intuition. We fix a set called the set of indices. In our case, we will only
use the set of integers $\mathbb{N}$.

Definition 5.2.1 A k-simplex is a set of $k + 1$ indices.

This very simple definition describes an abstract simplex. It can have an intuitive
geometric counterpart. Given a set of points numbered by indices, the geometric
k-simplex corresponding to a subset of indices is the convex hull of the subset of
points corresponding these indices. Within this geometric framework, a 0-simplex
is simply a point, a 1-simplex is an edge, a 2-simplex is a triangle, a 3-simplex is a
tetrahedron, and so on (see Fig. 5.1).

Definition 5.2.2 A simplicial complex X is a set of simplices such that for any $\sigma \in X$
and any $\sigma' \subset \sigma$, $\sigma' \in X$.

Therefore, a simplicial complex is a set of simplices with a very natural and simple
rule ensuring coherence. For example, if a triangle belongs to the simplicial complex
X, then the three edges that border it also belong to X as well as the three vertices.
Figure 5.2 illustrates this property. While the left object is a simplicial complex, the
middle one is not because the edge e is missing while the upper triangle exists. The
right one is also incorrect. A consequence of the definition is that the intersection
of two simplices is either empty or a simplex belonging to the simplicial complex.
Here p is the intersection of two simplices but it does not appear as a simplex. Note
that just adding p would not be sufficient to fix the construction.

We now introduce an algebraic notion of orientation to our simplices. Namely,
we fix an ordering on the indices.

Fig. 5.1 Example of
geometric simplices

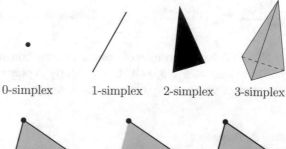

0-simplex   1-simplex   2-simplex   3-simplex

Fig. 5.2 One simplicial
complex (left) and some
objects that are not simplicial
complexes

Definition 5.2.3 Given a set of indices $\{v_1, \ldots, v_k\}$, we define the oriented simplex $\sigma = [v_1, \ldots, v_k]$ as an ordered set. The opposite simplex is obtained by permuting two indices: $[v_1, \ldots v_i, \ldots, v_j, \ldots, v_k] = -[v_1, \ldots, v_j, \ldots, v_i, \ldots, v_k]$.

We choose a field k in order to study the topology of simplicial complexes with the use of homology. Given a simplicial complex X, let $X^{(n)}$ be the set of all n-simplices of X. We use this set as the generating elements of the k-vector space $\Delta_n(X)$. This means that an element of $\Delta_n(X)$ is of the form $\sum_{\sigma \in X^{(n)}} \alpha_\sigma \sigma$ where $\{\alpha_\sigma\}$ are coefficient in k. The addition operation is naturally $\sum_{\sigma \in X^{(n)}} \alpha_\sigma \sigma + \sum_{\sigma \in X^{(n)}} \alpha'_\sigma \sigma = \sum_{\sigma \in X^{(n)}} (\alpha_\sigma + \alpha'_\sigma) \sigma$.

The next tool we need is to describe faces of a given simplex σ. We do so by indicating which vertex is opposite to it.

Definition 5.2.4 Given an ordered n-simplex $\sigma = [v_0, \ldots, v_n]$, we write $[v_0, \ldots \hat{v}_i, \ldots v_n]$ the $(n-1)$-simplex obtained by removing the index v_i.

Note that if an n-simplex σ belongs to a simplicial complex X, any one of its faces is a $(n-1)$-simplex and also belongs to X. We can hence define the following map.

Definition 5.2.5 Given a simplicial complex X, the boundary map $\partial_n : \Delta_n(X) \to \Delta_{n-1}(X)$ is defined on the basis elements by:

$$\partial_n([v_0, \ldots, v_n]) = \sum_{i=0}^{n} (-1)^i [v_0, \ldots \hat{v}_i, \ldots v_n].$$

For example, the definition for $n = 1, 2$ is given by $\partial_1([v_0, v_1]) = [v_1] - [v_0]$ and $\partial_2([v_0, v_1, v_2]) = [v_1, v_2] - [v_0, v_2] + [v_0, v_1]$. By extending this operator to all elements of $\Delta_n(X)$, we obtain a linear map. Geometrically, the boundary operator extracts the boundary of a chain while respecting the orientation (see Fig. 5.3).

By combining these operations for each dimension, we obtain the chain complex:

$$\cdots \xrightarrow{} \Delta_{n+1}(X) \xrightarrow{\partial_{n+1}} \Delta_n(X) \xrightarrow{\partial_n} \cdots \xrightarrow{} \Delta_1(X) \xrightarrow{\partial_1} \Delta_0(X) \xrightarrow{\partial_0} 0$$

Note that the composition of two consecutive boundary operators is zero. In other words, for any n, $\partial_{n-1}\partial_n = 0$. This property expresses the geometric fact that the boundary of the boundary of an object is empty.

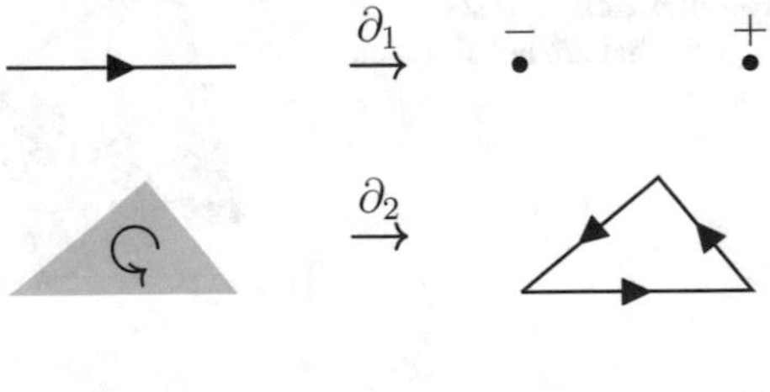

Fig. 5.3 Examples of boundaries

Let Ker $\partial_n = \{c \in \Delta_n(X): \partial_n c = 0\}$ and Im $\partial_{n+1} = \{c \in \Delta_n(X): c = \partial_{n+1} c', c' \in \Delta_{n+1}(X)\}$, be the kernel and the image of the boundary maps. From the above property, we have Im $\partial_{n+1} \subset$ Ker ∂_n. We can thus define homology by quotienting subspaces.

Definition 5.2.6 The n-dimensional homology of X is defined as $H_n(X) =$ Ker $\partial_n / $Im ∂_{n+1}.

Intuitively, homology describes holes of the structure. By counting generators of homology, we obtain the Betti numbers which count topological features. The Betti number in dimension 0 gives the number of connected components. In dimension 1, it corresponds to the number of holes and in dimension 2 to the number of cavities, and then generalizes to higher dimensions.

We now give an example of a simplicial complex with five vertices in Fig. 5.4 and compute its homology.

In this simplicial complex, the simplex of highest dimension is the 2-simplex, a.k.a. triangle, $[1, 2, 3]$. Therefore, $\Delta_2(X) = k[1, 2, 3]$. Looking at dimension 1 simplices, we can see five different edges. Therefore, $\Delta_1(X) = k[1, 4] \oplus k[4, 2] \oplus k[1, 2] \oplus k[2, 3] \oplus k[1, 3]$. Finally, we have 5 points and, therefore, $\Delta_0(X) = k[1] \oplus k[2] \oplus k[3] \oplus k[4] \oplus k[5]$.

First, remark that for any dimension $n \geq 3$, the boundary map $\partial_n = 0$ and, therefore, Ker $\partial_n = 0$ and $H_n(X) = 0$. We first need to consider the matrix associated with ∂_2. Writing the matrix M_2 associated with the boundary map ∂_2, we obtain,

$$M_2 = \begin{array}{c} [1,2,3] \\ \begin{pmatrix} 0 \\ 0 \\ 1 \\ 1 \\ -1 \end{pmatrix} \begin{array}{l} [1,4] \\ [4,2] \\ [1,2] \\ [2,3] \\ [1,3] \end{array} \end{array}$$

We can immediately deduce that Ker $\partial_2 = 0$ and Im $\partial_2 = k([1, 2] + [2, 3] - [1, 3])$. Hence $H_2(X) =$ Ker $\partial_2 / $Im $\partial_3 = 0$. To compute $H_1(X)$, we also need to consider the matrix M_1 associated with ∂_1.

Fig. 5.4 Example of a simplicial complex

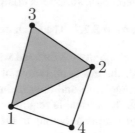

$$M_1 = \begin{array}{c} \begin{array}{ccccc} [1,4] & [4,2] & [1,2] & [2,3] & [1,3] \end{array} \\ \left(\begin{array}{ccccc} -1 & 0 & -1 & 0 & -1 \\ 0 & 1 & 1 & -1 & 0 \\ 0 & 0 & 0 & 1 & 1 \\ 1 & -1 & 0 & 0 & 0 \\ 0 & 0 & 0 & 0 & 0 \end{array} \right) \begin{array}{c} [1] \\ [2] \\ [3] \\ [4] \\ [5] \end{array} \end{array}$$

A simple computation yields that Ker $\partial_1 = k([1,4] + [4,2] - [1,2]) + k([1,2] + [2,3] - [1,3])$. Therefore, the homology $H_1(X) = $ Ker $\partial_1 /$Im $\partial_2 = k([1,4] + [4,2] - [1,2] + $ Im $\partial_2) \cong k$. In other words, the one-dimensional homology is isomorphic to k and, therefore, has dimension 1. It means that there exists one hole. Moreover, one possible representative of the class is the cycle $[1,4] + [4,2] - [1,2]$. Note that this representative is not unique as $[1,4] + [4,2] + [2,3] - [1,3]$ is also a representative of the same class. Intuitively, the quotient operation means that given a cycle in dimension d, we can add or remove the boundary of simplices of dimension $d + 1$ without changing the equivalence class. In our example, the cycle corresponding to the hole is equivalent to the one obtained by adding the boundary of the triangle $[1, 2, 3]$ to it.

To finish, remark that Im $\partial_1 = k([4] - [1]) + k([2] - [4]) + k([3] - [2])$ and that ∂_0 is a zero map. Therefore, Ker $\partial_0 = k[1] + k[2] + k[3] + k[4] + k[5]$ and $H_0(X) \cong k^2$ which indicates the presence of two connected components.

5.2.2 From Point Sets to Simplicial Complexes

The construction of simplicial homology relies on simplicial complexes. The first task is to build such a simplicial complex from our data. We consider here an input given as a set of points $P \subset \mathbb{R}^d$ in a Euclidean space. We want to build a geometric simplicial complex, id est a continuous space, from the point set P which is a discrete space. To do so, we consider balls around these points.

Given a radius r and a point x, we denote $B(x, r)$ the ball centered at x and of radius r. We consider the union $\cup_{x \in P} B(x, r)$ of all balls of radius r centered at points of P. We define the nerve of the union of balls also called the Čech complex, which is a geometric simplicial complex whose vertices are the points of P.

Definition 5.2.7 The Čech complex is defined as $C_r(P) = \{\sigma | \cap_{p \in \sigma} B(p, r) \neq \emptyset\}$.

Each point is associated with a ball. Note that all the balls are non-empty if $r > 0$ and, therefore, all points of P belong to the Čech complex. An edge belongs to the complex if and only if the two balls corresponding to its extremities intersect. Similarly, a triangle requires the common intersection of its three vertices' balls to be non-empty to belong to the Čech complex.

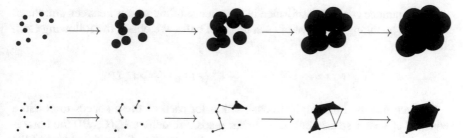

Fig. 5.5   Example of Čech complex

Considering the Čech complex is enough to study the topology of the union of balls as the Nerve Theorem [6, 4G.3] implies:

Proposition 5.2.8 *Given a set of points P in a Euclidean space and a radius r, the union of balls $\cup_{p\in P}B(p, r)$ and the Čech complex $C_r(P)$ are homotopy equivalent.*

Intuitively, two spaces are homotopy equivalent if we can deform continuously one into the other. Therefore, they have the same topological structure and studying the homology of one is equivalent to studying the homology of the other one. The construction is illustrated in Fig. 5.5.

It is important to note that the construction can be made with any union of balls. The Nerve Theorem is not limited to Čech complexes. From an applicative stand-point in material science, the notion of weighted Čech complexes is especially useful. When the input is a set of atomic positions with different type of atoms, we can reflect the size of each particular atom by modifying the radius accordingly. We obtain a union of balls with different radii, bigger atoms having larger balls.

5.2.3   Persistent Homology

A major problem that arises is the choice of the radius r. Choosing a radius gives a snapshot of the topology at the corresponding scale but does not encapture the whole topological structure. Persistent homology is a tool that allows multi-scale analysis. Instead of looking at one given radius, we can look at the evolution of topological features across scales.

In the context of material science, this allows to not only detect topological features but also to classify them depending on their scale. This is related to the diameter and the geometry of holes and cavities.

First, notice that the union of balls we considered previously possesses a natural inclusion when the radius increases. Given some radii $r_1 \leq \cdots \leq r_i \leq \cdots \leq r_l$, we have:

$$\cup_p B(p, r_1) \hookrightarrow \cup_p B(p, r_2) \hookrightarrow \cdots \hookrightarrow \cup_p B(p, r_i) \hookrightarrow \cdots \hookrightarrow \cup_p B(p, r_l).$$

This sequence can be transformed in a sequence of inclusions between simplicial complexes by taking the nerve of each union of balls. We obtain the following Čech filtration.

$$C_{r_1}(P) \hookrightarrow C_{r_2}(P) \hookrightarrow \cdots \hookrightarrow C_{r_i}(P) \hookrightarrow \cdots \hookrightarrow C_{r_l}(P).$$

We then use the homological construction for each of these spaces to obtain a sequence of vector spaces linked by linear maps. We denote $H_n(C_r(P))$ the homology vector space built using $C_r(P)$ for a given dimension n. Since the choice of the working dimension does not have an influence on the theoretical results, we indicate it by writing $H_*(C_r(P))$.

Definition 5.2.9 Given an ordered index set I and a field k, a persistence module H is a sequence $(\Phi_i)_{i \in I}$ of vector spaces and linear maps $(\phi_i^j)_{i \leq j}$ where $\phi_i^j : \Phi_i \to \Phi_j$ and for all $i \leq j \leq k$, $\phi_i^k = \phi_j^k \circ \phi_i^j$.

A persistence module is a sequence of vector spaces linked by linear maps. The condition on the linear maps is that they commute. Intuitively, this means that we can decompose and recompose them. Working on the previous chain sequence, we build at homology level the following persistence module.

$$H_*(C_{r_1}(P)) \to H_*(C_{r_2}(P)) \to \cdots \to H_*(C_{r_i}(P)) \to \cdots \to H_*(C_{r_l}(P))$$

The Persistent Nerve Lemma [7] guarantees that this persistent module is isomorphic to the one we can build using the union of balls. Therefore, studying the Čech filtered complex is equivalent to studying the filtered union of balls.

The critical property of the persistence module is its decomposability. Indecomposables, in other words, the building blocks, are called interval modules. They consist of a sequence of one-dimensional vector spaces linked by identity maps.

$$0 \to k \to k \to k \to 0 \to 0$$

In this example of an interval module, we have six values of indices we name $\{1, \ldots, 6\}$. The interval spans from the second to the fourth so we denote it $I[2, 4]$. All maps between the nonzero vector spaces are identity maps.

The following property ensures that the persistence modules we consider are uniquely decomposables into a direct sum of intervals.

Proposition 5.2.10 *A persistence module whose every vector space is finite dimensional is uniquely decomposable into a direct sum of interval modules.*

Note that in our setting, we build finite simplicial complexes from finite point sets. Therefore, everything is finite, especially the dimension of the vector spaces. Thus the Proposition applies. There exist various more general variants [8, 9] of this result but we limit ourselves to this one for the sake of simplicity.

Intuitively, intervals have a birth, the first index where the vector space is nonzero, and a death, the first index where the vector space is zero after having been nonzero. The first index for which a simplex σ belongs to the complex is called the apparition time of σ. Intervals correspond to the existence of topological features. In the case of a one-dimensional cycle, for example, the birth corresponds to the apparition time of the edge forming the cycle and the death corresponds to the apparition time of the triangle that fills it completely.

Formally, a persistence module H can be associated with a set of pairs (b_i, d_i) such that:

$$H = \bigoplus I[b_i, d_i]$$

We can represent each of the interval $I[b, d]$ as a bar starting at b and ending at d. We thus obtain a figure called barcode that describes the decomposition of the persistence module. Figure 5.6 shows an example of barcode.

There exists a natural bijection from barcodes to multi-sets of $\mathbb{R}^2$ denoted $D = \{(b, d)\}$. This multi-set is called a persistence diagram (PD for short) and is often represented as in Fig. 5.7.

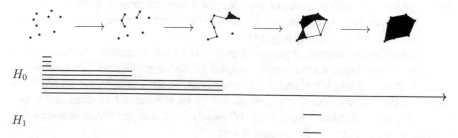

Fig. 5.6 Simplicial complex, topological features, and barcode for zero and one-dimensional homology

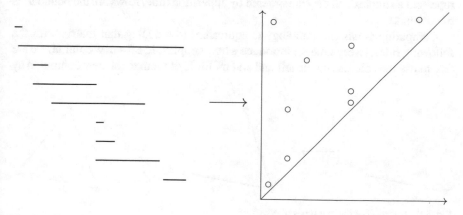

Fig. 5.7 From barcode to persistence diagram

Interpretation of persistence diagrams reveals two different kinds of information. First, it indicates, which features are probably relevant as they are those far away from the diagonal. Second, it can separate features according to a combination of size and shape that contribute to their lifespan.

5.2.4 Computation

From a computational point of view, persistent homology is very intuitive. Considering that we build the simplicial complex from scratch, we add one simplex at a time according to their apparition time. If multiple simplices are added at the same time, we can arbitrarily choose the order in which we insert them. This allows us to maintain a simplicial complex at all steps.

When a d-simplex is inserted, there are two possible cases. Either the simplex is *negative* which means that it destroys a $(d-1)$-dimensional feature, or it is *positive* and creates a d-dimensional feature. Figure 5.8 shows the two kinds of 1-simplices. Note that the object on the left has two connected components and no cycle. The first edge we introduce kills one of the connected component and, therefore, is negative. The second one has its two extremities in the same connected component and, therefore, is positive, creating a cycle.

To compute the barcode, a positive simplex is trivial to handle. We just need to create a new bar. However, a negative simplex is more complicated. We need to find which feature is being killed and that is nontrivial. In our example, we do not know which of the two connected component should be considered as dead and which one is still alive. Persistent homology follows the rule that the oldest one survives. Therefore, we kill the one that appeared last.

This very intuitive algorithm has an algebraic counterpart. We build a boundary matrix that contains the incidence information of all simplices. Each column and row represent a simplex and they are ordered by apparition time. Rows are the boundaries of columns.

Computing persistent homology is equivalent to reducing that matrix with the following rules. Every time we introduce a new simplex, id est a new column, we are free to use the columns on the left and add multiple of them to the new column. Any

Fig. 5.8 Insertion of the two types of edges

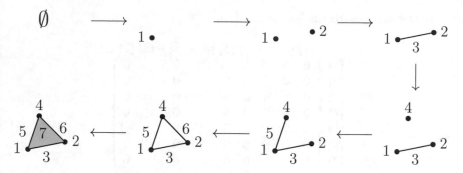

Fig. 5.9   Filtration on a triangle

zero column corresponds to a topological feature. A nonzero column corresponds to the death of the feature created at the time of the lowest nonzero index.

We now provide a simple example and do the whole computation. We build a complex containing a triangle, its edges and vertices filtered in the order shown in Fig. 5.9.

We fix an arbitrary orientation on every simplex by sorting indices in increasing order of apparition. Therefore, we consider the boundary of edge [3] to be [2] − [1]. We then obtain the following boundary matrix.

$$
\begin{array}{c}
\quad\quad\;\; [1]\;[2]\;[3]\;[4]\;[5]\;[6]\;[7] \\
\begin{array}{c}
[1] \\ [2] \\ [3] \\ [4] \\ [5] \\ [6] \\ [7]
\end{array}
\left(
\begin{array}{ccccccc}
0 & 0 & -1 & 0 & -1 & 0 & 0 \\
0 & 0 & 1 & 0 & 0 & -1 & 0 \\
0 & 0 & 0 & 0 & 0 & 0 & 1 \\
0 & 0 & 0 & 0 & 1 & 1 & 0 \\
0 & 0 & 0 & 0 & 0 & 0 & -1 \\
0 & 0 & 0 & 0 & 0 & 0 & 1 \\
0 & 0 & 0 & 0 & 0 & 0 & 0
\end{array}
\right)
\end{array}
$$

First, note that this matrix is upper triangular. This is a direct consequence of having a filtered complex. A simplex cannot appear before one of its faces.

We now do the computation for this example. First, we introduce columns [1] and [2] which are zero and corresponds to 0-simplices. Therefore, it creates two connected components. Then we add [3] which cannot be reduced by elements on its left and, therefore, kills a feature. The lowest nonzero entry corresponds to line [2] so [3] kills the feature created by [2]. In the same way, [4] creates a new connected component killed by [5].

The insertion of [6], however, introduces a column that can be reduced using columns located on its left. More precisely [6] = [5] − [3]. Note that it is easy to detect such a case as it suffices to look at the lowest nonzero entry, cancel it and then recurse. Hence [6] creates a cycle, id est a one-dimensional feature, which is then killed by the insertion of [7].

The resulting matrix can be expressed as:

$$
\begin{array}{c@{\quad}c}
& \begin{array}{ccccccc} [1] & [2] & [3] & [4] & [5] & [6]+[3]-[5] & [7] \end{array} \\
\begin{array}{c} [1] \\ [2] \\ [3] \\ [4] \\ [5] \\ [6] \\ [7] \end{array} &
\left(\begin{array}{ccccccc}
0 & 0 & -1 & 0 & -1 & 0 & 0 \\
0 & 0 & 1 & 0 & 0 & 0 & 0 \\
0 & 0 & 0 & 0 & 0 & 0 & 1 \\
0 & 0 & 0 & 0 & 1 & 0 & 0 \\
0 & 0 & 0 & 0 & 0 & 0 & -1 \\
0 & 0 & 0 & 0 & 0 & 0 & 1 \\
0 & 0 & 0 & 0 & 0 & 0 & 0
\end{array}\right)
\end{array}
$$

Note that the algorithm provides a few extra information for free. We obtain matches between positive simplices and negative ones. Moreover, we get a representant of each homology class being created. Here, the cycle can be represented by $[6] + [3] - [5]$. Beware that this representant is not necessarily the unique representant in its class nor looks good from a geometric point of view. Its structure is disconnected from the geometry.

This algorithm has a worst case running time that is cubic in the number of simplices. In practice, however, implementations work much faster, mostly because of the sparsity of the boundary matrix. There are numerous libraries that compute persistent homology and that are aimed at different public. Some of the most recent ones are the TDA package in R [10] intended for statisticians, DIPHA [11] and GUDHI [12] that are state-of-the-art approaches from the computational topology community or HomCloud [13] which aims at a more experimentalist public with additional tools and graphical output. This list is non-exhaustive and many more exist.

5.2.5 Digital Images

Until now we focused on point sets. We now look into what is different when we want to work with digital images.

By digital images, we mean a multidimensional array of value that can be either 0 or 1. For example, a two-dimensional array is a black and white image. The tabular structure is particular and our previous geometric construction using the Čech complex is not the most suitable here. We replace simplicial complexes by cubical complexes. The idea is similar but we use squares instead of triangles and cubes instead of tetrahedron and so on.

Taking the example of an image, we build the complex with the following rule. Every pixel is given a value α and the cubic complex at time α contains all pixels whose value is less than α. Moreover, two adjacent pixels are linked if both of them have values below α. Four pixels in a square shape corresponds to a square in the complex if all of them have value less than α. The construction extends naturally to higher dimensions. Note that the resulting object is indeed a complex in the sense that any element belonging to it has faces that also belong to it.

The next question is how to choose the value α for each pixel. We want to give a description of the topology of the areas, taking geometry into consideration. Note that if we just keep 0 and 1, we do only compare black and white areas. We thus put new values on each pixel depending on the distance to the other color. A black pixel adjacent to a white pixel is valued 0 and then the next black pixel is valued -1 and so on. Conversely, white pixels are valued increasingly depending on the distance to the nearest black pixel. Figure 5.10 shows the example of how to choose α and Fig. 5.11 shows the filtration by those α.

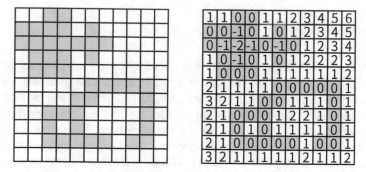

Fig. 5.10 Example of choosing α. The left figure shows an input digital image and the right figure shows the assignment of α on each pixel

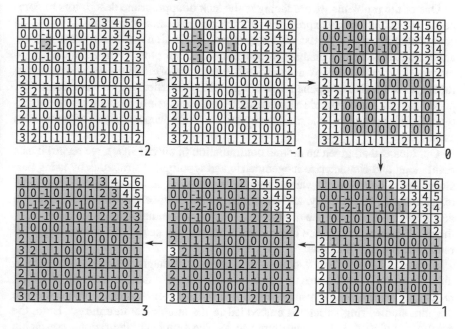

Fig. 5.11 Filtration for a digital image

Our construction provides a way to analyze digital images through the lense of persistent homology. It provides good insight into the structure of objects. Moreover, this simple approach to topological data analysis can be combined to machine learning to obtain interesting results [14].

5.3 Materials TDA

In this section, we briefly explain some applications of persistent homology to materials science. For details of each subject, we refer the readers to the original papers listed therein.

5.3.1 Silica Glass

Our first application is the structural analysis of silica glasses by using persistent homology [2]. There is a long history of trying to understand geometric structures of glass materials. From the experimental side, Xray/neutron scattering diffractions and the transmission electron microscopy (TEM) are often used to study the geometric structures of atomic configurations. On the other hand, from the computational side, molecular dynamics simulations, reverse Monte Carlo, and first-principles calculation based on density function theory are used to simulate atomic configurations. Although our understanding of glass structures is becoming richer, we have not yet reached a sufficient level.

One of the problems we are facing is the lack of appropriate descriptors to compactly and quantitatively express the geometry of glass atomic configurations. In the computational studies, we usually apply radial distribution functions, ring statistics, and Voronoi polyhedron analysis as conventional descriptors to the atomic configurations. However, those tools are restricted to the study of either the zero-dimensional topology (connected components) or single scale properties. As we have seen so far, persistence diagrams provide a tool for multi-scale analysis of higher topological features. This is presumably the most desired function for deeper analysis of amorphous structures.

Our idea is that, given an atomic configuration of silica (SiO_2), we regard it as a point cloud and characterize its geometric and topological structures by using persistent homology. Namely, we put balls with radius r_{Si} and r_O on silicon atoms and oxygen atoms, respectively, and gradually increase those radii to study birth and death events of holes in the atomic ball models in a multi-scale way. Technically, the initial radii r_{Si} and r_O are determined from the first peak positions of the partial radial distribution functions.

Figure 5.12 shows the one-dimensional persistence diagrams computed in the liquid, glass, and crystal states of silica, respectively. We denote them by $D_1(\mathcal{A}_{liq})$, $D_1(\mathcal{A}_{amo})$, and $D_1(\mathcal{A}_{cry})$, respectively. Recall that the one-dimensional persistence diagram studies ring structures embedded in the atomic configurations. Here, the color bar is plotted on the logarithmic scale. The atomic configurations, consisting

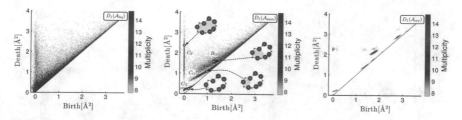

Fig. 5.12 Persistence diagrams of silica in liquid (left), glass (middle), and crystal (right) states (Reproduced from [2])

of 2,700 silicon atoms and 5,400 oxygen atoms, are prepared via the Beest-Kramer-Santen (BKS) model. We refer the readers to the original paper for details on preparing those atomic configurations by molecular dynamics simulations.

As we observe from Fig. 5.12, the persistence diagrams clearly distinguish these three states. Namely, the liquid, glass, and crystal states are characterized by planar (2-dim), curvilinear (1-dim), and island (0-dim) regions of the distributions, respectively. Here, the 0 and 2 dimensionality of the PDs result from the periodic and random atomic configurations of the crystal and liquid states, respectively. In particular, we emphasize that the presence of the curves in $D_1(\mathcal{A}_{\mathrm{amo}})$ clearly distinguishes the glass state from the others. This implies that specific geometric features of the rings generating these curves in $D_1(\mathcal{A}_{\mathrm{amo}})$ play a significant role to elucidate glass states.

Let us consider the meaning of curves. We first remark that, since our system consists of a large enough amount of atoms (8,100 atoms), statistical information is also embedded in each persistence diagram. From this respect, the presence of curve means that generators on each curve are restricted to that curve. Namely, each generator is not allowed to move in the normal direction of the curve, but possibly move to the tangential direction. We recall that generators in the persistence diagram are characterized by ring configurations of atoms. Hence, by pulling back these normal directions of curves, we obtain geometric constraints of local deformations to which atomic configurations are prohibited. In other words, a rigidity information with respect to small deformation of the atomic configuration is embedded in the persistence diagram. Actually, in the original paper, the relationship between persistence diagrams and rigidity based on the small deformation of atomic configurations induced by isotropic pressurization is studied in detail. From the same observation, we also remark that the persistence diagram of crystal state shows further geometric constraints.

The silica is a typical glass material classified as network forming glasses. In [2], we also studied another type of glass materials based on random packing structures. For instance, Fig. 5.13 shows the one-dimensional and two-dimensional persistence diagrams of the Lennard-Jones (LJ) system in crystal and glass states, denoted by $D_k(\mathcal{A}_{\mathrm{cry}}^{\mathrm{LJ}})$ and $D_k(\mathcal{A}_{\mathrm{amo}}^{\mathrm{LJ}})$ ($k = 1, 2$). In this case, not only the one-dimensional persistence diagrams but also the two-dimensional persistence diagrams show characteristic features. Similar to the silica case, a deviation of the persistence diagrams of the

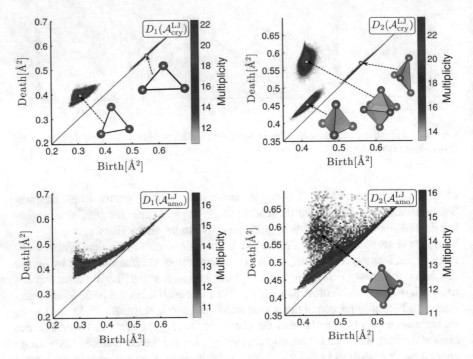

Fig. 5.13 Persistence diagrams of the Lennard-Jones system in crystal and glass states (Reproduced from [2])

glass state from those of the crystal state is observed. In particular, $D_2(\mathcal{A}_{\mathrm{amo}}^{\mathrm{LJ}})$ shows a peak corresponding to octahedral configurations.

As we see, the persistence diagrams clarify topological and geometric features embedded in atomic configurations, which cannot be characterized by other conventional methods. Note that those persistence diagrams are computed on atomic configurations given in a fixed system size. Therefore, we need to be careful about the dependence of the system sizes. The scaling properties of PDs with respect to the system size are computationally studied in [4]. Recently, the existence and uniqueness of limiting persistence diagram is mathematically solved in [15].

Starting from the research explained in this subsection, persistence diagrams are nowadays applied to a wide variety of structural analysis of materials.

5.3.2 Grain Packing

In the paper [5], crystallization mechanism of three-dimensional granular packings of frictional spheres is studied at the grain-scale using Xray tomography and persistent homology. Here, we briefly review some of the results.

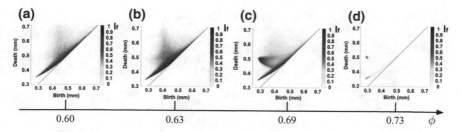

Fig. 5.14 Persistence diagrams of grain configurations for different packing ratios (Reproduced from [5])

In this study, three-dimensional images of granular packings with several packing ratio ϕ are obtained by using XCT, and these images provide precise positional coordinates of grains. Our interest is to characterize the skeleton deformation structures of grain configurations during the crystallization process. For experimental details, please see the original paper.

Figure 5.14 shows the two-dimensional persistence diagrams computed on the grain configurations for four packing ratios $\phi = 0.6, 0.63, 0.69$, and 0.73. Here, we note that the packing ratio $\phi = 0.64$ is known as the Bernal's density at which sharp structural transition to jamming is observed. As we observe from the figure, the persistence diagram (d) at the crystallized state consists of two strong peaks at $(0.288, 0.353)$ and $(0.288, 0.5)$, and they correspond to the regular tetrahedral and the regular octahedral configurations, respectively. We note that the persistence diagram (c) is similar to $D_2(\mathcal{A}_{amo}^{LJ})$ in Fig. 5.13 (the Lennard-Jones system), since both are classified as random packing systems.

The tetrahedral peaks are well preserved for all packing ratios, while the octahedral peaks only exist at (c) and (d). Actually, further studies show that the octahedral peaks are only observable for packing ratios $\phi > 0.64$.

Next, let us study the persistence diagram (c) at $\phi = 0.69$ in detail. Figure 5.15a is the same persistence diagram at $\phi = 0.69$, in which four curves (D1, D2, D3, and D4) corresponding to the boundaries are drawn. In the paper, we found the analytical expressions of the actual deformations of grain configurations corresponding to these curves. Figure 5.15b and c show those deformations. It follows from a discussion similiar to the silica glass case that distorted tetrahedra and octahedra are confined in the region bounded by D1-D4 and those deformations give geometric constraints during the crystallization process.

5.3.3 Craze Formation of Polymer

Craze formation has been intensively investigated by experiments such as electron microscopy, optical microscopy, atomic force microscopy, and so on. From these

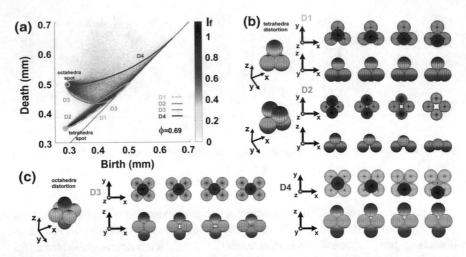

Fig. 5.15 Persistence diagram at $\phi = 0.69$ and the deformations of tetrahedra and octahedra generating the boundary curves D1–D4 (Reproduced from [5])

experimental observations, several kinetic models of craze formation have been proposed so far.

On the other hand, molecular dynamics (MD) simulations have also been applied to understand atomic-scale craze formation mechanisms, which are difficult to observe by experiments. However, the relation between the kinetic models and the MD simulations still remains unclear. This is partially due to the lack of definition of voids in the MD simulations. We note that, since MD simulations are based on the discretized systems, the definition of voids which are consistent with multi-scalability is not trivial. However, such a multi-scalable definition of voids is unavoidable to study the growing process of voids as continuum phenomena, where the kinetic models are discussed. As we now know, persistence diagrams provide an appropriate tool for this purpose.

In the paper [3], a persistent homology analysis is applied to investigate the behavior of nanovoids during the crazing process of glassy polymers. We carry out a coarse-grained molecular dynamics simulation of the uniaxial deformation of an amorphous polymer and analyze the results with persistent homology.

We first compute persistence diagrams of simulation results at each time snapshot. After yielding, several large voids appear, and we detect them from persistence diagrams as generators with large death values as these values measure the size of voids. Then, we reverse the time evolution of the simulation to investigate the initial configurations of those large voids. Then, we revealed that those large voids are created by the coalescences of small voids during craze formation. Figure 5.16 shows some

Fig. 5.16 Void percolation (Reproduced from [3])

of those coalescences during crazing, where gray voids correspond to large voids observed after yielding and other colored small voids coalesce to those gray voids. The results suggest that the yielding process should be regarded as the percolation of nanovoids created by deformation.

5.4 Discussions

In this paper, we summarized persistent homology and its applications to materials science. From these applications, we observed that persistence diagrams are significant descriptors for characterizing multi-scale disordered structures in materials. The next stage toward materials informatics is to combine TDA with machine learning.

Machine learning enables us to capture characteristic patterns from a large amount of data, and TDA enables us to summarize the shape of data quantitatively. Therefore, by combining these two data analysis methods, we can effectively capture the characteristic geometric patterns of the data. Since many machine learning methods accept vectors as input data, we need to convert a persistence diagram into a vector. Some vectorization methods are proposed, and here we introduce two methods with some applications.

One method is the persistence image (PI) [16], which uses a histogram on a finite mesh with smoothing and weighting applied. The histogram values are ordered consistently and we treat it as a finite dimensional vector. In [14], PI is used with logistic regression and linear regression to find a hidden relationship between a persistence diagram obtained from data and a parameter bound to data. In that paper, inverse analysis is effectively used to clarify the geometric origins of birth-death pairs important for the relationship. For materials informatics, we can apply the method to find the characteristic geometric patterns of materials data related to their physical properties such as Young's modulus and conductivity.

Another method is the persistence weighted Gaussian kernel (PWGK) [17, 18], a kind of kernel methods. PWGK maps a persistence diagram into a vector in an infinite dimensional Hilbert space. It is impossible to directly treat infinite dimensional vectors on a computer, but using the kernel trick technique, we can indirectly treat the vectors to apply various kinds of machine learning methods. This method shows good performance in some examples in [17] and is applied to practical problems in [17, 18], e.g., estimating the liquid-glass transition point by using changing point analysis and classifying proteins by using support vector machine.

Acknowledgements The authors appreciate all the collaborators relating materials TDA projects. This work is partially supported by JST CREST Mathematics15656429, JST Materials research by Information Integration Initiative (MI2I) project of the Support Program for Starting Up Innovation Hub, Structural Materials for Innovation Strategic Innovation Promotion Program D72, and New Energy and Industrial Technology Development Organization (NEDO).

References

1. G. Carlsson, Topology and data. Bull. Am. Math. Soc. **46**(2), 255–308 (2009)
2. Y. Hiraoka, T. Nakamura, A. Hirata, E.G. Escolar, K. Matsue, Y. Nishiura, Hierarchical structures of amorphous solids characterized by persistent homology. Proc. Nat. Acad. Sci. **113**(26), 7035–7040 (2016)
3. T. Ichinomiya, I. Obayashi, Y. Hiraoka, Persistent homology analysis of craze formation. Phys. Rev. E **95**(1), 012504 (2017)
4. T. Nakamura, Y. Hiraoka, A. Hirata, E.G. Escolar, Y. Nishiura, Persistent homology and many-body atomic structure for medium-range order in the glass. Nanotechnol. **26**(30), 304001 (2015)
5. M. Saadatfar, H. Takeuchi, V. Robins, N. Francois, Y. Hiraoka, Pore configuration landscape of granular crystallization. Nat. Commun. **8** (2017)
6. A. Hatcher, *Algebraic Topology*. (Cambridge University Press, 2002)
7. F. Chazal S.Y. Oudot, Towards persistence-based reconstruction in euclidean spaces, in *Proceedings of the 24th Annual Symposium on Computational Geometry*, (ACM, 2008), pp. 232–241
8. F. Chazal, V. De Silva, M. Glisse, S. Oudot, The structure and stability of persistence modules (2012), arXiv:1207.3674
9. S.Y. Oudot, *Persistence Theory: From Quiver Representations To Data Analysis*, vol. 209, (American Mathematical Society, 2015)
10. B.T. Fasy, J. Kim, F. Lecci, C. Maria, V. Rouvreau, *TDA: Statistical Tools For Topological Data Analysis* (2014)
11. U. Bauer, M. Kerber, J. Reininghaus, Distributed computation of persistent homology, in *2014 Proceedings of the 16th Workshop on Algorithm Engineering and Experiments (ALENEX)*, (SIAM, 2014), pp. 31–38
12. C. Maria, J.-D. Boissonnat, M. Glisse, M. Yvinec, The gudhi library: simplicial complexes and persistent homology, in *International Congress on Mathematical Software*, (Springer, 2014), pp. 167–174
13. http://www.wpi-aimr.tohoku.ac.jp/hiraoka_labo/homcloud-english.html
14. I. Obayashi, Y. Hiraoka, Persistence diagrams with linear machine learning models (2017), arXiv:1706.10082
15. T.K. Duy, Y. Hiraoka, T. Shirai, Limit theorems for persistence diagrams (2016), arXiv:1612.08371
16. H. Adams, T. Emerson, M. Kirby, R. Neville, C. Peterson, P. Shipman, S. Chepushtanova, E. Hanson, F. Motta, L. Ziegelmeier, Persistence images: a stable vector representation of persistent homology. J. Mach. Learn. Res. **18**(8), 1–35 (2017)
17. G. Kusano, K. Fukumizu, Y. Hiraoka, Kernel method for persistence diagrams via kernel embedding and weight factor (2017), arXiv:1706.03472

18. G. Kusano, Y. Hiraoka, K. Fukumizu, Persistence weighted gaussian kernel for topological data analysis, in *International Conference on Machine Learning* (2016), pp. 2004–2013

Chapter 6
Polyhedron and Polychoron Codes for Describing Atomic Arrangements

Kengo Nishio and Takehide Miyazaki

Abstract The arrangement of atoms can be represented as a tiling of Voronoi polyhedra by using the Voronoi tessellation. We can know how an atom is surrounded by its first nearest neighbour atoms by knowing the shape of the Voronoi polyhedron associated with that atom. Furthermore, by knowing how a Voronoi polyhedron is surrounded by other Voronoi polyhedra, we can know how an atom is surrounded by its first nearest neighbours, second nearest neighbours, third nearest neighbours, …. However, there existed no methods for describing the arrangements of polyhedra, or atomic arrangements. To overcome this problem, we have recently created the polyhedron and polychoron codes [Sci. Rep. 6, 23455, Sci. Rep. 7, 40269, and Bull. Soc. Sci. Form 32, 1 (2017)]. In this chapter, we review the methods.

Keywords Voronoi polyhedron · Amorphous · Glass · Atomic structure analysis · Molecular dynamics simulation

6.1 Introduction

Since the properties of materials depend on how atoms are arranged [1, 2], understanding the arrangement of atoms is essential for studying the material properties. When we perform molecular dynamics or Monte Carlo simulations, we obtain the xyz coordinates of all the atoms. However, knowing all the atomic coordinates does not mean understanding the atomic arrangements. To understand the atomic arrangements, the essence should be extracted from the raw data of the atomic coordinates.

When studying the atomic arrangements of materials, particularly amorphous materials, the Voronoi tessellation is often used [3–10]. By using this method, the

K. Nishio (✉) · T. Miyazaki
National Institute of Advanced Industrial Science and Technology (AIST), Central 2, Umezono 1-1-1, Tsukuba, Ibaraki 305-8568, Japan
e-mail: k-nishio@aist.go.jp

© The Author(s) 2018
I. Tanaka (ed.), *Nanoinformatics*, https://doi.org/10.1007/978-981-10-7617-6_6

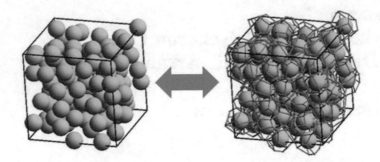

Fig. 6.1 Voronoi tessellation [11]. There is a one-to-one correspondence between the arrangement of atoms (left) and the tiling of Voronoi polyhedra (right)

arrangement of atoms can be represented as a tiling of Voronoi polyhedra (Fig. 6.1). Each Voronoi polyhedron contains one atom. We can know how an atom i is surrounded by its first nearest neighbour atoms by knowing the shape of the Voronoi polyhedron containing the atom i. For example, when the Voronoi polyhedron associated with the atom i is a dodecahedron, the atoms surrounding the atom i occupy the vertices of an icosahedron (Fig. 6.2). Therefore, we can reveal the dominant local atomic arrangements (short-range order) by identifying frequently found Voronoi polyhedra. Furthermore, by knowing how a Voronoi polyhedron is surrounded by other Voronoi polyhedra, we can know how the atom is surrounded by its first nearest neighbours, second nearest neighbours, third nearest neighbours, …. Therefore, we can reveal the long-range order by identifying frequently found assemblages of Voronoi polyhedra.

To classify Voronoi polyhedra, several methods have been proposed. For example, the Voronoi index $\langle n_3 n_4 n_5 n_6 \ldots \rangle$ [3] has often been used in studying amorphous materials. Here, n_i is the number of i-gons of a Voronoi polyhedron. However, different Voronoi polyhedra can accidentally have the same Voronoi index (Fig. 6.3). It is therefore impossible to study details of local atomic arrangements with the Voronoi index. To overcome this problem, Lazar et al. [13] used the Weinberg code [14, 15]. However, there arises a different problem. With this method, a dodecahedron, for example, is encoded as '1 2 3 4 5 1 5 6 7 8 1 8 9

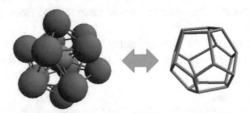

Fig. 6.2 Relation between the atomic arrangement and the shape of the Voronoi polyhedron [12]. First nearest neighbours of the pink atom are blue atoms, occupying the vertices of an icosahedron (left). The Voronoi polyhedron associated with the pink atom is a dodecahedron (right)

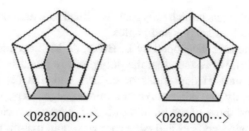

〈0282000⋯〉   〈0282000⋯〉

Fig. 6.3 Problem of Voronoi index [12]. Left and right polyhedra are composed of two squares, eight pentagons, and two hexagons. Therefore, both have the same Voronoi index. However, the left polyhedron is different from the right polyhedron. In fact, the hexagons of the left polyhedron adjoin each other, while the hexagons of the right polyhedron are separate from each other

10 2 10 11 12 3 12 13 14 4 14 15 6 15 16 17 7 17 18 9 18 19 11 19 20 13 20 16 20 19 18 17 16 15 14 13 12 11 10 9 8 7 6 5 4 3 2 1'. Such a long codeword is difficult for human to handle. Since our brain capacity is limited, a shorter codeword is desirable. More seriously, there existed no methods for classifying assemblages of Voronoi polyhedra. Although we might not become conscious, it has severely prevented our understanding of the long-range order of amorphous materials.

Considering that the knowledge of square pyramids was used to construct the ancient pyramids of Egypt at Gaza, the study of polyhedra has a more than 4500 years history [16, 17]. However, as described above, there existed no methods for briefly representing polyhedra and assemblages of polyhedra in a unified way. To overcome this problem, we have created the polyhedron code (p_3-code) and the polychoron code (p_4-code) [11, 12, 18, 19]. The p_3-code is a method for briefly representing polyhedra. It consists of (1) an encoding algorithm for converting a way of how polygons are arranged to form a polyhedron into a sequence of numbers, which we call a polyhedron codeword (p_3-codeword, or p_3 for short) and (2) a decoding algorithm for recovering the original polyhedron from its p_3. The p_4-code is a generalization of the p_3-code for representing assemblages of polyhedra. By using the p_4-code, a way of how polyhedra are arranged to form a polyhedral assemblage can be converted into a sequence of p_3s, which we call a polychoron codeword (p_4-codeword, or p_4 for short), from which the original polyhedral assemblage can be recovered. In this chapter, we review the p_3-code and p_4-code [11, 12, 18, 19].

6.2   Polyhedron Code

6.2.1   Our Way of Viewing a Polyhedron

We regard a polyhedron as a tiling by polygons of the surface of a three-dimensional object that is topologically the same as a sphere. We are interested in the relative arrangements of polygons (which polygons are glued to which polygons), while we ignore measures such as lengths and angles.

According to the idea developed by L. Euler, A. M. Legendre, F. Möbius, and P. R. Cromwell [16], we assume that the polygons are glued such that (1) any pair of polygons meet only at their sides or corners and that (2) each side of each polygon meets exactly one other polygon along an edge. Here, we stress that parts of a polyhedron and those of the building-block polygons are clearly distinguished (Fig. 6.4). Specifically, vertices and edges are zero- and one-dimensional parts of a polyhedron, respectively. On the other hands, corners and sides are zero- and one-dimensional parts of a polygon, respectively. Since this idea plays a central role in our theory, we need a verb to briefly describe the relation between parts of a polyhedron and those of polygons. For this purpose, we use the verb '*contribute*'. For example, when we say that corners contribute to a vertex or a vertex is contributed by corners, we mean that the vertex is a point of a polyhedron at which the corners of polygons meet. We also say that a polygon (side) contributes to a vertex if one of its corners (endpoints) contributes to that vertex. When we say that an edge is contributed by sides, we mean that the edge is a line segment of a polyhedron along which the sides of polygons meet. The face of a polyhedron is a polygon. But when we call a polygon, we regard it as a building block of a polyhedron. So, we may say the edge of a face. But we cannot say the edge of a polygon.

We first describe a method for simple polyhedra. By a simple polyhedron, we mean that every vertex is degree three. Here, the degree of a vertex is the number of edges incident to that vertex. We use the property that every vertex of a simple polyhedron is contributed by three corners in the method for simple polyhedra. After describing the method for simple polyhedra, we generalize it to non-simple polyhedra.

Fig. 6.4 Parts of polygons and parts of a polyhedron [18]

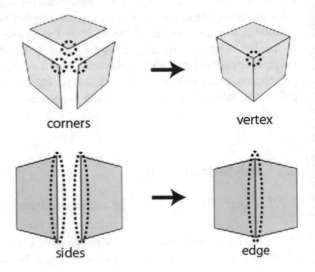

corners   vertex

sides   edge

6.2.2 Decoding Simple Polyhedra

The p_3-code consists of encoding and decoding algorithms. Since the decoding algorithm is incorporated in the encoding algorithm, we first describe the decoding algorithm. To formulate the decoding algorithm, simple but a lot of new ideas must be introduced. However, the completed algorithm can be described easily. We describe the completed algorithm below. See Ref. [18] for its formulation.

6.2.2.1 How to Recover a 34443-Polyhedron

In our theory, we refer to a polyhedron illustrated in Fig. 6.5 as a 34443-polyhedron. The number sequence 34443 is not only the name of the polyhedron, but also instructs how to construct the polyhedron from its building-block polygons. Each number in 34443 indicates a building-block polygon. Specifically, since the left most number is 3, the polygon 1 is a triangle. Similarly, the polygons 2, 3, and 4 are squares, and the polygon 5 is a triangle.

When recovering the 34443-polyhedron from 34443, we first convert each number to the building-block polygon (Fig. 6.6a). To instruct how to assemble the

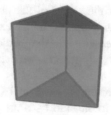

Fig. 6.5 34443-polyhedron [11]

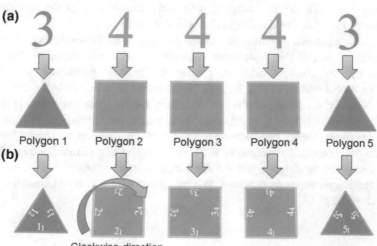

Fig. 6.6 Decoding procedures [11]

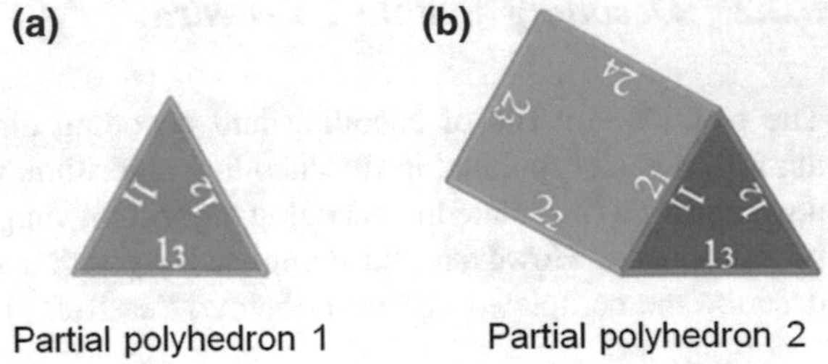

Fig. 6.7 Partial polyhedra 1 and 2 [11]

(a)   **(b)**

Partial polyhedron 1   Partial polyhedron 2

Fig. 6.8 Dangling sides [11]

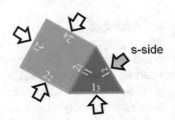

s-side

building-block polygons, we assign identification numbers (IDs) $i_1, i_2, i_3, \ldots$ to the sides of the polygon i in a clockwise direction (Fig. 6.6b). Here, the side i_j means the jth side of the polygon i or the side j of the polygon i. We assume that each symbol i_j has a lexicographical number. In an example shown in Fig. 6.6, the lexicographical number increases in the order of $1_1, \ldots, 1_3, 2_1, \ldots, 2_4, 3_1, \ldots, 3_4, 4_1, \ldots, 4_4, 5_1, \ldots, 5_3$. In general, we define that $i_j < m_n$ when $i < m$, and $i_j < i_k$ when $j < k$.

We call the polygon 1 the partial polyhedron 1 (Fig. 6.7a). By glueing the side 2_1 (the side 1 of the polygon 2) to the side 1_1 of the partial polyhedron 1, we obtain a structure illustrated in Fig. 6.7b, which we call the partial polyhedron 2. Here, we introduce a term '*dangling side*'. The dangling side is a side that is not glued to another side. In the example of Fig. 6.8, the sides $1_2, 1_3, 2_2, 2_3,$ and 2_4 are the dangling sides. We call the dangling side with the smallest ID the *s-side*. In the example of Fig. 6.8, the dangling side 1_2 is the s-side.

By glueing the side 3_1 (the side 1 of the polygon 3) to the s-side 1_2 of the partial polyhedron 2, we obtain a structure illustrated in Fig. 6.9. When a vertex contributed by two dangling sides are contributed by three polyhedra, we call that vertex an illegal vertex. In Fig. 6.9, the illegal vertex is indicated by an open circle. Every vertex of a simple polyhedron is contributed by three polygons. If we proceed decoding with leaving the illegal vertex, then the number of polygons that contribute to that vertex can increase to four, five, six, ..., and we cannot construct a simple polyhedron. Therefore, when an illegal vertex is generated, we *rectify* it by glueing together the dangling sides contributing to it as illustrated in Fig. 6.10. As a result, the illegal vertex is removed. We call the structure thus obtained the partial polyhedron 3.

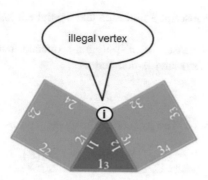

Fig. 6.9 Illegal vertex of a partial polyhedron [11]

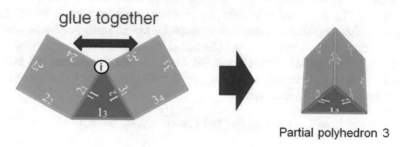

Partial polyhedron 3

Fig. 6.10 How to rectify an illegal vertex [11]

We then repeat procedures described above. Specifically, we glue the side 4_1 (the side 1 of the polygon 4) to the s-side 1_3 of the partial polyhedron 3. As a result, two illegal vertices are generated. We, therefore, rectify them, and obtain the partial polyhedron 4. Then, we glue the side 5_1 (the side 1 of the polygon 5) to the s-side 2_3 of the partial polyhedron 4, and rectify illegal vertices. As a result, the 34443-polyhedron is completed.

In addition to the number sequence 34443, the polyhedron illustrated in Fig. 6.5 can be constructed from 43434 or 44343. Note that the number sequence 34443 is the sequence of numbers three, four, four, four and three. To give only one unique number sequence to the polyhedron, we regard the number sequences as numbers. Since the 34443, thirty-four thousand four hundred forty-three, is the smallest of three, we define it as the unique number sequence of the polyhedron. Therefore, we call the polyhedron the 34443-polyhedron.

6.2.2.2 Polyhedron Codeword

As we have seen, the polyhedron illustrated in Fig. 6.5 can be represented by 34443. We refer to the number sequence 34443 that represents the polyhedron as the p_3-codeword. The subscript 3 indicates that a polyhedron is a three-dimensional object.

The p_3 formally consists of a polygon-sequence codeword (ps_2) and a side-pairing codeword (sp), and is denoted as

$$p_3 = ps_2; sp.$$

Here, ';' is a separator. The ps_2-codeword is denoted as

$$ps_2 = p_2(1)p_2(2)p_2(3)\ldots p_2(F).$$

Here, $p_2(i)$ is the number of sides of the polygon i. F is the number of faces of the polyhedron, in other words the number of polygons of the polyhedron. Although the formal form is $p_3 = ps_2; sp$, the p_3-codeword of the polyhedron illustrated in Fig. 6.5 consists of only ps_2. In other words, $p_3 = ps_2 = 34443$. There are many polyhedra whose p_3 does not have sp. However, some polyhedra need sp. For example, Tutte's polyhedron illustrated in Fig. 6.11 is represented by $p_3 = 4555A4559554AA55555454555; E_6 9_6$. Here, A and E indicate 10 and 13, respectively. $sp = E_6 9_6$ instructs that the side E_6 should be glued to the side 9_6. The sp-codeword is formally denoted as

$$sp = y(1)x(1)y(2)x(2)y(3)x(3)\ldots y(N_{na})x(N_{na}).$$

Here, we refer to the pair of $y(i)$ and $x(i)$ as a *necessary additional pair* (*necessary a-pair*). Note that a necessary a-pair is identical with a non-curable a-pair of Ref. [18]. To stress that the a-pair is necessary, we call a non-curable a-pair a necessary a-pair in this chapter. The necessary a-pair $y(i)x(i)$ instructs that the sides $y(i)$ and $x(i)$ should be glued together. N_{na} is the number of the necessary a-pairs. Note that $y(i) > x(i)$ and $y(i) < y(i+1)$.

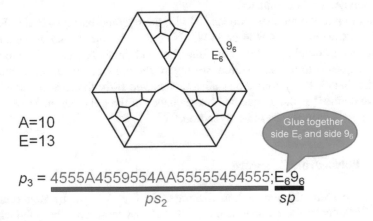

Fig. 6.11 Tutte's polyhedron [12]

6.2.2.3   Algorithm for Recovering the Original Polyhedron from p_3

In Sect. 6.2.2.1, we have described how to recover the 34443-polyhedron. Here, we describe how to recover the original polyhedron from its $p_3 = ps_2; sp$.

Algorithm A (Fig. 6.12)

1. $i = 1$

   (a) Polygon α is a $p_2(\alpha)$-gon $(1 \leq \alpha \leq F)$.
   (b) Assign IDs $(\alpha_1, \alpha_2, \alpha_3, \ldots, \alpha_{p_2(\alpha)})$ to sides of polygon α in a clockwise direction.
   (c) Polygon 1 is partial polyhedron 1.

2. $i = i + 1$

   (a) Glue side i_1 to the s-side of partial polyhedron $i - 1$.
   (b) When side $y(\beta)$ $(1 \leq \beta \leq N_{na})$ is a side of polygon i, glue side $y(\beta)$ to side $x(\beta)$.
   (c) Rectify illegal vertices.
   (d) Resultant structure is partial polyhedron i.

3. Repeat procedure 2 until all the polygons are placed.

Fig. 6.12 Decoding algorithm [12]

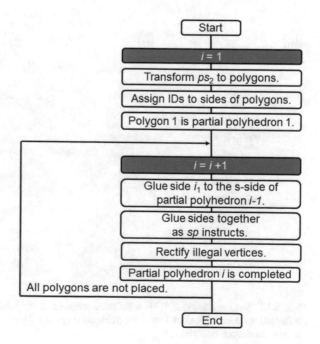

6.2.3   Encoding Simple Polyhedra

6.2.3.1   Schlegel Diagram

So far, we have dealt with three-dimensional polyhedra. For convenience, we use
Schlegel diagrams [17, 20] to illustrate polyhedra from now on. The Schlegel
diagram is the projection of a polyhedron onto a plane. The Schlegel diagram of the
34443-polyhedron is illustrated in Fig. 6.13a. Here, there are two things we should
note. First, the outside polygon *abc* of the Schlegel diagram corresponds to the
interior of the polygon *abc* of the polyhedron. Second, counterclockwise directions
around inside polygons of a Schlegel diagram correspond to clockwise directions
around the corresponding polygons of the polyhedron, while a clockwise direction
around the outside polygon of a Schlegel diagram corresponds to a clockwise
direction around the corresponding polygon of the polyhedron. For example, a
travel $z \rightarrow x \rightarrow a \rightarrow c$ in the Schlegel diagram is in a counterclockwise direction.
However, the corresponding travel in the polyhedron is in a clockwise direction. On
the other hand, a travel $a \rightarrow b \rightarrow c$ around the outside polygon of the Schlegel
diagram and the corresponding travel in the polyhedron are both in clockwise
directions. Figure 6.13b illustrates the decoding process of the 34443-polyhedron
by using Schlegel diagrams.

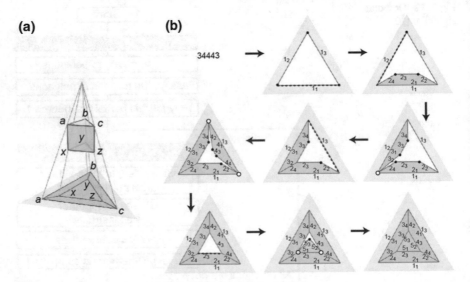

Fig. 6.13 Schlegel diagram [18]. **a** Relation between a polyhedron and its Schlegel diagram.
b Decoding process illustrated by using Schlegel diagrams. Open circles are illegal vertices. Filled
circles are degree two vertices

6.2.3.2 Polygon-Sequence Codeword

When encoding a polyhedron, we first choose a polygon and its side as a seed. Different seeds yield different p_3s. To assign one unique p_3 to that polyhedron, we introduce the lexicographical number $\text{Lex}(p_3)$. We define p_3 with the smallest lexicographical number as the unique p_3 of that polyhedron. We have described the lexicographical numbers of 34443, 43434, and 44343 in Sect. 6.2.2.1. We will describe how to deal with $p_3 = ps_2; sp$ in Sect. 6.2.3.7.

The p_3-codeword consists of ps_2 and sp. We first describe how to generate ps_2. The ps_2-codeword is the sequence of $p_2(i)$ s. Generating ps_2 is, therefore, assigning IDs to polygons. To distinguish between polygons to which IDs have already been assigned and polygons to which IDs will be assigned later, we use colours. We first colour all the polygons. When an ID is assigned, we make the polygon transparent.

We explain how to generate ps_2 by using the 34443-polyhedron as an example. In Fig. 6.14, polyhedra are expressed by using Schlegel diagrams. We choose the outside polygon and its side indicated by the arrow as a seed (Fig. 6.14a). The polygon chosen as the seed is the polygon 1. Since the outside polygon is a triangle, $p_2(1) = 3$. We then assign IDs to the sides of the polygon 1 from the side chosen as the seed in a clockwise direction, and make the polygon 1 transparent (Fig. 6.14b).

When decoding, we have defined a dangling side as a side that is not glued to another side. In encoding, we define a dangling side as a side of a transparent polygon that is glued to a coloured polygon. In Fig. 6.14b, the sides $1_1, 1_2$, and 1_3 are dangling sides. Since 1_1 is the lexicographically smallest, the side 1_1 is the

Fig. 6.14 Encoding process [18]. The dashed lines are s-sides

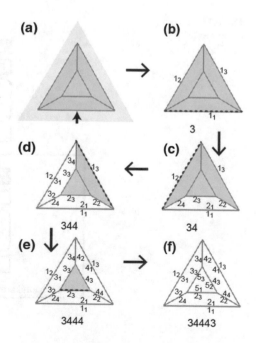

s-side. The polygon 2 is the one that is glued to the s-side 1_1. Since the polygon 2 is a square, $p_2(2) = 4$. We assign IDs to the sides of the polygon 2 from the side glued to the s-side 1_1 in a clockwise direction, and then make the polygon 2 transparent (Fig. 6.14c). We note that the side IDs of the polygon 2 are assigned in a counterclockwise direction in Fig. 6.14c. This is because a counterclockwise direction around any polygon of the Schlegel diagram other than the outside polygon corresponds to a clockwise direction around the corresponding polygon of a polyhedron.

All that is left now is to repeat above described procedures. Specifically, since the polygon that is glued to the s-side 1_2 is the polygon 3, $p_2(3) = 4$. After assigning IDs to its sides, we make the polygon 3 transparent (Fig. 6.14d). The polygon 4 is the one that is glued to the s-side 1_3, and $p_2(4) = 4$. After assigning IDs to its sides, we make the polygon 4 transparent (Fig. 6.14e). The polygon 5 is the one that is glued to the s-side 2_3, and $p_2(5) = 3$. After assigning IDs to its sides, we make the polygon 5 transparent (Fig. 6.14f). All the polygons have become transparent, and $ps_2 = 34443$ has been completed.

To summarize, for a given seed, ps_2 can be generated as follows.

Algorithm B (Fig. 6.15)

1. $i = 1$

    (a) Polygon chosen as a seed is polygon 1.
    (b) Assign IDs $(1_1, 1_2, 1_3, \ldots, 1_{p_2(1)})$ to its sides in a clockwise direction from the side chosen as a seed.

Fig. 6.15 Encoding algorithm [12]

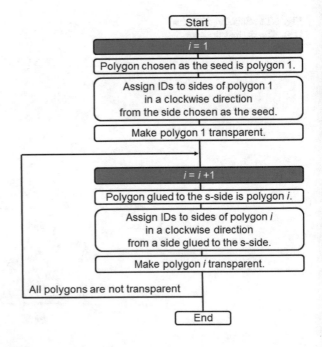

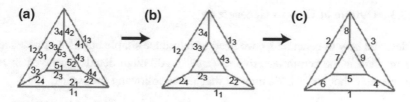

Fig. 6.16 How to assign edge IDs [18]

(c) Make polygon 1 transparent.

2. $i = i + 1$

  a. Polygon glued to the s-side is polygon i.

  b. Assign IDs $(i_1, i_2, i_3, \ldots, i_{p_2(i)})$ to its sides in a clockwise direction from the side that is glued to the s-side.

  c. Make polygon i transparent.

3. Repeat the procedure 2 until all the polygons get transparent.

Face and side IDs are assigned by generating ps_2. By using the side IDs, we can assign edge and vertex IDs. The edge IDs will be used when generalizing the p_3-code to the p_4-code for polyhedral assemblages, while vertex IDs will be used when dealing with non-simple polyhedra.

We first describe how to assign edge IDs (Fig. 6.16). Since two side IDs are associated with every edge, we tentatively assign the smaller side ID to the edge (Fig. 6.16b). Since the tentative IDs thus assigned are not in a sequential order, we relabel the IDs so that the edge i is the one with the ith smallest tentative ID (Fig. 6.16c). To assign vertex IDs, we first assign IDs to corners as illustrated in Fig. 6.17. We note that i_1 is assigned to the corner shared by the sides i_1 and $i_{p_2(i)}$. For $1 < j \leq p_2(i)$, i_j is assigned to the corner shared by the sides i_{j-1} and i_j. Since three corner IDs are associated with every vertex, we tentatively assign the smallest corner ID to the vertex, and then relabel the IDs so that the vertex i is the one with the ith smallest tentative ID.

Fig. 6.17 Corner and edge IDs [18]

6.2.3.3 Outline of How to Generate *sp*

To describe how to generate *sp*, we need to introduce simple but a lot of new ideas. One of them is the zeroth tentative *sp* ($tsp^{(0)}$). Although details are given in Ref. [18], we note that $tsp^{(0)}$ is defined so that it has following properties;

1. The original polyhedron can be recovered from ps_2; $tsp^{(0)}$ by using Algorithm A (Sect. 6.2.2.3),
2. but $tsp^{(0)}$ can contain information that is not needed for recovering the original polyhedron.

The *sp*-codeword is obtained by reducing the redundancy in $tsp^{(0)}$.

To describe a little bit more about the relation between *sp* and $tsp^{(0)}$, we introduce the partial polyhedron $D(i)$, polyhedron $P(i)$, and partial polyhedron $E(i)$. When decoding, polygons are glued together one by one. $D(i)$ is the assemblage of polygons obtained when the polygon i is attached. On the other hand, when generating ps_2, the polygons get transparent one by one. $P(i)$ is the polyhedron obtained when the polygon i becomes transparent. We note that encoded polygons of $P(i)$ are transparent, but the others are coloured. $E(i)$ is the assemblage of polygons obtained by removing the coloured polygons from $P(i)$. $P(4)$ of Fig. 6.14e is reproduced in Fig. 6.18a. $E(4)$ is obtained from $P(4)$ by removing the coloured polygon (Fig. 6.18b). Since recognizing transparent polygons is difficult, we coloured polygons of $E(4)$ (Fig. 6.18c). As for partial polyhedra, we do not distinguish between transparent polygons and coloured polygons. We therefore consider that coloured $E(4)$ is identical with $E(4)$.

Now we look at the sequence of partial polyhedra obtained in encoding $E(1)E(2)E(3)\ldots E(F)$ and the sequence of partial polyhedra obtained in decoding $D(1)D(2)D(3)\ldots D(F)$. When decoding from ps_2; $tsp^{(0)}$, $E(i) = D(i)$ for $1 \leq i \leq F$ (Fig. 6.19a). But, what we need is $E(F) = D(F)$. We therefore admit $E(i) \neq D(i)$ for $i < F$, and reduce redundancy from $tsp^{(0)}$ to obtain *sp* (Fig. 6.19b).

To describe details of $tsp^{(0)}$, we need to describe a-pairs. For this purpose, we first describe a term '*plot*'.

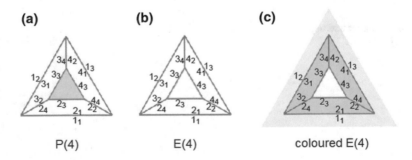

(a) P(4)   **(b)** E(4)   **(c)** coloured E(4)

Fig. 6.18 Relation between polyhedron $P(i)$ and partial polyhedron $E(i)$ [18]

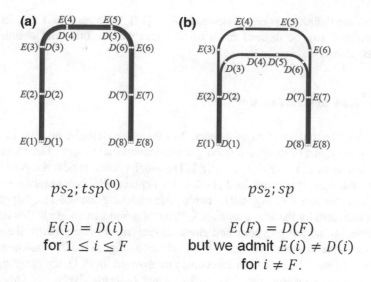

(a)   **(b)**

$$ps_2; tsp^{(0)}$$ $$ps_2; sp$$

$$E(i) = D(i)$$ $$E(F) = D(F)$$
$$\text{for } 1 \leq i \leq F$$ but we admit $E(i) \neq D(i)$
                                                $$\text{for } i \neq F.$$

Fig. 6.19 Comparison between $tsp^{(0)}$ and sp [12]

6.2.3.4 Plot

When two dangling sides of different polygons adjoin each other, we consider that they are *chained*. We call the chain of dangling sides the plot. We also call a separate dangling side the plot. We assign the smallest ID of the dangling sides constituting a plot to that plot. In an example shown in Fig. 6.20, the dangling sides 1_2 and 2_4 of different polygons adjoin each other, so that they are chained. On the other hand, since the dangling sides 1_2 and 1_3 belong to the same polygon, they are not chained. Similarly, the dangling sides 2_3 and 2_4 are not chained. Therefore, the dangling sides 1_2 and 2_4 constitute the plot 1_2. Similarly, the dangling sides 1_4 and 2_2 constitute the plot 1_4. The separate dangling side 1_3 constitutes the plot 1_3 by itself. Similarly, the separate dangling side 2_3 froms the plot 2_3. Here, we point out that the dangling sides of the same plot are all glued to the same polygon.

In Ref. [18], we defined 'chained' as follows: two dangling sides are chained when they contribute to the same vertex contributed by two transparent polygons.

Fig. 6.20 Plot. The pair of Dangling sides contributing to a circle are chained [18]

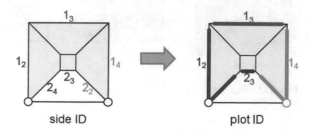

side ID   plot ID

However, the definition is complicated. In Ref. [12], we have found that the same term 'chained' can be defined more briefly, and we use this brief definition as described above.

6.2.3.5 How to Generate $tsp^{(0)}$

The $tsp^{(0)}$-codeword consists of a-pairs. We therefore describe a-pairs. In generating ps_2, polygons of the polyhedron get transparent one by one. The process can be represented as $P(1)P(2)P(3)\ldots P(F)$. The a-pairs relate to how the polygon i is glued to the other polygons in $P(i-1)$. To explain this, we generate ps_2 of a polyhedron illustrated in Fig. 6.21a twice with choosing the outside polygon and the side indicated by the arrow as a seed. When the first generation is finished, IDs are assigned to all the polygons and sides. Therefore, we can perform the second generation with knowing all the IDs in advance. Figure 6.21b illustrates $P(1)$, which is obtained when polygon 1 becomes transparent. In $P(1)$, the dangling sides 1_1, 1_2, 1_3 and 1_4 constitute the plots 1_1, 1_2, 1_3 and 1_4, respectively. The smallest ID plot (*s-plot* for short) is the plot 1_1, to which the polygon 2 is glued. In general, the polygon i is glued to the s-plot of $P(i-1)$. This is because, by definition, the polygon that is glued to the s-side of $P(i-1)$ is the polygon i. Figure 6.21c illustrates $P(7)$, where the polygon 8 is glued to the s-plot 3_4. In addition to the s-plot, the polygon 8 is glued to the plot 5_6, which we call an additional plot. In general, the additional plots of $P(i-1)$ are plots other than the s-plot to which the polygon i is glued. By definition, the smallest ID of dangling sides constituting the additional plot 5_6 is 5_6. The side 5_6 is glued to the side 8_5 of the polygon 8, and we refer to the pair of the sides 8_5 and 5_6 as the a-pair 8_55_6. Note that the lexicographically larger 8_5 proceeds 5_6. As is illustrated in Fig. 6.21d, 10_45_4 is also an a-pair. When generating ps_2 of the polyhedron illustrated in Fig. 6.21a, 8_55_6 and 10_45_4 are a-pairs. By collecting the a-pairs, $tsp^{(0)} = 8_55_610_45_4$.

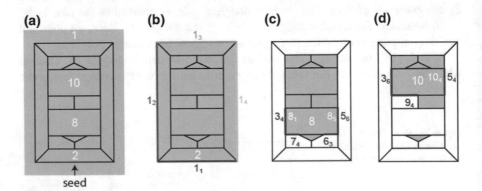

(a)   **(b)**   **(c)**   **(d)**

Fig. 6.21 Explanation of $tsp^{(0)}$ [18]

The $tsp^{(0)}$-codeword is formally denoted as

$$tsp^{(0)} = y_a(1)x_a(1)y_a(2)x_a(2)y_a(3)x_a(3) \ldots y_a(N_a)x_a(N_a).$$

Here, $y_a(i)x_a(i)$ is the ith a-pair, where $y_a(i) > x_a(i)$ and $y_a(i) < y_a(i+1)$. N_a is the number of a-pairs.

6.2.3.6 How to Generate sp

As described above, when we encode the polyhedron illustrated in Fig. 6.21a, we obtain $ps_2; tsp^{(0)} = 458585574755433; 8_5 5_6 10_4 5_4$. The original polyhedron can be recovered from $ps_2; tsp^{(0)}$ using Algorithm A described in Sect. 6.2.2.3 (See Ref. [18] for its proof). But $tsp^{(0)}$ can contain information that is not needed for recovering the original polyhedron.

We now examine whether $10_4 5_4$ is necessary or not. To do so, we try to decode from $458585574755433; 8_5 5_6$ which is obtained by removing $10_4 5_4$ from $ps_2; tsp^{(0)}$ (Fig. 6.22). Since it does not have the information of the a-pair $10_4 5_4$, the sides 10_4 and 5_4 are not glued together in $D(10)$, which is obtained when the polygon 10 is placed. And then the polygon 13 is glued to the side 5_4 in $D(13)$. This means that we cannot recover the original polyhedron without $10_4 5_4$. On the other hand, when we decode $458585574755433; 10_4 5_4$ which is obtained by removing $8_5 5_6$ from $ps_2; tsp^{(0)}$, the sides 8_5 and 5_6 are separate in $D(8)$, but they are glued together in $D(13)$, and the original polyhedron can be recovered (Fig. 6.23). This means that $8_5 5_6$ is not necessary. By removing the unnecessary $8_5 5_6$ from $tsp^{(0)}$, sp is obtained as $10_4 5_4$.

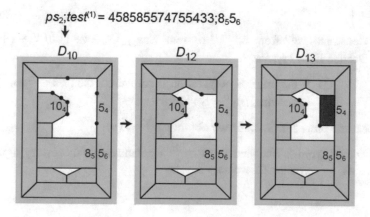

$ps_2; test^{(1)} = 458585574755433; 8_5 5_6$

D_{10} D_{12} D_{13}

Fig. 6.22 Necessary a-pair [18]

$ps_2; test^{(2)} = 458585574755433; 10_4 5_4$

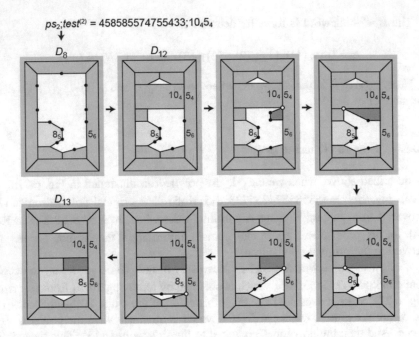

Fig. 6.23 Unnecessary a-pair [18]

For a given $tsp^{(0)}$, sp can be generated as follows.
Algorithm C

1. $i = 0$

   (a) $tsp^{(0)} = y_a(1)x_a(1)y_a(2)x_a(2)y_a(3)x_a(3) \ldots y_a(N_a)x_a(N_a)$.

2. $i = i + 1$

   (a) Construct $test^{(i)}$ from $tsp^{(i-1)}$ by removing $y_a(N_a - i + 1)x_a(N_a - i + 1)$.
   (b) Decode from $ps_2; test^{(i)}$.

           ① If the original polyhedron is recovered, then $tsp^{(i)} = test^{(i)}$.
           ② Otherwise, $tsp^{(i)} = tsp^{(i-1)}$.

3. Repeat the procedure 2, until we obtain $tsp^{(N_a)} = sp$.

   The sp-codeword is obtained from $tsp^{(0)}$ by removing unnecessary a-pairs. In other words, sp consists of necessary a-pairs.

6.2.3.7 Lexicographical Number of p_3

Different seeds yield different p_3s. To assign one unique p_3 to a polyhedron, we describe the lexicographical number $\text{Lex}(p_3)$. We regard $\text{Lex}(p_3)$ as a base-n number, where n is any sufficiently large number as described below. Since p_3 consists of ps_2 and sp, $\text{Lex}(p_3)$ consists of $\text{Lex}(ps_2)$ and $\text{Lex}(sp)$. We first describe $\text{Lex}(ps_2)$. Since ps_2 is the sequence of F numbers, we define $\text{Lex}(ps_2)$ as a F-digit base-n number $\text{Lex}(ps_2) = p_2(1)p_2(2)p_2(3)\ldots p_2(F)$, where $p_2(i)$ is the value of the $(F-i+1)$th digit. Note that $p_2(i)$ in the number sequence $ps_2 = p_2(1)p_2(2)p_2(3)\ldots p_2(F)$, is the ith number. Similarly, $\text{Lex}(sp)$ is a $2N_{\text{na}}$-digit base-n number $y(1)x(1)y(2)x(2)y(3)x(3)\ldots y(N_{\text{na}})x(N_{\text{na}})$. $\text{Lex}(p_3)$ is the concatenation of $\text{Lex}(ps_2)$ and $\text{Lex}(sp)$. For reference, the concatenation of 24 expressed in base-10 (twenty-four) and 5 expressed in base-10 (five) is 245 (two hundred forty-five).

Note that, to regard $\text{Lex}(p_3)$ as a base-n number, n should be larger than $p_2(i)$, $y(i)$ and $x(i)$. Since the number of sides of a polyhedron is $2E = 6(F-2)$, n should be larger than $2E$. Here, E is the number of edges of a polyhedron.

Since there are $2E$ different selections of seeds, $2E$ different p_3s can be generated from a polyhedron. If we consider that the polyhedron is identical with its mirror image, additional $2E$ different p_3s can be generated from the mirror image. By selecting the smallest of $4E$ different p_3s, we can assign one unique p_3 to the polyhedron. By considering that a polyhedron is identical with its mirror image, the unique p_3s can be used to determine the isomorphism of polyhedral graphs. A polyhedral graph is a planer triply connected graph that has no multiple edges. If we regard a region enclosed by two edges as a 2-gon, p_3s can be used to determine the isomorphism of planer triply connected graphs. When we want to distinguish the polyhedron from its mirror image, we may select the smallest of $2E$ different p_3s. But the unique p_3s thus generated cannot be used to determine the isomorphism of polyhedral graphs.

6.2.3.8 Solving the Problem of Voronoi Index

As is shown in Fig. 6.24, two different polyhedra have the same Voronoi index $\langle 0282000\ldots\rangle$. By using p_3s, we can say that 455665555455- and

Fig. 6.24 Solving the problem of Voronoi index [12]

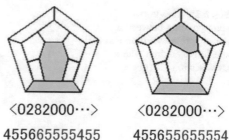

$\langle 0282000\cdots\rangle$ $\langle 0282000\cdots\rangle$

455665555455   455655655554

455655655554-polyhedra have the same Voronoi index $\langle 0282000\ldots\rangle$. In other words, different polyhedra can be distinguished by using our method.

6.2.4 Non-simple Polyhedron

6.2.4.1 Cut-and-Dot Method

So far, we have assumed that polyhedra are simple. The theory for simple polyhedra can be easily generalized to non-simple polyhedra that have one or more vertices of degree more than three. Figure 6.25 illustrates a pentagonal pyramid. Since the apex is degree five, the pentagonal pyramid is a non-simple polyhedron. But when we cut the apex, a simple polyhedron can be obtained. By distinguishing the cross section from other faces, we can establish a one-to-one correspondence between a non-simple polyhedron and a simple polyhedron with a cross section. Using this relation, the non-simple polyhedron can be represented by 5444445. The dot over '5' indicates that the pentagon is a cross section which should be shrunk to a vertex. Note that this approach was inspired by Kempe's patch method for the four colour problem [17].

When dealing with polyhedra without cross sections, we have defined the s-side as the smallest ID dangling side. As for polyhedra with cross sections, we define the s-side as the smallest ID dangling side of polygons that are not cross sections. The reason is given later. A non-simple polyhedron can be encoded as follows:

1. Cut vertices of degree more than three.
2. Choose a polygon that is not a cross section and its side that does not contribute to any edge of the cross sections as a seed.
3. Generate p_3 using Algorithm B.
4. Put dots over numbers of ps_2 corresponding to the cross sections.

By encoding, IDs can be assigned to faces, edges and vertices of a non-simple polyhedron as follows (Fig. 6.26):

1. Assign IDs to a polyhedron with cross sections.
2. Shrink cross sections to vertices. As a result, some IDs disappear.

Fig. 6.25 One-to-one corresponding between a non-simple polyhedron and a simple polyhedron with a cross section [18]

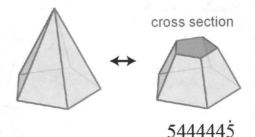

cross section

$5444\dot{4}\dot{4}\dot{5}$

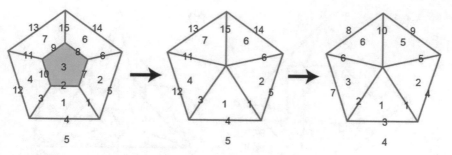

Fig. 6.26 How to assign IDs to a non-simple polyhedron [18]

3. Relabel IDs so that they become sequential orders.

We have modified the definition of the s-side. As a result, IDs assigned using the method described above conform to IDs assigned by directly applying Algorithm B to the non-simple polyhedron. Note that a codeword can be generated by directly applying Algorithm B to a non-simple polyhedron. However, the original polyhedron cannot be recovered from the codeword using Algorithm A. For example, a pentagonal pyramid can be encoded as 533333. But it cannot be recovered from 533333.

To assign one unique codeword to a non-simple polyhedron, we define $\text{Lex}(p_3)$ for p_3 with dots. For this purpose, we define $\text{Lex}(ps_2)$ as the concatenation of $\text{Lex}\left(ps_2^{(1)}\right)$ and $\text{Lex}\left(ps_2^{(2)}\right)$. The $ps_2^{(1)}$-codeword is obtained from ps_2 by replacing every number without a dot by 0 and then removing all dots, while $ps_2^{(2)}$ is obtained by removing all dots from ps_2. For example, when $ps_2 = 544444\dot{5}$, $ps_2^{(1)} = 0000005$ and $ps_2^{(1)} = 5444445$. Therefore, $\text{Lex}(ps_2) = 00000055444445$.

6.2.4.2   Using Duality

The cut-and-dot method described in the previous section is applicable to all non-simple polyhedra, but is sometimes inefficient. For example, the octahedron is encoded as $66\dot{4}6\dot{4}6\dot{4}6\dot{4}6\dot{4}6\dot{4}6$. Since 6 is repeated twice and then $\dot{4}6$ is repeated six times, $66\dot{4}6\dot{4}6\dot{4}6\dot{4}6\dot{4}6\dot{4}6$ can be shortened to $6^2(\dot{4}6)^6$. However, the representation of the octahedron with beautiful symmetries is not beautiful. We think that it is a problem. To overcome this problem, we use the duality of polyhedra [17, 21]. Since the octahedron is the dual of the hexahedron, we represent the octahedron as $\star 4^6$. Here, $\star$ represents the dual, 4^6 is p_3 of the hexahedron, and $\star 4^6$ means the dual of the 4^6-polyhedron. We describe the details of this method below.

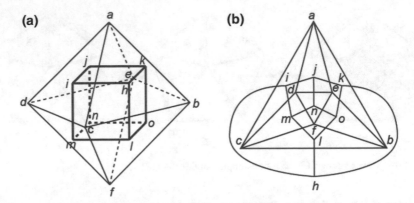

Fig. 6.27 Duality. **a** The dual of the octahedron is the hexahedron. **b** Graph representation of (**a**). The octahedron and hexahedron are dual to each other [18]

For any polyhedron, its dual is constructed as follows (Fig. 6.27):

1. Put a vertex v_i^* of the dual polyhedron on the face f_i of the given polyhedron.
2. Link v_i^* and v_j^* by the edge e_{ij}^*, when f_i and f_j share an edge e.

The dual of the octahedron is the hexahedron, and the dual of the hexahedron is the octahedron. Thus, the octahedron and hexahedron are dual to each other. By using ★, a polyhedron composed of triangles only can be briefly represented, for its dual is a simple polyhedron.

Since there is a one-to-one correspondence between an original polyhedron and its dual, we can determine the edge and face IDs of the original from those of its dual. For example, in an example shown in Fig. 6.27b, the face dcf of the octahedron corresponds to the vertex m of the hexahedron. We, therefore, assign the ID of the vertex m to the face dcf. Similarly, the edge dc of the octahedron corresponds to the edge mi. We, therefore, assign the ID of the edge mi to the edge dc.

When we encode a simple polyhedron, we first choose a seed, and then generate p_3 from the seed. The side chosen as a seed contributes to the edge 1, and the polygon chosen as a seed becomes the polygon 1. Therefore, choosing a seed is determining the edge 1 and face 1. When we encode a non-simple polyhedron using the duality, we also choose a side and polygon of the non-simple polyhedron as a seed. We then choose a seed for its dual so that the edge of the dual corresponding to the edge of the original contributed by the side chosen as a seed becomes the edge 1 and that the vertex of the dual corresponding to the polygon of the original chosen as a seed becomes the vertex 1. For example, when we encode the octahedron with choosing the polygon dcf and its side dc as a seed, the polygon $mihl$ and its side mi is a seed for its dual. Then 1 is assigned to the vertex m of the dual, and therefore to the polygon dcf of the original. Similarly, 1 is assigned to the edge mi of the dual, and therefore to the edge dc of the original contributed by the side dc.

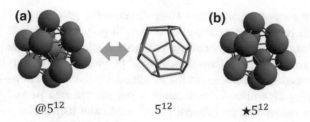

Fig. 6.28 Relation between a local atomic arrangement and a Voronoi polyhedron [12]. **a** The Voronoi polyhedron associated with the pink atom is 5^{12}. The pink and its neighbouring atoms form a $@5^{12}$ cluster. **b** The atoms adjacent to the pink atom occupy the vertices of a $\star 5^{12}$-polyhedron

Note that in Ref. [18], we determine the seed for the dual in a different way. But for simplicity, we have modified the method in Ref. [12], and the modified version is described above.

To assign one unique p_3 to any non-simple polyhedron, we assume $\text{Lex}(\star) = 1$ and define $\text{Lex}(\star p_3)$ as the concatenation of $\text{Lex}(\star)$ and $\text{Lex}(p_3)$. By doing so, $\text{Lex}(\star 4^6) = 1444444$, while $\text{Lex}\left(6^2(46)^6\right) = 00404040404040406646464646464646$. Therefore, the unique p_3 of the octahedron is $\star 4^6$.

6.2.5 Relation Between an Atomic Arrangement and a Voronoi Polyhedron

To represent atomic arrangements, we introduce the symbol @ that relates a Voronoi polyhedron and its corresponding atomic arrangement as follows. We refer to the Voronoi polyhedron associated with the atom i as the Voronoi polyhedron i. In other words, the atom i and its nearest neighbour atoms define the Voronoi polyhedra i. When the Voronoi polyhedron i is a p_3-polyhedron, we represent the arrangement of atoms defining the Voronoi polyhedron, namely the atom i and its first nearest neighbour atoms, as an $@p_3$-cluster (Fig. 6.28a). Note that, in the $@p_3$-cluster, first nearest neighbour atoms of the atom i occupy the vertices of a $\star p_3$-polyhedron and the atom i locates at the centre of the $\star p_3$-polyhedron (Fig. 6.28b).

6.3 Polychoron Code

We can study the short-range order of amorphous materials by classifying the Voronoi polyhedra with the p_3-code. We can study the long-range order by classifying assemblages of Voronoi polyhedra. A polyhedral assemblage can be regarded as a part of a polychoron (four-dimensional polytope). The p_3-code for polyhedra can be easily generalized to deal with polychora, for it is based on the hierarchy of structures of polytopes: a polyhedron (three-dimensional polytope) is an assemblage of polygons (two-dimensional polytopes). In this section, we generalize the p_3-code for polyhedra to the p_4-code for polychora. The p_4-code consists of the encoding algorithm for converting a polychoron into p_4 and the decoding algorithm for recovering the original polychoron from its p_4. The p_4-code can be used to study the long-range order of amorphous materials.

Since we are living in the three-dimensional world, understanding four-dimensional objects is not easy. But understanding Schlegel diagrams of polychora is not difficult. As shown in Fig. 6.13a, a polyhedron can be represented as a two-dimensional object by using a Schlegel diagram. Similarly, a polychoron can be represented as a three-dimensional object by using a Schlegel diagram. The Schlegel diagram of a polychoron $abcdefgh$ is illustrated in Fig. 6.29. We can see that the polychoron is an assemblage of two 3333-polyhedra and four 34443-polyhedra, and polyhedra are glued face to face. We note that the outside of the polyhedron $abcd$ of the Schlegel diagram corresponds to the inside of the polyhedron $abcd$ of the polychoron. By using the p_4-code, the polychoron $abcdefgh$ is represented by $p_4 = 3333\ 34443\ 34443\ 34443\ 34443\ 3333$. The p_4-codeword is the sequence of p_3s and instructs how to construct the polychoron from its building-block polyhedra. Since the left most p_3 is 3333, the polyhedron 1 is a 3333-polyhedron. Similarly, the polyhedra 2, 3, 4, and 5 are 34443-polyhedra, and the polyhedron 6 is a 3333-polyhedra. To describe the p_4-code, we first describe the relations between parts of polyhedra and parts of a polychoron.

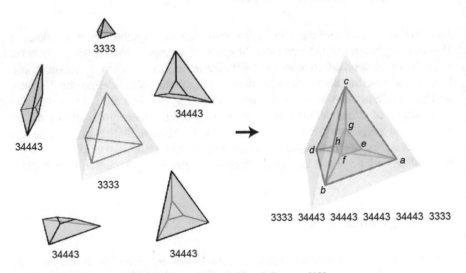

Fig. 6.29 Polychoron represented by using a Schlegel diagram [19]

6.3.1 Our Way of Viewing a Polychoron

We regard a polychoron as a tiling by polyhedra of the surface of a four-dimensional object that is topologically the same as a 3-sphere. We assume that the polyhedra are glued together such that (1) any pair of polyhedra meet only at their faces, edges, or vertices and that (2) each face of each polyhedron meets exactly one other polyhedron along a ridge. We distinguish parts of a polychoron and parts of its building-block polyhedra (Fig. 6.30). The 0-face is a point of a polychoron where the vertices of polyhedra meet; the peak is a line segment of a polychoron where the edges of polyhedra meet; the ridge is an area of a polychoron where the faces of polyhedra meet. The cell of a polychoron is a polyhedron.

6.3.2 1-Simple Polychoron

Polyhedra can be classified into simple and non-simple polyhedra according to the degrees of the vertices. As described above, we have first created the method for simple polyhedra, and then generalized it for non-simple polyhedra. Polychora can be classified into simple and non-simple polychora according to the types of the

Fig. 6.30 Parts of polyhedra and parts of a polychoron [18]

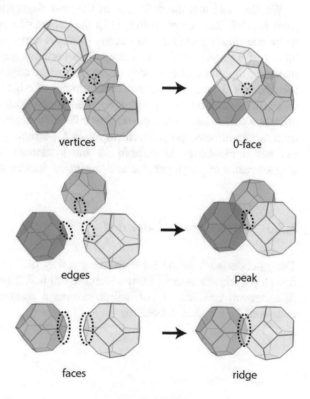

vertices   0-face

edges   peak

faces   ridge

0-faces as well. A polychoron whose 0-faces are all degree four is called a simple polychoron. Here, the degree of a 0-face is the number of peaks incident to that 0-face. However, to describe the p_4-code, we need to classify polychora according to the types of the peaks. We, therefore, generalize the concept of 'simple'. For this purpose, we first define the degree of a peak as the number of ridges incident to that peak. We then call a polychoron whose peaks are all degree three a *1-simple* polychoron. The '1' indicates the simplicity regarding one-dimensional parts of a polychoron, namely peaks or 1-faces. We first describe the method for 1-simple polychora, and then briefly describe the generalization of the method for non-1-simple polychora. In the method for 1-simple polychora, we use the property that every peak of a 1-simple polychoron is contributed by three edges.

Note that, using our generalized notation, a simple polychoron is called a 0-simple polychoron. Here, '0' means the simplicity regarding zero-dimensional parts of a polychoron, namely 0-faces. Similarly, a simple polyhedron is a 0-simple polyhedron. If a polychoron is 0-simple, then it is also 1-simple, for its peaks are all degree three. However, a 1-simple polychoron is not always 0-simple. In other words, a set of 0-simple polychora is a subset of 1-simple polychora. For example, the 24-cell composed of 24 octahedra is 1-simple, but is not 0-simple. The polyhedra of a 0-simple polychoron are all 0-simple. On the other hand, the polyhedra of a 1-simple polychoron can be non-0-simple. For example, the octahedra of the 1-simple 24-cell are non-0-simple.

We also note that the definition of the peak degree described above is different from the definition given in Ref. [18]: the degree of a peak is the number of polyhedra contributing to that peak. In the previous definition, the polyhedra (building blocks of a polychoron) are focused. The previous definition, therefore, does not match the definition of the vertex degree where the edges (parts of a polyhedron) are focused: the degree of a vertex is the number of edges incident to that vertex. To match the definition of the vertex degree, namely to focus on the parts of a polychoron, we have introduced the new definition. In addition, in Ref. [18], we used the term 'a non-affected polychoron' to mean a 1-simple polychoron. However, since our new terminology is suitable for the systematic description of fundamental characteristics of polytopes, we use '1-simple' instead of 'non-affected'.

6.3.3 Polychoron Codeword

The p_4-code is a method for converting a way of how polyhedra are arranged to form a polychoron into p_4 from which the original structure can be recovered. The p_4-codeword consists of polyhedron-sequence codeword (ps_3) and face-pairing codeword (fp), and is denoted as

$$p_4 = ps_3; fp.$$

Here, ps_3 is denoted as

$$ps_3 = p_3(1)p_3(2)p_3(3)\ldots p_3(C),$$

where $p_3(i)$ is p_3 of polyhedron i, C is the number of cells of a polychoron, in other words the number of polyhedra of a polychoron. The fp-codeword is denoted as

$$fp = w(1)z(1)v(1)w(2)z(2)v(2)w(3)z(3)v(3)\ldots w(N_{\text{na}})z(N_{\text{na}})v(N_{\text{na}}).$$

Here, $w(i)z(i)v(i)$ is the necessary a-pair for the polychoron, instructing that the face $w(i)$ and face $v(i)$ should be glued together in such a way that the edge of the face $w(i)$ contributed by the side $z(i)$ is glued to the smallest ID edge of the face $v(i)$. N_{na} is the number of necessary a-pairs. Note that $w(i) > v(i)$ and $w(i) < w(i+1)$.

6.3.4 How to Generate ps_3

To encode a polychoron, we first chose a polyhedron, face of the polyhedron, and edge of the face as a seed. Depending on how we choose the seed, p_4 changes. By using lexicographical numbers described in Sect. 6.3.8, one unique p_4 can be assigned to each polychoron. The p_4-codeword consists of ps_3 and fp. We first describe how to generate ps_3.

Generating ps_3 is assigning IDs to polyhedra. We use colours to distinguish between polyhedra to which IDs have already been assigned and polyhedra to which IDs will be assigned later. We first colour all the polyhedra. When an ID is assigned, we make the polyhedron transparent (Fig. 6.31). We refer to the face of a transparent polyhedron glued to a coloured polyhedron as a dangling face. To instruct how to assign IDs to polyhedra, we assign IDs to the faces (edges) of every polyhedron by encoding the polyhedron. Specifically, we assign i_j to the jth face (edge) of the polyhedron i. The dangling face with the smallest ID is called the s-face.

For a given seed, ps_3 can be generated as follows.

Algorithm D

1. $i = 1$

   a. Polyhedron chosen as the seed is polyhedron 1.
   b. Generate p_3 of polyhedron 1 and assign IDs to its faces (edges) by encoding polyhedron 1 in such a way that 1_1 is assigned to the face (edge) chosen as the seed.
   c. Make polyhedron 1 transparent.

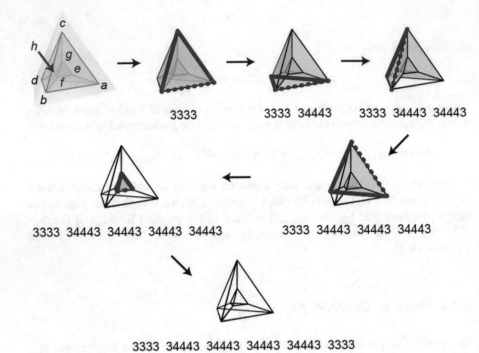

Fig. 6.31 Encoding a polychoron. The edges of each s-face are indicated by red lines. The smallest ID edge of each s-face is indicated by a dotted line [18]

2. $i = i + 1$

    a. Polyhedron glued to the s-face is polyhedron i.

    a. Generate p_3 of polyhedron i and assign IDs to its faces (edges) by encoding polyhedron i in such a way that i_1 is assigned to the face glued to the s-face (the edge glued to the smallest ID edge of the s-face).

    b. Make polyhedron i transparent.

3. Repeat the procedure 2 until all the polyhedra get transparent.

6.3.5 How to Generate $tfp^{(0)}$

To describe fp, we introduce a zeroth tentative fp ($tfp^{(0)}$). For this purpose, we introduce some ideas. When two dangling faces of different polyhedra adjoin each other, we consider that they are chained. The chained dangling faces constitute a plot. A separate dangling face also constitutes a plot by itself. We assign the smallest face ID of all the dangling faces constituting a plot to that plot.

In generating ps_3, polyhedra of the polychoron get transparent one by one. We consider a polychoron $P_4(i-1)$ obtained when the $(i-1)$th polyhedron gets transparent. The polyhedron i is glued to the s-plot of $P_4(i-1)$. When the polyhedron i is glued to plots other than the s-plot, those plots are the a-plots for the polychoron. Suppose that the face w_a of the polyhedron i is glued to the face v_a of the a-plot v_a in such a way that the edge contributed by the side z_a of the polygon w_a is glued to the smallest ID edge of the face v_a. Then the pair of $w_a z_a v_a$ is the a-pair for the polychoron. The $tfp^{(0)}$-codeword is obtained by collecting the a-pairs in such a way that $w_a(i) < w_a(i-1)$;

$$tfp^{(0)} = w_a(1)z_a(1)v_a(1) \ldots w_a(N_a)z_a(N_a)v_a(N_a).$$

Here, N_a is the number of a-pairs.

6.3.6   How to Recover a Polychoron from ps_3; $tsp^{(0)}$

To describe how to recover a polychoron from its ps_3; $tsp^{(0)}$, we first describe the dangling face for decoding and an illegal peak. In decoding, we call a face that is not glued to another polyhedron a dangling face. When a peak contributed by two dangling faces is also contributed by three polyhedra, we call that peak an illegal peak. The illegal peak can be rectified by glueing the two dangling faces together.

The polychoron can be recovered from its ps_3; $tsp^{(0)}$ as follows:
Algorithm E

1. $i = 1$

   a. Polyhedron α is a $p_3(\alpha)$-polyhedron $(1 \le \alpha \le C)$.
   b. Assign α_j to the jth face (edge) of polyhedron α.
   c. Polyhedron 1 is partial polychoron 1.

2. $i = i + 1$

   a. Glue face i_i (face 1 of polyhedron i) to the s-face of partial polychoron $i - 1$.
   c. When $w_a(\beta)$ is the face ID of polyhedron i $(1 \le \beta \le N_a)$, glue together faces $w_a(\beta)$ and $v_a(\beta)$ in such a way that the edge contributed by side $z_a(\beta)$ is glued to the smallest ID edge of the face $v_a(\beta)$.
   d. Rectify illegal peaks.
   e. Structure thus obtained is partial polychoron i.

3. Repeat the procedure 2 until all the polyhedra are placed.

For reference, recovering a polychoron from 3333 34443 34443 34443 34443 3333 is illustrated in Supplemental Information of Ref. [18].

6.3.7 How to Generate fp

The original polychoron can be recovered from $ps_3; tfp^{(0)}$ by using Algorithm E. However, $tfp^{(0)}$ can contain information that is not necessary for recovering the original polychoron. The fp-codeword is obtained by reducing the redundancy in $tfp^{(0)}$ as follows;

Algorithm F

1. $i = 0$

    a. $tfp^{(0)} = w_a(1)z_a(1)v_a(1) \ldots w_a(N_a)z_a(N_a)v_a(N_a)$.

2. $i = i + 1$

    a. Construct $test^{(i)}$ by removing $w_a(N_a - i + 1)z_a(N_a - i + 1)v_a(N_a - i + 1)$ from $tfp^{(i-1)}$.

    b. Decode from $ps_3; test^{(i)}$.

        i. If the original polychoron is recovered, then $tfp^{(i)} = test^{(i)}$.
        ii. Otherwise, $tfp^{(i)} = tfp^{(i-1)}$.

3. Repeat the procedure 2 until $fp = tfp^{(N_a)}$ is obtained.

6.3.8 Lexicographical Number of p_4

Different p_4s are generated from different seeds. To determine one unique p_4, we define $\text{Lex}(p_4)$ as follows. Since $p_4 = ps_3; fp$, $\text{Lex}(p_4)$ is the concatenation of $\text{Lex}(ps_3)$ and $\text{Lex}(fp)$. Since ps_3 is the sequence of $p_3(i)$s, $\text{Lex}(ps_3)$ is a C-digit number $\text{Lex}(p_3(1))\text{Lex}(p_3(2))\text{Lex}(p_3(3)) \ldots \text{Lex}(p_3(C))$, where $\text{Lex}(p_3(i))$ is the value of the $(C - i + 1)$th digit. Similarly, $\text{Lex}(fp)$ is a $3N_a$-digit number $\text{Lex}(w(1))\text{Lex}(z(1))\text{Lex}(v(1)) \ldots \text{Lex}(w(N_{na}))\text{Lex}(z(N_{na}))\text{Lex}(v(N_{na}))$. A total of $12P$ p_4s are obtained from a polychoron and its mirror image, where P is the number of peaks. By choosing the lexicographically smallest one, we can assign one unique p_4 to a polychoron.

6.3.9 Non-1-Simple Polychora

A 1-simple polychoron has one or more peaks of degree more than three. By cutting such peaks and distinguishing cross-section polyhedra from other polyhedra, a one-to-one correspondence can be established between a non-1-simple polychoron and a 1-simple polychoron with cross-section polyhedra. By using this correspondence, the p_4-code can be generalized to deal with non-1-simple poly-chora (See [18] for details). However, this approach is not always efficient. So, we also use the duality of polychora. By using the duality, the 5-cell can be represented by T^4, the 8-cell by H^8, the 16-cell by $\star H^8$, the 24-cell by O^{24}, the 120-cell by D^{120}, and the 600-cell by $\star D^{120}$. Here, $T = 3333 = 3^4$ represents the tetrahedron, $H = 444444 = 4^6$ represents the hexahedron, $O = \star H$ represents the octahedron, and $D = 555555555555 = 5^{12}$ represents the dodecahedron.

6.3.10   Ridge-Sequence Codeword

A polychoron relating to an amorphous material is 0-simple. Since they are all 0-simple, the polyhedra of a 0-simple polychoron can be represented without '·' and '$\star$'. In other words, all the numbers of sides of polygons of a 0-simple polychoron are recorded in p_4. We introduce a ridge-sequence codeword (rs) to briefly represent a 0-simple polychoron below.

We first describe tentative ridge IDs and ridge IDs. Since two face IDs are associated with every ridge, we tentatively assign the smaller face ID to the ridge. Since the tentative IDs thus assigned are not in a sequential order, we relabel the IDs so that the ridge i is the one with the ith smallest tentative ID.

In a polychoron, polyhedra are glued together face to face. For example, in the polychoron illustrated in Fig. 6.29, the first face of the polyhedron 2 (34443-polyhedron) is glued to the first face of the polyhedron 1 (3333-polyhedron). Since the face 2_1 is glued to the face 1_1, $p_2(2_1) = p_2(1_1) = 3$. Here, $p_2(i_j)$ is the value of p_2 of jth number of ps_2 of the polyhedron i. In general, when the face a_b is glued to the face x_y, $p_2(a_b) = p_2(x_y)$. The ridge is a part of a polychoron where two faces of polyhedra meet. Suppose that $a < x$ and $r_t(a_b)$ is the number of peaks of the ridge with a tentative ID a_b. Then $p_2(a_b) = p_2(x_y) = r_t(a_b)$. Since $r_t(a_b)$ is recorded twice, p_4 is redundant. This originates in that we regard a polychoron as an assemblage of polyhedra. If we regard a polychoron as an assemblage of ridges and use rs, we can reduce the redundancy in p_4 (See Ref. [19] for details). The rs-codeword is denoted as,

$$rs = r(1)r(2)r(3)\dots r(R).$$

Here, $r(i)$ is the number of peaks of the ridge i. R is the number of ridges of a polychoron. By using rs, we can represent a polychoron whose p_4 is 3333 34443 34443 34443 34443 3333 by a briefer codeword $p_4^{(rs)} = rs = 33334443443433$ (Fig. 6.32).

Fig. 6.32 Ridge-sequence
codeword [19]

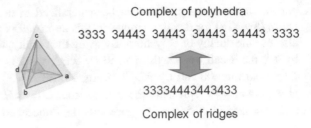

Complex of polyhedra

3333 34443 34443 34443 34443 3333

33334443443433

Complex of ridges

Fig. 6.33 Relation between a
local atomic arrangement and
an assemblage of Voronoi
polyhedra. Here, $D = 5^{12}$ is p_3
of the dodecahedron

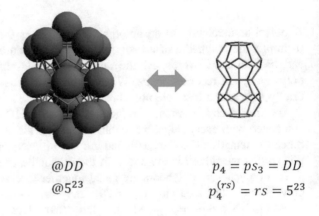

$@DD$

$@5^{23}$

$p_4 = ps_3 = DD$

$p_4^{(rs)} = rs = 5^{23}$

6.3.11 Relation Between a Local Atomic Arrangement and an Assemblage of Voronoi Polyhedra

An assemblage of polyhedra can be regarded as a partial polychoron, and is represented by p_4. For example, as is illustrated in Fig. 6.33, an assemblage of two dodecahedra is represented by $p_4 = ps_3 = DD$ ($p_4^{(rs)} = rs = 5^{23}$). The arrangement of atoms that define the polyhedral assemblage is represented by $@DD$ (@ 5^{23}). The $@DD$-cluster (@ 5^{23}-cluster) can be regarded as overlapping two $@D$-clusters.

6.4 Summary

We have reviewed the p_4- and p_3-codes, which we have created as a method for studying the structures of amorphous materials [11, 12, 18, 19]. The p_3-code is a method for briefly representing polyhedra. It consists of (1) an encoding algorithm for converting a way of how polygons are arranged to form a polyhedron into p_3 and (2) a decoding algorithm for recovering the original polyhedron from its p_3. We can study the short-range order of amorphous materials by classifying Voronoi polyhedra according to their p_3s. The p_4-code is a generalization of the p_3-code for representing assemblages of polyhedra. By using the p_4-code, a way of how polyhedra are arranged to form a polyhedral assemblage can be converted into p_4, from which the original polyhedral assemblage can be recovered. We can study the long-range order of amorphous materials by classifying assemblages of Voronoi polyhedra according to their p_4s.

References

1. K. Nishio, J. Kōga, T. Yamaguchi, F. Yonezawa, Theoretical study of light-emission properties of amorphous silicon quantum dots. Phys. Rev. B **67**, 195304 (2003). https://doi.org/10.1103/PhysRevB.67.195304
2. K. Nishio, A.K.A. Lu, G. Pourtois, Low-strain Si/O superlattices with tunable electronic properties: ab initio calculations. Phys. Rev. B **91**, 165303 (2015). https://doi.org/10.1103/PhysRevB.91.165303
3. J. Bernal, Bakerian lecture 1962—structure of liquids. Proc. R. Soc. Lond. A-Math. Phys. Sci. **280**, 299+ (1964). https://doi.org/10.1098/rspa.1964.0147
4. J. Finney, Random packings and structure of simple liquids. 1. Geometry of random close packing. Proc. R. Soc. Lond. A **319**, 479–493 (1970). https://doi.org/10.1098/rspa.1970.0189
5. J. Finney, Random packings and structure of simple liquids. 2. Molecular geometry. Proc. R. Soc. Lond. A **319**, 495–507 (1970). https://doi.org/10.1098/rspa.1970.0190
6. F. Yonezawa, Glass transition and relaxation of disordered structures. Solid State Phys. **45**, 179–254 (1991)
7. Y.Q. Cheng, E. Ma, Atomic-level structure and structure–property relationship in metallic glasses. Prog. Mater. Sci. **56**, 379–473 (2011). https://doi.org/10.1016/j.pmatsci.2010.12.002
8. K. Nishio, J. Kōga, T. Yamaguchi, F. Yonezawa, Confinement-induced stable amorphous solid of Lennard-Jones Argon. J. Phys. Soc. Jpn. **73**, 627–633 (2004). https://doi.org/10.1143/JPSJ.73.627
9. K. Nishio, T. Miyazaki, H. Nakamura, Universal medium-range order of amorphous metal oxides. Phys. Rev. Lett. **111**, 155502 (2013). https://doi.org/10.1103/PhysRevLett.111.155502
10. A. Hirata, L.J. Kang, T. Fujita et al., Geometric frustration of icosahedron in metallic glasses. Science **341**, 376–379 (2013). https://doi.org/10.1126/science.1232450
11. Creation of a mathematical method to express irregular atomic arrangements of amorphous materials, http://www.aist.go.jp/aist_e/list/latest_research/2017/20170201/en20170201.html. Accessed 28 Aug 2017
12. K. Nishio, T. Miyazaki, Polyhedron code—rule hidden in polyhedra (in Japanese). Bull. Soc. Sci. Form **32**, 1 (2017)
13. E.A. Lazar, J.K. Mason, R.D. MacPherson, D.J. Srolovitz, Complete topology of cells, grains, and bubbles in three-dimensional microstructures. Phys. Rev. Lett. **109**, 095505 (2012). https://doi.org/10.1103/PhysRevLett.109.095505
14. L. Weinberg, On the maximum order of the automorphism group of a planar triply connected graph. SIAM J. Appl. Math. **14**, 729–738 (1966). https://doi.org/10.1137/0114062
15. L. Weinberg, A simple and efficient algorithm for determining isomorphism of planar triply connected graphs. IEEE Trans. Circuit Theory **13**, 142–148 (1966). https://doi.org/10.1109/TCT.1966.1082573
16. P.R. Cromwell, *Polyhedra* (Cambridge University Press, 1999)
17. I. Stewart, R. Wilson, *Four Colors Suffice: How the Map Problem Was Solved*, Color edn. (Princeton University Press, Princeton, New Jersey, 2013)

18. K. Nishio, T. Miyazaki, How to describe disordered structures. Sci. Rep. **6**, 23455 (2016). https://doi.org/10.1038/srep23455
19. K. Nishio, T. Miyazaki, Describing polyhedral tilings and higher dimensional polytopes by sequence of their two-dimensional components. Sci. Rep. **7**, 40269 (2017). https://doi.org/10.1038/srep40269
20. G.M. Ziegler, *Lectures on Polytopes* (Springer, New York, 2012)
21. R.J. Wilson, *Introduction to Graph Theory* (Pearson, Harlow, England; New York, 2012)

Part II
Nanoscale Analyses and Informatics

Chapter 7
Topological Data Analysis for the Characterization of Atomic Scale Morphology from Atom Probe Tomography Images

Tianmu Zhang, Scott R. Broderick and Krishna Rajan

Abstract Atom probe tomography (APT) represents a revolutionary characterization tool for materials that combine atomic imaging with a time-of-flight (TOF) mass spectrometer to provide direct space three-dimensional, atomic scale resolution images of materials with the chemical identities of hundreds of millions of atoms. It involves the controlled removal of atoms from a specimen's surface by field evaporation and then sequentially analyzing them with a position sensitive detector and TOF mass spectrometer. A paradox in APT is that while on the one hand, it provides an unprecedented level of imaging resolution in three dimensions, it is very difficult to obtain an accurate perspective of morphology or shape outlined by atoms of similar chemistry and microstructure. The origins of this problem are numerous, including incomplete detection of atoms and the complexity of the evaporation fields of atoms at or near interfaces. Hence, unlike scattering techniques such as electron microscopy, interfaces appear diffused, not sharp. This, in turn, makes it challenging to visualize and quantitatively interpret the microstructure at the "meso" scale, where one is interested in the shape and form of the interfaces and their associated chemical gradients. It is here that the application of informatics at the nanoscale and statistical learning methods plays a critical role in both defining the level of uncertainty and helping to make quantitative, statistically objective interpretations where heuristics often dominate. In this chapter, we show how the tools of Topological Data Analysis provide a new and powerful tool in the field of nanoinformatics for materials characterization.

Keywords Atom probe tomography · Topological data analysis
Persistent homology

T. Zhang · S. R. Broderick · K. Rajan (✉)
Department of Materials Design and Innovation, University at Buffalo:
The State University of New York, Buffalo, NY 14260, USA
e-mail: krajan3@buffalo.edu

© The Author(s) 2018
I. Tanaka (ed.), *Nanoinformatics*, https://doi.org/10.1007/978-981-10-7617-6_7

133

7.1   Introduction

The modern development of Atom Probe Tomography (APT) has opened new exciting opportunities for material design due to its ability to experimentally map atoms with chemistry in a 3D space [1–7]. However, the challenges exist to accurately reconstruct the 3D atomic structure and to more precisely identify features (for example, precipitates and interfaces) from the 3D data. Because data is in the format of discrete points in some metric space, i.e., a point cloud, many data mining algorithms which have been developed are applicable to extract the geometric information embedded in the data. Nevertheless, those geometric-based methods have certain limitations when being applied to solve the problems in atom probe data. We summarize below the limitations of geometric-based methods and present a data-driven approach to address significant challenges associated with massive point cloud data and data uncertainty at sub-nanoscales which can be generalized to many other applications.

7.1.1   Atom Probe Tomography Data and Analysis

In APT, atoms are removed from a region on a specimen's surface (the area may be as large as 200 nm × 200 nm) and are then spatially mapped (see Fig. 7.1). When combined with depth resolution of one inter-planar atomic layer for depth profiling, APT provides the highest spatial resolution of any microanalysis technique. This capability provides a unique opportunity to study experimentally with atomic resolution, chemical clustering, and 3D distributions of atoms, and to directly test and refine atomic and molecular-based modeling studies. While APT has its origins in FIM, originally developed by Erwin W. Müller in 1955, and the atom probe microscope dates back to ca. 1968, it is only fairly recently that highly sophisticated and reliable instruments have become commercially available.

Improvements in data collection rates, field-of-view, detection sensitivity (at least one atomic part per million), and specimen preparation have advanced the atom probe from a scientific curiosity to a state-of-the-art research instrument [9–18]. While APT is a powerful technique with the capacity to gather information containing hundreds of millions of atoms from a single specimen, the ability to effectively use this information has significant challenges. The main technological bottleneck lies in handling the extraordinarily large amounts of data in short periods of time (e.g., giga- and terabytes of data). The key to successful scientific applications of this technology in the future will require that handling, processing, and interpreting such data via informatics techniques be an integral part of the equipment and sample preparation aspects of APT.

As applies to APT, two main phases are involved in the data processing and analysis. The first one is the reconstruction of the 3D image, which identifies the 3D coordinate and chemistry for each collected atom. The second phase is to extract

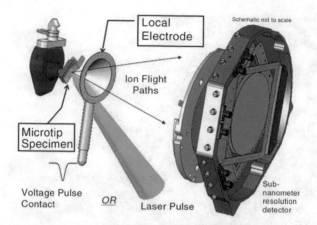

Fig. 7.1 In APT, the specimen is inserted into a cryogenically cooled, UHV analysis chamber. The analysis chamber is cryogenically cooled to freeze out atomic motion. It is at ultrahigh vacuum (UHV) to allow individual atoms to be identified without interference from the environment. A positive voltage is applied to the specimen via a voltage/laser pulse. The positive voltage attracts electrons and results in the creation of positive ions. These ions are repelled from the specimen and pulled toward a position sensitive detector. The location of the atom in the specimen is determined from the ion's hit position on the detector. This configuration magnifies the specimen by a million times and in due course, atoms from the surface ionize, exposing another layer of atoms under them. This process of field ionization continues until the specimen has been fully analyzed, and provides a 3D image of the entire specimen. The difference in APT with other characterization techniques is that the image is mathematically a point cloud, as opposed to a traditional gray scale voxelized image. Reproduced from Ref. [8] with permission

useful information from the reconstructed image; for example, to identify crystalline structures, clusters, and precipitates. There are two parameters of interest here which need to be determined during the 3D image reconstruction: the voxel size [19–21] and the elemental concentration threshold for the voxels. Normally these two parameters are determined empirically by trial and error—i.e., a value is set for the parameter and if the expected features are visible then the image is considered to be correct. Once the parameters are set, they are treated as fixed values and all the subsequent analyses are done based on these set values. There are two issues with this approach (1) the determination of the values for the parameters is largely subjective, and (2) once the values are chosen, the results of the subsequent analyses are biased toward those particular values.

In the following, we use a practical case to elaborate the issues. Because the number of atoms being imaged is very large, using visual inspection to detect the existence of crystalline structure is very difficult. This is a particular problem which we address later on in this chapter for defining interfaces and precipitates. That is, by identifying where there is a change in crystal structure, we can identify phase transitions. A popular way to detect the crystalline structure in a set of atoms is to find repetitive patterns formed by local subsets of atoms [21]. The local subset of an atom is defined as the set of neighboring atoms within the nth coordinate shell together with the atom itself. The nth coordinate shell of an atom is defined as the

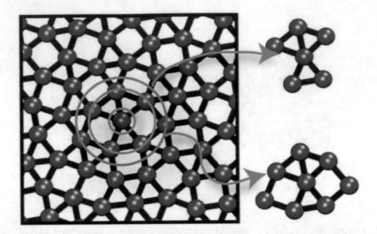

Fig. 7.2 Example of how the definition of a cluster or atomic scale feature is largely dependent on user selection of parameters. In this case, based on a difference in nearest neighbor distances, two very different clusters are defined. This is a significant problem in APT, where this same issue can result in two totally different microstructural characterizations. For example, multiple boundaries of a precipitate can be reasonably defined. Through the use of topological methods, we propose to address this issue and define a bias-free approach to reconstruction and data analysis in APT and, therefore, provide believable and sufficiently robust results not provided with geometry-based approaches. Reproduced from Ref. [22] with permission from The Royal Society of Chemistry

distance from the atom to the nth peak of the radial distribution function. Figure 7.2 shows the 1st and 2nd coordinate shells of a point/atom. As a result, every neighboring point is either in or not in the local subset. Although this is an effective method, the results of the 1st and 2nd coordinate shells are relatively independent, and there is no collective way of summarizing the results for all the coordinate shells. On the other hand, there have been many algorithms developed for the detection of precipitates and atomic clustering. Almost all of these algorithms require some parameter inputs, including bin size, chemical threshold, and number of neighbors. As discussed in terms of defining reconstruction parameters, the end result is largely user biased. Therefore, in both aspects (reconstruction and data analysis), a protocol which removes this bias is necessary if we are to trust the results from an APT experiment. Such a protocol is described in the following section, with application of the approach demonstrated in the results section.

7.1.2 Characteristics of Geometric-Based Data Analysis Methods

The modern development of (APT) has opened new exciting opportunities for material design due to its ability to experimentally map atoms with chemistry in a 3D space. However, the challenges exist to accurately reconstruct the 3D atomic structure and to more precisely identify features (for example, precipitates and interfaces) from the 3D data [23–30]. Because data is in the format of discrete points in some metric space, i.e., a point cloud, many data mining algorithms, which have been developed, are applicable to extract the geometric information embedded in the data [31–34]. Nevertheless, those geometric-based methods have certain limitations when being applied to solve the problems in atom probe data. We summarize below the limitations of geometric-based methods.

In the category of supervised learning [35], many methods require prior knowledge about the data. In the case when the prior knowledge is not available, assumptions need to be made and a bias could be introduced. For example, regression usually assumes a mathematical function between the variables, which means the conclusion we draw from the regression would bias the function that is chosen. On the other hand, for unsupervised learning methods [34], there is usually some parameter(s) that needs to be determined for the algorithm. For example, clustering methods usually require the number of clusters (or some equivalent parameter) to be manually determined; in the case of dimensionality reduction, a common assumption is that the data resides on a lower dimensional manifold, which will sufficiently represent the data, although the dimension of the manifold may not be something that can be determined by the algorithm.

Due to the wide range of applications, there is hardly a universal rule to determine the values of the parameters required by the geometric-based methods. For a particular task, the parameters can be determined either empirically based on the constraints of the situation at hand, or by some algorithm [36]. In these cases, the hidden assumption is that the number of the parameter is fixed once chosen. In some scenario, it would be worthwhile to make those fixed parameters variables. This is not equivalent to giving a set of values to the parameters and collecting all of the results, since the results are independent from each other. What is needed is a scheme that can summarize the results as the parameter changes value. The lack of variability also exists on another level, that is, geometric-based approaches have the property of being exact, i.e., two points in a space are geometrically distinguishable as long as they do not share the same coordinates. As a result of this, for example, classification algorithms determine the classes by using a set of hyper boundaries which are fixed once obtained by training the algorithm.

Topological-based methods have certain properties that are not available for the geometrical-based methods [37]:

(i) topology focuses on the qualitative geometric features of the object, which themselves are not sensitive to coordinates, which means that the data can be studied without having to use some algebraic function, and thus no prior assumption or parameter needs to be dealt with;

(ii) instead of using a metric for distance, topology uses a less clear metric, i.e., "proximity" or "neighborhood", since "proximity" is less absolute than the actual metric, topology is capable of dealing with the scenarios where information is less exact;

Table 7.1 Comparison of geometric-based and topological-based methods

	Requirement for model/Assumption	Requirement for coordinate	Parametric flexibility
Geometric methods	Based on algebraic model or assumptions such as data has certain algebraic property	Need to have coordinate information since metric(s) is used	Parameter is fixed once the value is determined; result cannot reflect the impact of different parameter values
Topological methods	No model required, no algebraic assumption on the data	Use neighborhood (proximity) as metric, possible for coordinate-free applications	Parameter can be variable; result is integrated with the parameter of different values

(iii) the qualitative geometric features can be associated to some algebraic structure through homology, so changes in the topology can be tracked by these algebraic structures, which can be useful when assessing the impact of a parameter on the result of a given analysis. All these properties make the topological-based methods good candidates for dealing with APT data. Table 7.1 summarizes the main differences between the geometric- and topological-based methods.

7.2 Persistent Homology

Persistent Homology [38–55] is a means of topological data analysis. Now let us use an example to show how the topological data analysis methods can overcome the limitations of geometrical methods. Given, as shown in Fig. 7.3, is a set of points

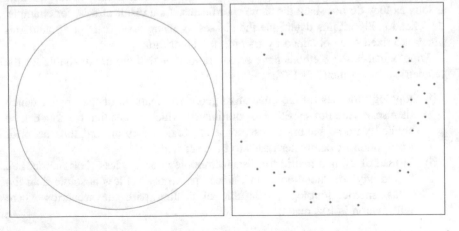

Fig. 7.3 Circle and sampled points. Left: a perfect circle as the object of interest; right: the set of points obtained by sampling the circle at random intervals with noise

Table 7.2 Summary of homology classes and their corresponding qualitative geometric features for the first few dimensions

0th homology class	1st homology class	2nd homology class	3rd homology class
Connected components	2D holes (enclosed area)	3D cavities (enclosed volume)	4D Hyper-voids (enclosed hyper-volume)

(a point cloud) obtained by randomly sampling a circle with some added noise. Let us assume that we are assigned with the task to infer what kind of object the set of points is sampled from. There are several ways we can approach this, with the most straight-forward one being to make a judgment based on how the points visually appear. Because the points appear to be located close to the rim of a disk, one can intuitively connect those points along the outskirt of the set, which would give us a zig-zag version of a circle, and thus one may conclude that the points are sampled from a circle. However, not only is this visual observation-based intuition very subjective, but also it is not feasible when the dimension of the data space is higher than three. Alternatively, because we know the points are sampled from some unknown object, we can assume that two or more points are from the same portion of the object if they are close. With this assumption, we can connect two points with a line if they are close. However, this means we need to choose a distance threshold, and it is obvious that choosing different thresholds will result in different conclusions.

Persistent Homology provides a concise way to deal with the above question. First, homology is, generally speaking, a link between topology and algebra, which associates the qualitative geometric features like connected components and n-dimensional holes with algebraic objects (homology groups). Table 7.2 shows the summary of the homology class with the corresponding qualitative geometric features for the first few dimensions. The homology classes of the same dimension in a topological space form a group and the rank of the group (the Betti number) equals the number of distinct qualitative geometric features of that dimension. Thus by using homology, different topological spaces can be distinguished by comparing the number of distinct homology classes. In the case of a circle versus an annulus, homology cannot distinguish them because both the circle and the annulus have one connected component and one hole, i.e., the number of distinct homology classes for both 0th and 1st dimensions are the same. Thus, in the above case, the ambiguity between the circle and the annulus does not affect the result. Next, as the points are sampled from some object, adding continuum to the subsets of discrete points is necessary to recover the shape of the object. Persistent Homology accomplishes this by associating every point with a disk (or hyper-disk in higher dimensional cases) and increasing the diameter of the disks from 0 (growing the size of the disks). During the process, the topology of the space defined by the union of all of the disks will change. The changes will include two or more isolated disks merging together and/or hole(s) being formed or covered by a set of disks. Through homology, these changes will be seen as the change in the number of distinct homology classes and all this will be recorded by a so-called "barcode" representation, which serves as the result of the persistent homology. Because all the disks have the same size at any given time, and the homology classes are recorded for a series of continuous disk sizes, the conclusion can be made without biasing toward a particular value of the disk size.

Figure 7.4 demonstrates the process with the barcode. The left column shows the points and their associated disks at different sizes, and the right column is the barcode which is a summary of the qualitative geometric features (different homology classes) of the space defined by the intersection of the disks. The horizontal axis of the barcode is the diameter of those disks, and the vertical axis is the ordering of the bars. At the beginning when the diameter of all of the 30 disks is 0, no disk overlaps with others, so there are 30 connected components in the space. These 30 connected components are represented by 30 horizontal bars (in orange color) of length 0, while the left end of the bars line up to the horizontal value 0 since they begin to exist when the diameter is zero. As the disk sizes increase, the bars also extend horizontally toward the right, as can be seen in the first row of Fig. 7.4. During this process, whenever there are two disks whose union changes from an empty set to a non-empty set with the two disks originally belonging to two different connect components, then there are two originally connected components becoming one connected component. At this point, one of the two bars representing the two isolated connected components will stop extending, while the other bar will continue extending and represent the union of both connected components. This dynamic is shown in the 1st row through the 3rd row of Fig. 7.4 (notice that the color is lighter for the bars that are still extending). At some large diameter value all the disks merge together and become one single connected component, so there remains one bar representing the union of all the disks. Also at a certain diameter value, an enclosed hole will be formed by the union of the disks, this enclosed hole represents a 1st homology class, and a new bar (green color) is added to the barcode as shown in the fourth row of the Fig. 7.4. Notice this bar did not start from the horizontal value of 0, but it starts from when the diameter value corresponds to when the enclosed hole is just formed. As the disks size increases, this bar also extends horizontally to the right, and at the same time the area of the enclosed hole decreases. When this enclosed hole is totally covered by the disks, it is considered to be no longer "live" and the corresponding bar stops extending at the value of the disk diameter such that the hole is just covered by the disks. We call the diameter value when the enclosed hole is formed the "birth time" of that hole/homology class and the diameter value when the region is just covered by the disks the "death time" of the hole/homology class. Correspondingly, all of the 0th homology classes have a birth time of zero and a death time equal to the diameter when their corresponding bar is no longer extending. By examining the barcode, we can read off the birth and death time of every homology class. This provides an idea of all the qualitative geometric features with their relative birth and death time, and thus how persistent they are within the range of the diameter values being considered. In general, the

Fig. 7.4 Demonstration of Persistent Homology. Each row of subfigures corresponds to one value of disk diameter. The left column of subfigures shows the points (blue point) with their associated disks in purple color; the right column is the corresponding barcode. The 0th homology classes are represented by the orange bars and the 1st homology class is represented by the green bar. All the bars for the 0th homology class are sorted in the order of their death time. The lighter shaded bars indicate that the bar is not terminated at that point

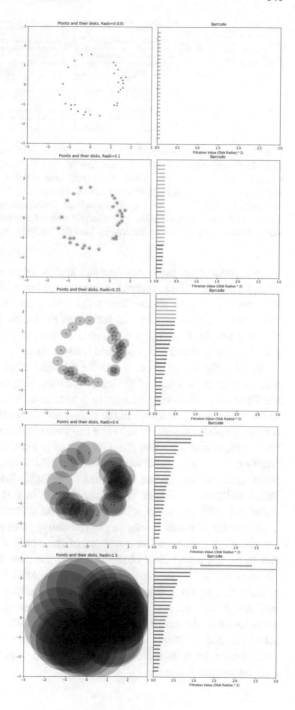

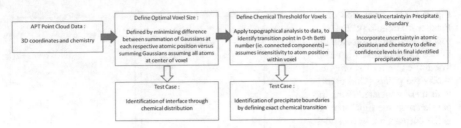

Fig. 7.5 Process followed in this chapter for defining optimal reconstruction parameters in APT, and identifying microstructural features free of bias. In Sect. 7.3, we introduce an approach for defining optimal voxel size, where the optimal size is defined as that which is least sensitive to atomic positioning with the respective voxels. This makes the data most applicable to topological analyses. In Sect. 7.4, we then apply a topological analysis to the data after voxelized to identify the optimal chemical thresholds which reflect microstructural phase transitions. In Sect. 7.5, we then introduce application of uncertainty into the analysis, reflecting the experimental conditions

long-lived or more persistent features are the ones of greater significance than the short-lived ones. Here, except for one orange bar which keeps extending forever, all the orange bars have shorter length than the green bar, so our conclusion is that the 2D hole is the main topological feature of the space. It is worth pointing out that the above example demonstrated the idea of persistent homology in 2D space, but in practice, the data and the geometric features are not restricted to 2D (see Table 7.2).

In fact, the above process can be thought of as viewing the points through a telescope with the focus of the telescope changing continuously, and thus we name the Persistent Homology on a set of points as a "Data Telescope". When the image in the telescope is clear and all the points are sharp, it corresponds to the case which the diameter of the disks is 0. When the focus is detuned and the points in the image are vague, it indicates that the diameter of the disks is no longer zero. This tuning ability can be very useful when processing APT data. One point to clarify: in atom probe data the raw output is a point cloud with information associated with each individual atom. We then define the voxels in order to visualize the data and to make the analyses manageable through data reduction. In APT data analyses, the counterpoint of the disk diameter in the above example can be the voxel size and/or the chemical concentration threshold of the voxel. By tuning these parameters, we can track the topological changes within the raw data space, and based on these changes we select the critical transition point(s) in reconstruction parameters to capture the useful information which is otherwise hidden. The exact procedure developed and applied in this chapter is described schematically in Fig. 7.5, encompassing the entire "Data Telescope" for APT data by focusing the voxel size selection and chemical threshold for each voxel (the adjustable reconstruction parameters) in a bias-free manner, and for which we apply for identifying microstructural features.

7.3 Voxel Size Determination: Identification of Interfaces

In visualizing APT data, the large number of data points can obscure the underlying structures. Therefore, the reconstructed APT data is first sectioned into voxels (i.e., 3D boxes), which encompass a collection of atoms (Fig. 7.6) and a local density value is assigned to each voxel based on the chemical composition of the voxel. By tracking the variation in the local density associated within each voxel across the sample, one can detect underlying features such as grain boundaries or precipitates. The following section describes the process we outlined in a previous report [56]. Figure 7.6 graphically denotes the process of voxelization for random points scattered in 3D, which represent the atoms in a material. The volume encompassing the data is initially sectioned into voxels of edge length of 0.2 nm (chosen arbitrarily for illustration). The data was then binned into voxels. The voxels were classified by the number of data points they contained and a density value was assigned. The density value of voxels is useful for pinpointing regions of high chemical density, potentially indicating the presence of precipitates, or capturing regions of different densities delineating different phases.

The procedure which we have developed and which is described below roughly follows these steps:

(i)   for a given voxel size, apply a Gaussian kernel to each atom at its exact position;

(ii) sum all of the Gaussian kernels to define a estimated density across the voxel;

(iii) define another Gaussian assuming every atom in the voxel is located within the center of the voxel;

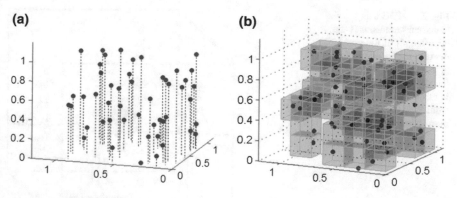

Fig. 7.6 Voxelization of random data points scattered in 3D, representative of atoms in a material. **a** Original data set within a 1 nm³ box. **b** The atoms are grouped into uniform voxels of edge length 0.2 nm. The numbers of atoms contained within the different voxels represent the local density of that voxel volume. Reproduced from Ref. [56] with permission

(iv) calculate the difference between the estimated density and the central Gaussian which approximates the true density; and

(v) define the optimal voxel size as that which has the minimum difference.

The specifics of each step are expanded below.

Kernel density estimation (KDE) methods can capture the contribution of each atom toward the voxel and to obtain a smooth overall density function representing the voxel. Each atom is represented by a kernel, which is a symmetric function which integrates to one and contributes to a value at the center of the voxel. The center is considered representative of the region encompassed by the voxel. The contribution of the atoms within the voxel to the voxel center is a function of the atom's location and is determined by a sampling formula [57]. In the simplest case, the contribution of each atom, located at position "x", toward a voxel of unit length centered around the origin can be represented by a Parzen window [58] given by:

$$K(x) = \frac{1}{2}, \quad |x| \leq 1$$

$$K(x) = 0, \quad |x| > 1 \tag{9.1}$$

which indicates that all atoms within the voxel contribute equally, independent of their location within the voxel. There are several other sampling functions and merits of each are discussed elsewhere [59, 60]. In this work, we make use of Gaussian kernels as shown in Fig. 7.7.

If "x" is the atom position and X is the center of the voxel of edge length "h" where the individual kernel contribution is measured, the contribution of the Gaussian kernel is defined by the weighting function

Fig. 7.7 KDE using Gaussian kernels. h is the edge length of the voxel. x_i denotes the location of atom "i". X denotes the center of the voxel at which the density is calculated. Reproduced from Ref. [56] with permission

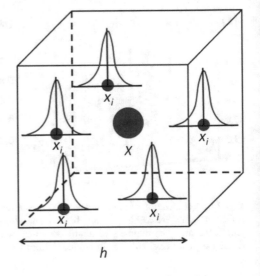

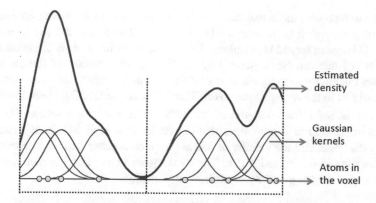

Fig. 7.8 Illustration of density estimation through kernel density function (red lines) representing atom positions (yellow circles) in different voxels. The blue line is the estimated density of atoms in each voxel obtained by summing up the contributions of the various kernels within the voxel. Reproduced from Ref. [56] with permission

$$w_g(x - X, h_d) = \frac{1}{\sqrt{2\pi\sigma^2}} e^{\frac{-(x-X)^2}{2\sigma^2}} = \frac{1}{h_d} K\left(\frac{x-X}{h_d}\right) \tag{9.2}$$

where the value h_d is known as the bandwidth of the kernel and in the case of the Gaussian kernel: $h_d = \sigma$ where σ is the standard deviation. The standard Gaussian kernel (with zero mean and unit variance) is given by $K(t) = \frac{1}{\sqrt{2\pi}} e^{-\frac{1}{2}t^2}$.

The estimated density function at any point x within the voxel (Fig. 7.8) is defined by the average of the different kernel contributions (Eq. 9.2) as

$$\hat{f}(x) = \frac{1}{nh_d} \sum_{j=1}^{n} K\left(\frac{x-X_j}{h_d}\right) \tag{9.3}$$

where $h_d > 0$ is the window width, smoothing parameter or bandwidth.

To automate the voxel size, an error function is defined to compute the difference between the kernel estimated density of the data $\hat{f}(x)$ and its true density f(x). A typical measure of the accuracy over the entire voxel is obtained by integrating the square of the error computed given by:

$$MISE(\hat{f}) = E\left[\int_{-\infty}^{\infty} (\hat{f}(x) - f(x))^2 dx\right] \tag{9.4}$$

where *MISE* is the mean integrated square error. Since the distribution of atoms does not follow any known pattern, especially at the region of interest such as the interface, the true density $f(x)$ is not known. The approach followed here to approximate the true density as closely as possible to the estimated density consists of the following sequence: $f(x)$ is first assumed to be a Gaussian distribution, assumed to represent the actual distribution of atoms within the voxel, although the atoms may very well be non-normally distributed. The mean and variance of this assumed Gaussian spread is calculated for the atoms within and on the boundary of the voxel of interest. Next, depending on the real distribution of the atoms, the Gaussian function may peak either at or off center in the voxel and in the latter case it is translated to the center of the voxel. The difference of this Gaussian distribution with $\hat{f}(x)$ is used for computation of *MISE*. For the cases where the initial assumption of f(x) is a poor one, it will results in a high MISE. By gradually varying the voxel size the validity of this assumption reaches a most probable value corresponding to minimized MISE. The total squared error (E_{tot}) is then computed for the entire dataset given by the following equation

$$E_{tot} = \sum_{j=1}^{V} (MISE)_j \qquad (9.5)$$

where V is the total number of voxels. E_{tot} is then minimized with respect to varying voxel size. The kernel density estimation was carried out on the Ni–Al–Cr dataset comprising ~8.72 million atoms. For each atom in a voxel, a Gaussian kernel was fit at the atom location and its amplitude was set at 1 with full width at half maximum set to the voxel edge length. The kernel contributions of atoms to the voxel were calculated at the voxel center for all atoms within and on the boundary of a particular voxel. These values were then added giving the amplitude of density at the voxel center. The error was then calculated between the actual density and estimated density using the procedure explained in the previous section. This procedure is repeated for the voxel size varied from 0.5 to 2.5 nm in steps of 0.1 nm. A minimum error was obtained for 1.6 nm voxel size (Fig. 7.9) providing a tradeoff between the noise and data averaging. This voxel size of 1.6 nm reduces the atomic data set into a representation of 83,754 voxels at 1.6 nm^3 each.

Fig. 7.9 Dependence of the normalized mean integrated square error (MISE) on the voxel size. The error is minimum for a voxel edge length of 1.6 nm. Reproduced from Ref. [56] with permission

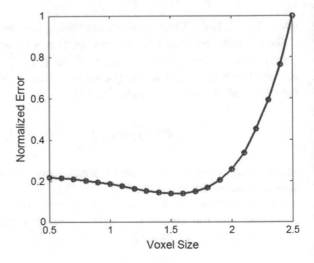

Fig. 7.10 Feature topology for voxel size of 1.6 nm. Only those voxels where the Al = Cr concentrations are imaged, reflecting the γ/γ' interface. This elucidates the effect of voxel edge length on capturing the precise interface between the γ/γ' region of the Ni–Al–Cr sample, with the approach applicable to all APT data for defining voxel size and for any metallic samples for defining microstructural features. Adapted from Ref. [56] with permission

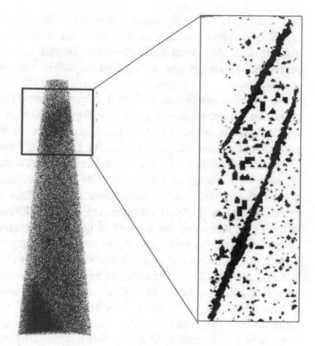

In the case of 1 nm, the voxel size is too small to accurately estimate the density. Due to statistical fluctuation in the distribution of atoms, there are many pockets of 1 nm^3 throughout the sample where the concentration of Al and Cr are almost equal. As the voxel size is increased, the increase in volume averages out the noise and a clear interface starts emerging. At 1.6 nm most of the statistical noise vanishes and a very sharp interface is obtained, with nanometer scale fluctuations visible on the isosurface representing the interface (Fig. 7.10). As the voxel size is increased beyond this value, over smoothing of data starts occurring. The interface starts becoming diffuse and the graininess in the image disappears. At this stage there is ideally no statistical noise and the residual clusters scattered throughout the volume could potentially be capturing the presence of nanoclusters.

7.4 Topological Analysis for Defining Morphology of Precipitates

Interfaces and precipitate regions are typically identified from APT data by representing them as isoconcentration surfaces at a particular concentration threshold, thereby making the choice of concentration threshold critical. The popular approach to selecting the appropriate concentration threshold is to draw a proximity histogram [61], which captures the average concentration gradient across the interface and visually identifies a concentration value that is the best representative of an interface or phase change occurrence. This makes the choice of concentration gradient user dependent and subjective. In this section, we will showcase how persistent homology can be applied to better recover the morphology of the precipitates.

As we have mentioned in Sect. 7.2, metric properties such as the position of a point, the distance between points, or the curvature of a surface are irrelevant to topology. Thus, a circle and a square have the same topology although they are geometrically different. Such qualitative geometric features can be represented by simplicial complexes, which are combinatorial objects that can represent spaces and separate the topology of a space from its geometry [62]. Simplicial homology is a process that provides information about the simplicial complex by the number of cycles (a type of hole) it contains. One of its informational outcomes are Betti numbers which record the number of qualitative geometric features such as connected components, holes, tunnels, or cavities. A microstructural features such as a nanocluster can have only limited topological features depending on its dimension. For example, in 3D, a structure can be simply connected, or it can be connected such that a tunnel passes through it, or it can be connected to itself such that it encloses a cavity, or it can remain unconnected. Thus, we can characterize the topology of a structure by counting the number of simply connected components, number of tunnels and number of cavities denoted by Betti numbers β_0, β_1, and β_2. The relationship between the Betti numbers, the data topology, and the concept of barcodes as described in the introduction is summarized in Fig. 7.11.

As discussed earlier and expanded upon in our prior work [62], the persistence of different topological features can be recorded as barcodes, which we now group according to each Betti number. The horizontal axis represents the parameter ε or the range of connectivity among points in the point cloud while the vertical axis captures the number of topological components present in the point cloud at each interval of ε. There has to be some knowledge of the appropriate range for ε, such as the interatomic distance when dealing with raw atom probe data or voxel length if the data has been voxelized. The persistence of features is a measure of whether these features are actually present in the data or if they are artifacts appearing at certain intervals.

Having voxelized the APT data following the approach discussed in Sect. 7.3, each voxel represents a certain value of local concentration. By varying the concentration threshold as our filtration parameter, our underlying dataset provides a different set of voxels corresponding to each concentration threshold. We vary the concentration threshold of each element independently in Fig. 7.12 to show how the process evolves.

The top panel shows the evolution of Betti numbers for varying Sc concentration. At each value of Sc concentration threshold "δ", those voxels having a concentration of $\delta \pm 0.02$ were chosen. Consider β_0: at a high concentration threshold, beyond 0.5, a very small number of simply connected components are observed. This is because very few voxels have concentration value equal or more than this threshold. As concentration threshold is decreased, more voxels qualify to be

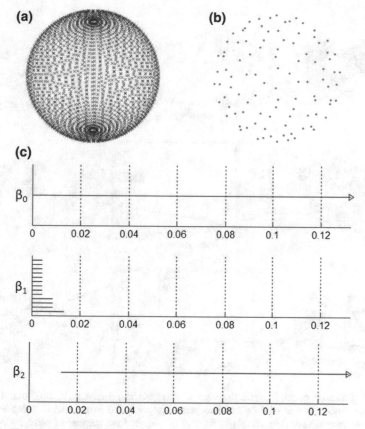

Fig. 7.11 Persistent homology and barcodes as described in the introduction, and their relationship with Betti numbers, as applied in this section. **a** the original point cloud data consisting of 2000 points. **b** Extracting 1% of the original data points as landmark points. **c** The barcode using the witness complex [21] achieves the identification of $(\beta_0, \beta_1, \beta_2) = (1, 0, 1)$, representing 1 connected component and 1 cavity. Reproduced from Ref. [62] with permission

included in the group leading to an increase in β_0. The value of β_0 remains constant for a certain range indicating that these are real features. A plot of the voxels at $\delta = 0.3$ shows that it indeed captures real clusters of Sc. With further decrease of concentration threshold, a decrease in β_0 is observed. This is because every voxel outside the Sc clusters has some minimal content of Sc and the inclusion of all exterior voxels results in one single connected component. We also observe a peak in the value of β_1 at a low concentration of $\delta = 0.03$. When we plot the isoconcentration surface for those voxels we find that these represent cavities. These voxels with very low Sc concentration sit on the edge of the Sc clusters, and thereby, enclose Sc clusters within themselves. A similar trend is observed with Mg where for low concentration we see Mg Isosurface containing cavities that enclose Sc clusters, whereas for high concentration there are few voxels.

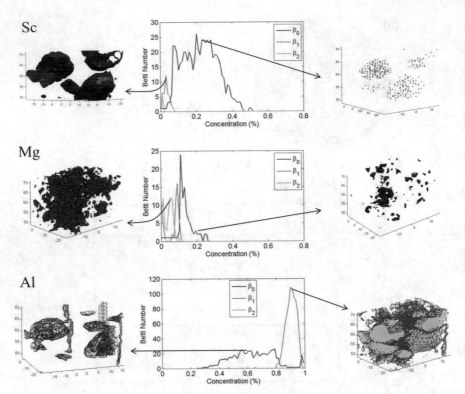

Fig. 7.12 Filtration of an AlMgSc structure with respect to concentration threshold. The three panels show the evolution of Betti numbers with changes in the concentration threshold for Sc, Mg, and Al respectively. The set of (β_0, β_1, and β_2) captures the number of precipitates and the persistence of β_0 denotes appropriate concentration thresholds for the different elements Isoconcentration Surfaces obtained at different concentration threshold are shown corresponding to each figure to denote regions of interest. Adapted from Ref. [62] with permission

7.5 Spatial Uncertainty in Isosurfaces

APT data is a point cloud data and in order to study hidden features like precipitates or grain boundaries, isosurfaces are often used. These isosurfaces are drawn at a particular concentration threshold. We calculate the uncertainty in spatial location of isosurfaces here and use visualization techniques that lead to the incorporation of uncertainty information in the final image (Fig. 7.13). Isosurfaces were drawn by joining voxels which have the same value of density or concentration, as defined in Sect. 7.4. For uncertainty calculations in the APT data, we followed the approach described in Sect. 7.3 for calculating the error and difference from ideal density. Consider x_i as the atom in a voxel with coordinates (x_i^x, x_i^y, and x_i^z), the mean μ_x and standard deviation σ_x along the x-axis will be given as follows:

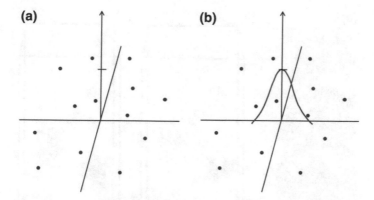

Fig. 7.13 The atoms are distributed throughout a voxel, as discussed in Sect. 7.3. When isosurfaces are drawn, only the net value is included in the representation (**a**). We have added another quantity to include spatially scattered data. We model the uncertainty as Gaussian noise (**b**). Parameters of Gaussians such as FWHM or variance are calculated based on spatial distribution in the region. We have net value, as before, but also have another parameter which includes information about neighborhood. However, this also adds another dimension in visualization. We represent the uncertainty through selective blurring of the region where blur intensity is mapped to values at the Gaussian distribution

$$\mu_x = \frac{\sum_{i=1}^{N} x_i^x}{N} \tag{9.6}$$

$$\sigma_x = \sqrt{\frac{1}{N} \sum_{i=1}^{N} \left(x_i^x - \mu_x\right)^2} \tag{9.7}$$

With the above information, a full definition of Gaussian distribution at the voxel center was obtained, following the logic described in Fig. 7.7. We have added the concept of uncertainty in isosurfaces. As we have seen, the calculation for voxels and chemical thresholds involves averaging of data points, while averaging involves some variability in the final result. We may not be able to remove the variability but we can attempt to quantify it. At present, density values are calculated at the centers and we get one net value at each point. Further interpolation across voxels provides uncertain isosurfaces. The above equations were used to study the APT data described in Sect. 7.4. Voxel data was used to convert the data into a structured grid format. The data was then visualized. As a first step, crisp isosurfaces were drawn. In a second step, uncertainty information was added by assigning a shaded region around it. The intensity of the shade was dependent on the uncertainty value. In the present study, an uncertainty of ±1% was used. Figure 7.14 shows an isosurface drawn at a concentration threshold of 12%. Here, each voxel was assigned a value which was assumed to be constant throughout the voxel. Further, all of the voxels

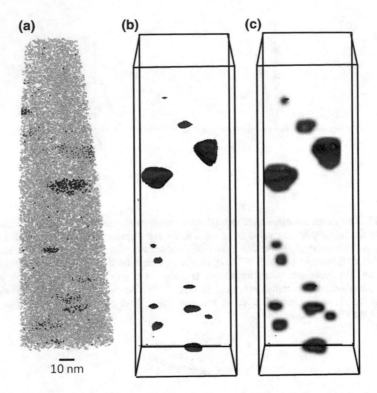

Fig. 7.14 a Output from APT experiment. **b** Isoconcentration surface obtained at 12% concentration threshold, following the approaches described in Sects. 3 and 4 for defining voxel size and chemical threshold, respectively. Crisp and bias free definition of precipitate boundaries is provided. **c** Isoconcentration surface shown with inclusion of uncertainty

with values equal to the threshold were joined to give crisp isosurfaces. Figure 7.14b shows the same isosurface with the inclusion of uncertainty. Uncertainty is a function of spatial distribution of atoms. Distribution of atoms is less at distances away from the isosurface and thus no effect was observed. Near the surface, the uncertainty of the isosurface decreases. From the image, it is observed that there is an increased level of intensity as the surface is approached.

7.6 Summary

Atom probe tomography is a chemical imaging tool that produces data in the form of mathematical point clouds. Unlike most images which have a continuous gray scale of voxels, atom probe imaging has voxels associated with discrete points that are associated with individual atoms. The informatics challenge is to assess nano and sub-nanoscale variations in morphology associated with isosurfaces when clear physical models for image formation do not exist given the uncertainty and sparseness in noisy data. In this chapter, we have provided an overview of the application of topological data analysis and computational homology as powerful new informatics tools that address such data challenges in exploring atom probe images.

Acknowledgements We gratefully acknowledge support from NSF DIBBs Project OAC-1640867 and NSF Project DMR-1623838. KR acknowledges support from the Erich Bloch Endowed Chair at the University at Buffalo-State University of New York.

References

1. M.K. Miller, R.G. Forbes, *Atom-Probe Tomography* (Springer, New York, 2014)
2. D.J. Larson, T.J.T.J. Prosa, R.M. Ulfig, B.P. Geiser, T.F. Kelly, *Local Electrode Atom Probe Tomography: A User's Guide* (Springer, New York, 2013)
3. B. Gault, M.P. Moody, J.M. Cairney, S.P. Ringer, *Atom Probe Microscopy* (Springer, New York, 2012)
4. S.K. Suram, K. Rajan, Microsc. Microanal. **18**, 941 (2012)
5. B. Gault, S.T. Loi, V.J. Araullo-Peters, L.T. Stephenson, M.P. Moody, S.L. Srestha, R.K.W. Marceau, L. Yao, J.M. Cairney, S.P. Ringer, Ultramicroscopy **111**, 1619 (2011)
6. E.R. McMullen, J.P. Perdew, Solid State Commun. **44**, 945 (1982)
7. E.R. McMullen, J.P. Perdew, Phys. Rev. B **36**, 2598 (1987)
8. M.K. Miller, T.F. Kelly, K. Rajan, S.P. Ringer, Mat. Today **15**, 158 (2012)
9. M.K. Miller, E.A. Kenik, Microsc. Microanal. **10**, 336 (2004)
10. K. Hono, Acta Mat. **47**, 3127 (1999)
11. T.F. Kelly, D.J. Larson, K. Thompson, J.D. Olson, R.L. Alvis, J.H. Bunton, B.P. Gorman, Ann. Rev. Mat. Res. **37**, 681 (2007)
12. T.F. Kelly, M.K. Miller, Rev. Sci. Instr. **78**, 1101 (2007)
13. M.K. Miller, *Atom-probe tomography* (Kluwer Academic/Plenum Publishers, New York, 2000)
14. M.K. Miller, A. Cerezo, M.G. Heatherington, G.D.W. Smith, *Atom-probe field-ion microscopy* (Clarendon Press, Oxford, 1996)
15. J. Rsing, J.T. Sebastian, O.C. Hellman, D.N. Seidman, Microsc. Microanal. **6**, 445 (2000)
16. D.N. Seidman, R. Herschitz, Acta Metall. **32**, 1141 (1985)
17. D.N. Seidman, R. Herschitz, Acta Metall. **32**, 1155 (1985)
18. D.N. Seidman, B.W. Krakauer, D. Udler, J. Phys. Chem. Sol. **55**, 1035 (1994)
19. M. Hetherington, M. Miller, J. Phys. Colloq. **50**, C8 (1989)
20. M. Hetherington, M. Miller, J. Phys. Colloq. **48**, C6–559 (1987)
21. K. Torres, M. Daniil, M.A. Willard, G.B. Thompson, Ultramicroscopy **111**, 464 (2011)
22. C. Phillips, G. Voth, Soft Matter **9**, 8552 (2013)
23. M.P. Moody, L.T. Stephenson, P.V. Liddicoat, S.P. Ringer, Contingency table techniques for three dimensional atom probe tomography. Micros. Res. Tech. **70**, 258 (2007)
24. F. Vurpillot, F. De Geuser, G. Da Costa, D. Blavette, J. Microscopy **216**, 234 (2004)
25. J.M. Hyde, A. Cerezo, T.J. Williams, Ultramicroscopy **109**, 502 (2009)
26. M.P. Moody, B. Gault, L.T. Stephenson, D. Haley, S.P. Ringer, Ultramicroscopy **109**, 815 (2009)

27. B. Gault, X.Y. Cui, M.P. Moody, A.V. Ceguerra, A.J. Breen, R.K.W. Marceau, S.P. Ringer, Scripta Mat. **131**, 93 (2017)
28. K. Hono, D. Raabe, S.P. Ringer, D.N. Seidman, MRS Bull. **41**, 23 (2016)
29. F. Vurpillot, W. Lefebvre, J.M. Cairney, C. Obderdorfer, B.P. Geiser, K. Rajan, MRS Bull. **41**, 1 (2016)
30. J.M. Cairney, K. Rajan, D. Haley, B. Gault, P.A.J. Bagot, P.-P. Choi, P.J. Felfer, S.P. Ringer, R.K.W. Marceau, M.P. Moody, Ultramicroscopy **159**, 324 (2015)
31. O. Hellman, J. Vandenbroucke, J. Blatz du Rivage, D.N. Seidman, Mater. Sci. Eng. A **327**, 29 (2002)
32. L.T. Stephenson, M.P. Moody, P.V. Liddicoat, S.P. Ringer, Microsc. Microanal. **13**, 448 (2007)
33. A. Shariq, T. Al-Kassab, R. Kirchheim, R.B. Schwarz, Ultramicroscopy **107**, 773 (2007)
34. F.D. Geuser, W. Lefebvre, Microsc. Res. Tech. **74**, 257 (2011)
35. L. Ericksson, T. Byrne, E. Johansson, J. Trygg, C. Vikstrom, *Multi- and Megavariate Data Analysis: Principles, Applications* (Umetrics Ab, Umea, 2001)
36. H.-P. Kriegel, P. Krger, J. Sander, A. Zimek, Wiley Interdisciplinary Reviews. Data Min. Knowl. Disc. **1**, 231 (2011)
37. G. Carlsson, Bull. Am. Math. Soc. **46**, 255 (2009)
38. H. Edelsbrunner, D. Letscher, A. Zomorodian, Discret. Comput. Geom. **28**, 511 (2002)
39. S. Bhattacharya, R. Ghrist, V. Kumar, IEEE Trans. Rob. **31**, 578 (2015)
40. G. Carlsson, T. Ishkhanov, V.d. Silva, A. Zomorodian, Int. J. Comput. Vision **76**, 1 (2008)
41. P.G. Cmara, *Current Opinion in Systems Biology Future of Systems Biology Genomics and Epigenomics*, vol. 1, p. 95 (2017)
42. I. Donato, M. Gori, M. Pettini, G. Petri, S. De Nigris, R. Franzosi, F. Vaccarino, Phys. Rev. E **93**, 052138 (2016)
43. H. Edelsbrunner, D. Letscher, A. Zomorodian, Discr. Comput. Geom. **28**, 511 (2002)
44. Y. Hiraoka, T. Nakamura, A. Hirata, E.G. Escolar, K. Matsue, Y. Nishiura, Proc. Natl. Acad. Sci. **113**, 7035 (2016)
45. D. Horak, S. Maletic, M. Rajkovic, J. Stat. Mech Theor. Exp. **2009**, P03034 (2009)
46. H. Liang, H. Wang, PLoS Comput. Biol. **13**, e1005325 (2017)
47. N. Otter, M.A. Porter, U. Tillmann, P. Grindrod, H.A. Harrington, arXiv:1506.08903 (2015)
48. B. Rieck, H. Leitte, Comput. Graph. Forum **34**, 431 (2015)
49. B. Rieck, H. Leitte, Comput. Graph. Forum **35**, 81 (2016)
50. B. Rieck, H. Mara, H. Leitte, IEEE Trans. Visual Comput. Graph. **18**, 2382 (2012)
51. C.M. Topaz, L. Ziegelmeier, T. Halverson, PLoS One **10**, e0126383 (2015)
52. B. Wang, G.-W. Wei, J. Comput. Phys. **305**, 276 (2016)
53. K. Xia, G.-W. Wei, International journal for numerical methods. Biomed. Eng. **30**, 814 (2014)
54. K. Xia, G.-W. Wei, J. Comput. Chem. **36**, 1502 (2015)
55. A. Zomorodian, G. Carlsson, Discr. Comput. Geom. **33**, 249 (2005)
56. S. Srinivasan, K. Kaluskar, S. Dumpala, S. Broderick, K. Rajan, Ultramicroscopy **159**, 381 (2015)
57. V. Epanechnikov, Theor. Probab. Appl. **14**, 153 (1969)
58. E. Parzen, Ann. Math. Stat. **33**, 1065 (1962)
59. M. Rosenblatt, Ann. Math. Stati. **27**, 832 (1956)
60. D.W. Scott, Biometrika **66**, 605 (1979)
61. O.C. Hellman, J.A. Vandenbroucke, J. Rusing, D. Isheim, D.N. Seidman, Microsc. Microanal. **6**, 437 (2000)
62. S. Srinivasan, K. Kaluskar, S. Brodeick, K. Rajan, Ultramicroscopy **159**, 374 (2015)

Chapter 8
Atomic-Scale Nanostructures by Advanced Electron Microscopy and Informatics

Teruyasu Mizoguchi, Shin Kiyohara, Yuichi Ikuhara
and Naoya Shibata

Abstract Interfaces dramatically affect the properties of materials because their atomic configurations often differ from the bulk material. A determination of the atomic structure of the interface is, therefore, one of the most significant tasks in materials research. Electron microscopy and theoretical calculations have been effectively used to accomplish this important task. In addition, an informatics approach has recently been combined with theoretical calculations to efficiently determine the atomic structures of interfaces. This chapter introduces the determination of interface structures using an informatics approach (Bayesian optimization and virtual screening) along with advanced electron microscopy. In the informatics approach, calculation acceleration on the order of 10^6 can be achieved. Determination of the interface structure with resolution better than ~45 pm is now possible using advanced electron microscopy. In this way, nanostructures at grain boundaries and heterointerfaces can be qualified. We will introduce these state of the art methods to investigate nanostructures.

Keywords Interface · Electron microscopy · Scanning transmission electron microscopy · Bayesian optimization · Virtual screening

8.1 Atomic Structures of Interfaces

Interfaces are a kind of lattice defect inside materials and can have significant effects on the overall material properties. For instance, interfaces in polycrystalline materials, i.e., grain boundaries (GB), determine the ion transportation properties and high temperature mechanical properties [1–4]; an atomically controlled interface in a thin film often provides unique properties such as the formation of

T. Mizoguchi · S. Kiyohara · Y. Ikuhara · N. Shibata (✉)
The University of Tokyo, Tokyo, Japan
e-mail: shibata@sigma.t.u-tokyo.ac.jp

© The Author(s) 2018
I. Tanaka (ed.), *Nanoinformatics*, https://doi.org/10.1007/978-981-10-7617-6_8

two-dimensional gases [5, 6]. The fact that interfaces have different properties from the bulk is a consequence of the fact that they have different atomic structures from the bulk. Thus, for a comprehensive understanding of interface properties, determination of the atomic structure of the interface is crucial.

Since the atomic structure of the interface is strongly dependent on the crystal orientation, lattice planes, and terminations, a systematic study of the interface structure is indispensable for achieving a comprehensive understanding. Thus, the atomic structures of interfaces have already been extensively investigated. Because these characteristic atomic configurations at the interface appear within a very limited area (below 10 nm), high spatial resolution observations using transmission electron microscopy (TEM) and theoretical calculations using atomistic simulations have been effectively applied to investigate interfaces.

Similar to TEM, aberration-corrected scanning transmission electron microscopy (STEM) has achieved sub-0.45 Å spatial resolution [7], and direct atom-by-atom imaging is now routinely possible via annular dark-field (ADF) imaging. In addition, owing to the rapid improvement in detectors, interface chemical analysis using energy-dispersive X-ray spectroscopy (EDS) and electron energy loss spectroscopy (EELS) can also achieve atomic resolution. In short, atomic-resolution STEM has become a very powerful tool for characterizing the atomic structures of interfaces [8, 9].

In terms of calculations, extensive calculations are usually necessary to determine even one interface structure because of the geometrical freedom of the interface. Nine degrees of freedom (five macroscopic and four microscopic) are present in an interface. The number of atomic configurations to be considered often reaches 10^4 in even the simplified coincidence site lattice (CSL) grain boundary, namely Σ grain boundaries [10, 11]. In a straightforward manner, as schematically illustrated in Fig. 8.1, structure and energy calculations for all candidates must be performed, and leading to optimized configurations and energies of these are obtained ($E_{i,j}$ in Fig. 8.1). The most stable configuration with the minimal energy ($E_{i, \text{mim}}$ in Fig. 8.1) can then be determined from the density functional theory/molecular dynamics (DFT/MD) simulation of the interface. Furthermore, the same "brute force" computation is necessary to determine other types of interfaces because the interface structure is dependent on the type of interface (ΣGB_1, ΣGB_2, ... ΣGB_N in Fig. 8.1).

If the structure and energy of unknown interfaces could be determined more efficiently and accurately, the investigation of interfaces would be dramatically accelerated, which could lead to a deeper understanding of the mechanisms that give rise to interface properties. To more efficiently determine interface structures, a genetic algorithm method and a random structure searching algorithm method have been proposed [13, 14]. However, many trial calculations are still necessary to determine a single grain boundary structure. More recently, much more efficient methods based on machine learning techniques, including virtual screening and Bayesian optimization have been proposed by the present authors [12, 15–17]. Those methods are described below.

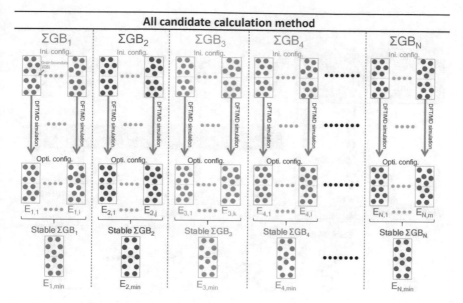

Fig. 8.1 Schematics of all-candidate calculation methods [12]

8.2 Informatics Approach for Interfaces

8.2.1 Virtual Screening

In this section, a virtual screening method for interface structure determination is described. Virtual screening is an effective method in time-critical problems and was applied to determine the structure and energy of an interface. This virtual screening technique has been used in drug discovery, where a prediction model was constructed using machine learning from a relatively small dataset and a large database consisting of the actual data and data predicted by the prediction model. Then, the candidate drug that is most likely to have the intended effectiveness is selected from the larger constructed database. More recently, this virtual screening method has been applied to discover new molecules for organic electro-luminescence (EL) applications and has succeeded in its discovery aims [18]. We have applied this virtual screening technique to predict the structure and energy of certain interfaces [12].

The idea of our virtual screening method is illustrated in Fig. 8.2. A prediction model (predictor) is constructed via regression analysis of the training data, in this case ΣGB_1 and ΣGB_2. Once the predictor is constructed, the grain boundary energies can be predicted from the initial configurations. Then, the candidate configuration that is most likely to give the minimal energy $E_{i, \text{ mim } (i=3, 4, \ldots N)}$ can be determined. Next, the promising initial configuration is optimized using the

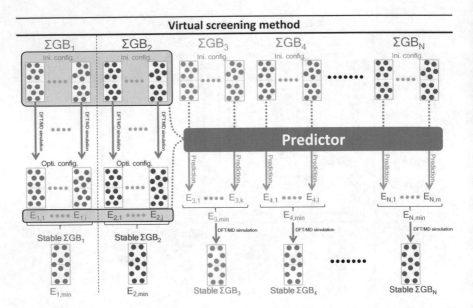

Fig. 8.2 Schematic illustration of virtual screening method for interface structure searching [12]

structure and subsequent energy calculations. Finally, the accurate energy and stable structure are obtained (Stable $\Sigma GB_{3 \sim N}$ in Fig. 8.2).

Seventeen [001] axis-symmetric tilt CSL grain boundaries of Cu were considered in this chapter: $\Sigma 5[001]/(210)$, $\Sigma 5[001]/(310)$, $\Sigma 13[001]/(230)$, $\Sigma 17[001]/(410)$, $\Sigma 17[001]/(350)$, $\Sigma 25[001]/(430)$, $\Sigma 25[001]/(710)$, $\Sigma 29[001]/(520)$, $\Sigma 29[001]/(730)$, $\Sigma 37[001]/(610)$, $\Sigma 37[001]/(750)$, $\Sigma 41[001]/(910)$, $\Sigma 41[001]/(540)$, $\Sigma 53[001]/(720)$, $\Sigma 53[001]/(950)$, $\Sigma 61[001]/(11\ 1\ 0)$, and $\Sigma 125[001]/(11\ 2\ 0)$. To obtain stable structures for these grain boundaries, approximately 1,000,000 configurations must be considered. Namely, structure and energy calculations (such as DFT and MD) must be performed 1,000,000 times to determine the structures of these grain boundaries. To construct the predictor, $\Sigma 5[001]/(210)$, $\Sigma 5[001]/(310)$, $\Sigma 17[001]/(350)$, and $\Sigma 17[001]/(410)$ were selected as the training data, corresponding to ΣGB_1 and ΣGB_2 in Fig. 8.2. Those grain boundaries were selected as the training data based on the variance of their tilt angles and computational costs for their calculations. Structure and energy calculations for a total of 150,000 configurations, corresponding to approximately 15% of all possible configurations, were performed. We can confirm that the calculated structures are almost identical to the previously reported structures [19, 20], indicating that these training data are suitable for constructing the predictor.

The selection of descriptors for regression analysis is important when predicting the grain boundary energy of non-calculated structures. In this study, geometrical

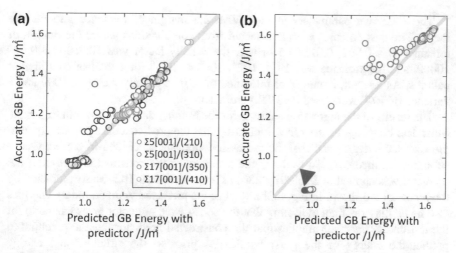

Fig. 8.3 **a** Results of the regression and **b** results for the test data for Σ13[001]/(230) [12]

Table 8.1 List of descriptors

List of descriptors	
tan(θ/2)	Number of longer bond length
sin(θ/2)	Average shorter bond length
Atomic density around GB	Average longer bond length
Average 1st NN (Near Neighbor) bond length	Shortest bond length
Average 2nd NN bond length	Number of dangling bond around GB
Average 1st NN bond length around GB	Relative translation distance along x direction
Average 2nd NN bond length around GB	Relative translation distance along y direction
Number of shorter bond length	Relative translation distance along z direction

data for the "initial atomic configurations" are used as the descriptors. This choice enables one to predict the grain boundary energy without performing structure and energy calculations. The selected descriptors, such as the minimum and maximum bond lengths are listed in Table 8.1.

In addition to these descriptors, their square, inverse, exponential and exponential inverse values were considered. As a result, 83 descriptors were obtained, which were standardized to align their average and variance to zero and one, respectively.

The nonlinear support vector machine (SVM) method was used for regression analysis. In this study, the most stable structures and metastable structures of Σ5 [001]/(210), Σ 5[001]/(310), Σ17[001]/(410), and Σ17[001]/(350) were considered for construction of the prediction model. We have selected those grain boundaries as the training data based on the variance of tilt angles and computational costs for their calculations.

There are two parameters in the SVM, the margin of tolerance and penalty factor. The best parameters were selected from combinations where the margin of tolerance was 0.001, 0.01, 0.05 or 0.1, the penalty factor was 10, 100, 1000 or 10000, and the variance was 10^{-2}, 10^{-3}, 10^{-4} or 10^{-5}, for a total of 64 different patterns. As a result, a margin of tolerance of 0.01, a penalty factor of 1000 and a variance of 10^{-4} were used as SVR parameters.

The results of the regression analysis for the training data are shown in Fig. 8.3a. Most data lie along the grey line, indicating that the predicted energies are equal to the accurate energies and that the regression analysis succeeded in correctly constructing the predictor. To evaluate the accuracy of the constructed predictor, the predictor was applied to $\Sigma 13[001]/(230)$ as a test situation. The results predicted by the predictor are shown in Fig. 8.3b. Most of the predicted grain boundary energies also lie on the grey line, indicating that the constructed predictor is also suitable for the test data. This result implies that the constructed predictor has the potential to predict the energy of the grain boundaries prior to the structure and energy calculations.

Here, we focus on the blue data point marked by the blue arrow in Fig. 8.3b. Based on the constructed predictor, the blue data point was predicted to provide the minimum grain boundary energy. It should be mentioned that the virtual screening method and the calculations of all candidates give the minimum grain boundary energy at the same blue data point. The predicted grain boundary energy is 0.96 J/m^2, which is only 10% larger than that the minimum grain boundary energy obtained from all-candidate calculations. It is also noteworthy that the predicted rigid body translation state (X = 5.0 Å, Y = 1.0 Å, and Z = 0.0 Å) is identical to the most stable rigid body translation state determined by all-candidate calculations.

We succeeded in screening all possible candidates and selecting the most promising candidate configuration for accurately provide the most stable structure. By performing the structure and energy calculation once for this rigid body translation state, a grain boundary energy and structure identical to those obtained from all-candidate calculations can be obtained. Namely, the stable grain boundary structure and energy can be determined with only a one-time calculation using the present virtual screening method, which is significantly more efficient than previously reported methods.

Since the constructed prediction model (the predictor shown in Fig. 8.2) was established, this predictor was also applied to other GBs. Here, based on the constructed predictor, the structures and energies of 12 other [001]-axis-symmetric tilt CSL grain boundaries, $\Sigma 25[001]/(430)$, $\Sigma 25[001]/(710)$, $\Sigma 29[001]/(520)$, $\Sigma 29 [001]/(730)$, $\Sigma 37[001]/(610)$, $\Sigma 37[001]/(750)$, $\Sigma 41[001]/(910)$, $\Sigma 41[001]/(540)$, $\Sigma 53[001]/(720)$, $\Sigma 53[001]/(950)$, $\Sigma 61[001]/(11\ 1\ 0)$, and $\Sigma 125[001]/(11\ 2\ 0)$, are predicted.

Figure 8.4 shows the results of the predicted grain boundary energies and a comparison with previously reported grain boundary energies [19, 20]. Based on previous studies, the grain boundary energy exhibits a convex profile in relation to the misorientation angle θ. Small cusps are also present, namely energy drops at 16.26°, 28.07°, 36.87°, 53.13°, and 67.38° corresponding to $\Sigma 25[001]/(710)$, $\Sigma 17$

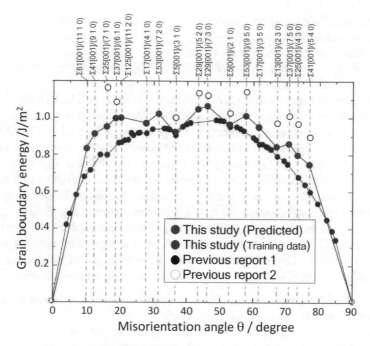

Fig. 8.4 Predicted GB energies using the constructed predictor [15]. Reported values in previous studies are also plotted [19, 20]

[001]/(410), Σ5[001]/(310), Σ5[001]/(210), and Σ13[001]/(230) respectively. The predicted grain boundary energies of all grain boundaries obtained using the predictor are also plotted in the same figure. Although the absolute value is not identical to the previous studies due to the differences in empirical potential, the overall profile of the grain boundary energy is in good agreement with previous reports. Notably, small cusps at 16.26° and 67.38° are also reproduced by the prediction model (other cusps at 28.07°, 36.87°, and 53.13° were used for training). In addition to the GB energy, it was also confirmed that those predicted models fit well to the other calculation and TEM observations. The above results clearly demonstrate that the presented virtual screening method based on machine learning is sufficiently robust and powerful for predicting stable interface structures and energies from initial atomic configurations. The success of this method implies that the initial atomic configuration is correlated to the grain boundary energy, and its correlation is studied by machine learning.

8.2.2 Bayesian Optimization (Kriging) [15]

In this section, we demonstrate an alternative and powerful method that can be used to search for stable interface structures with the aid of a geostatistics approach called kriging. Kriging is an effective interpolation method based on a Bayesian optimization and Gaussian process governed by prior covariances. This Kriging method has been previously used to predict the optimum access points for geological mining operations. Here, we apply this Kriging technique to determine the stable structures of interfaces.

To demonstrate the performance of the Kriging method, the $\Sigma 5[001](210)$ CSL GB of fcc-Cu was again selected as the test case. The three-dimensional translations were considered with 0.1 Å steps, resulting in the generation of a total of 17,983 configurations. The data space that must be searched to determine the most stable structure can be visualized as shown in Fig. 8.5a. In the conventional approach, namely all-candidate calculations, one must calculate the interface energies of all configurations and determine the most stable point within this space. In other words, the search space is occupied by the calculated results as shown in Fig. 8.5b.

To accelerate this search process, a Kriging method based on a Gaussian process was applied. The Gaussian process is a nonparametric regression analysis based on Bayesian statistics. This method allows for the prediction of values and uncertainties of a random field at a point. The steps of this Kriging process are as follows:

(1) Several initial configurations (20 configurations here) are randomly selected from the search space shown in Fig. 8.5a.
(2) The structure optimization and energy calculations for the selected configurations are performed, and a joint probability distribution is calculated with the Gaussian process considering the three rigid body translations along each direction as the descriptors. The grain boundary energy, E_{GB}, is then estimated.
(3) Based on the joint probability distribution, the possible GB energies are estimated at all points in the search space.
(4) Z-scores, calculated by the following equation, are estimated at all points in the search space.

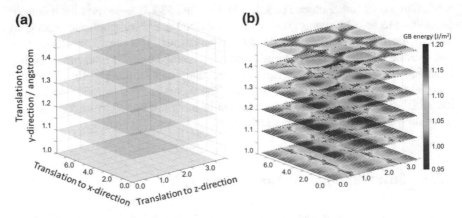

Fig. 8.5 **a** Data space for searching and **b** calculated data space. All data points are calculated [15]

$$Z - score_i = (GB\,Energy_{current\,min} - GB\,Energy(x_i)) \,/\, \sqrt{\sigma(x_i)}$$

where *GB Energy*$_{current\ min}$ is the minimum GB energy at this moment, while *GB Energy(x_i)* and $\sigma(x_i)$ are the mean and standard deviation at the point x_i in the search space respectively.

(5) The point that has the maximum Z-score is selected as the next search point because that point is most likely to have the least GB energy at the moment.

(6) It is confirmed whether or not the acquired GB energies meet the convergence conditions, such as energy difference between the *i*th calculation and the *i + 1*th calculation.

The cycle of above operations ((2)–(6)) is repeated until the convergence criteria have been satisfied. In the structure optimization and energy calculation for (2), we have performed a static lattice calculation using an empirical potential method with the general utility lattice program (GULP) code [21]. The embedded atom potential method reported by Cleri et al. was employed [22].

First, using the conventional approach, all configurations were calculated and the most stable point was determined from the search space shown in Fig. 8.5b. The obtained stable structure is shown in Fig. 8.6a, where the calculated GB energy was 0.96 J/m². As can be seen here, the GB is composed of an array with a six-membered structure unit, in agreement with previously reported structures [23, 24]. However, 17,983 complete calculations were necessary to reach this stable structure using the conventional all-candidate calculation.

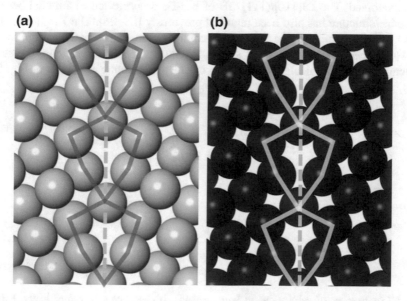

Fig. 8.6 **a** Calculated structure obtained using the all-candidate calculation and **b** using the Kriging method [15]

On the other hand, in the Kriging approach, the search space was interpolated based on the Gaussian process. We found that this Kriging approach greatly decreased the data necessary for calculations. In this case, the most stable point was determined after only 69 trials (including the initial 20 trials). The most stable structure obtained is shown in Fig. 8.6b; the GB is composed of a six-membered structure unit that is very similar to the stable structure obtained by comprehensive data searching. Furthermore, the calculated GB energy is $0.96 \, \text{J/m}^2$, which is identical to that determined by the conventional method, indicating that our present method can accurately determine the most stable structure.

The convergence processes for the all-candidate calculation and the Kriging method are displayed in Fig. 8.7. Although the conventional method requires the calculation of all 17,983 configurations, the present Kriging method requires only 69 calculations (Fig. 8.7a). Figure 8.7b shows the calculation trajectory in the Kriging method. The red numbers show the position in the random sampling, and the pink triangle shows the most stable structure found by the Kriging method. As can be seen in Fig. 8.7b, data space was randomly selected at the beginning of the Kriging method, and it gradually concentrates to neighboring points around the most stable point.

We repeated the Kriging operation 74 times for the same GB and found that the 43 Kriging operations were completed using fewer than 50 time calculations. As a result, the average number of calculations for determining the stable structures is 70. Based on this, the Kriging method has succeeded in accelerating the process of interface structure determination by ~150 times.

Finally, to confirm the applicability of the Kriging method, a different GB was also examined. The $\Sigma3[110]/(111)$ GB of bcc-Fe was selected as a model because its stable structure has also been reported previously [25, 26]. The Kriging method was applied to search for the most stable configuration, just as in the case of the $\Sigma5$ [001]/(210) copper GB described above. We succeeded in determining the stable structure after 105 calculations, and the calculated structure is shown in Fig. 8.8.

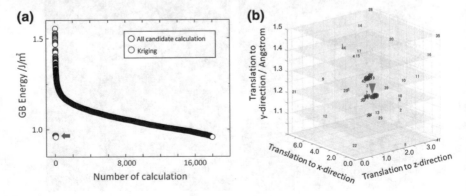

Fig. 8.7 **a** Number of calculations in both methods. **b** Calculation trajectory in the Kriging method. Red numbers indicate the position of the initial random sampling and the pink triangle shows the position of the most stable point found by the Kriging method [15]

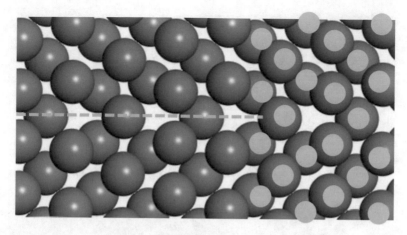

Fig. 8.8 Calculated structure of Fe $\Sigma 3$ GB using the Kriging method [15]. The dashed line represents the position of the GB and the yellow circles show the structure reported previously [25]

The stable structure determined by the Kriging approach agrees well with that of the previous study [25, 26] (yellow circles in Fig. 8.8).

Since 17,466 configurations are present for this $\Sigma 3[110]/(111)$ GB bcc-Fe grain boundary, the Kriging method again achieves two orders of magnitude better efficiency than the conventional method. This clearly indicates that the Kriging method is a very powerful technique to determine the stable interface structure.

8.2.3   Kriging Method for Oxide Interfaces [16]

Comparing the virtual screening method and the Kriging method, the efficiency of the virtual screening is superior to that of the Kriging method. However, one has to construct a predictor in order to maintain this great efficiency. The most important advantage of the Kriging method is its wide applicability. No training is needed for the Kriging method, and thus it can be easily applied to other GBs in other materials. To show the wider applicability of the Kriging method, we used it to conduct similar studies on oxide interfaces. In particular, we applied the Kriging approach to grain boundaries of metal oxides including MgO, TiO_2, and CeO_2 which commonly exhibit more complex structures than metals.

Four kinds of metal oxide grain boundaries, namely rock-salt-MgO $\Sigma 5[001]/(210)$ and $\Sigma 5[001]/(310)$, rutile-TiO_2 $\Sigma 5[001]/(210)$, and fluorite-CeO_2 $\Sigma 3[110]/(111)$ were selected to test the applicability of the present method. These grain boundaries have different complexities; the number of termination planes for MgO $\Sigma 5[001]/(210)$ and $\Sigma 5[001]/(310)$ is one (Fig. 8.9a, b), whereas that for TiO_2 $\Sigma 5[001]/(210)$ and CeO_2 $\Sigma 3[110]/(111)$ is two (Fig. 8.9c, d).

The same Kriging method was applied to these oxides GBs. Two hyper-parameters, pre-distribution and kernel parameter, were set to 0 and 3.0

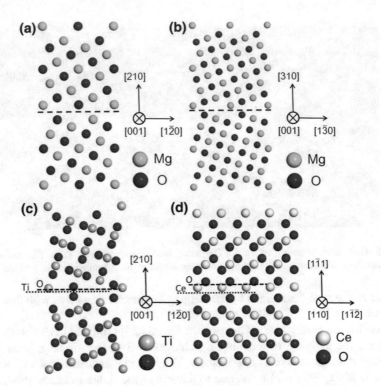

Fig. 8.9 Atomic structure of **a** Σ5[001]/(210), **b** Σ5[001]/(310) GBs of MgO, **c** Σ5[001]/(210) GB of TiO₂, and Σ3[110]/(111) GB of CeO₂ [16]

respectively so that the kernel is not biased to 0 or 1. The random selection number for the initial calculation was set to 5, with the actual size of the three-dimensional rigid body translations in each xyz-direction acting as descriptors. Namely, smaller numbers than the above metal cases were used due to the higher computational cost of the oxide simulations.

For structure optimization and energy calculation, static lattice calculations with an empirical potential were performed using a general utility lattice program (GULP) code [21]. Buckingham-type potentials were used for MgO (Catlow et al. [27]), TiO₂ (Bandura et al. [28]), and CeO₂ (Minervini et al. [29]).

Regarding the convergence criteria in the Kriging method, the structure searching continues until five structures which have the identical lowest grain boundary energy are found. In this case, the grain boundary energies within 0.005 J/m² were judged to be the same. Until these convergence criteria are met, the Kriging algorithm continues searching for the lowest energy configuration.

Figure 8.10a, b show the obtained Σ5[001](210) and Σ5[001](310) grain boundaries of rock-salt-MgO, which has a single grain boundary termination plane as shown in Fig. 8.9a, b. Previously reported structures are overlaid on the structures calculated herein using black or white circles [30, 31]. The number of candidate configurations for Σ5[001](210) and Σ5[001](310) structures equals 28,896 and 40,635 respectively. In

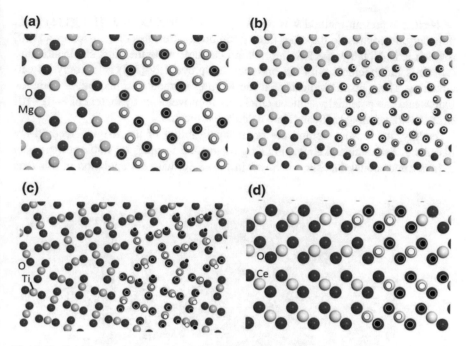

Fig. 8.10 Calculated structures of MgO **a** Σ5(210), **b** Σ5(310), **c** TiO$_2$ Σ5(210), and **d** CeO$_2$ Σ3 (111) GBs [16]

the conventional method, structure optimization and energy calculations for all candidates need to be performed to determine the most stable structure. Conversely, using the Kriging method, we have succeeded in determining the most stable structures for Σ5[001](210) and Σ5[001](310) by performing only 18 and 15 calculations respectively, including initial random calculations.

We performed the Kriging method several times and confirmed that the number of calculations needed to reach convergence in these cases is 14–22, including the initial random sampling. This variation comes from the selection of the initial sampling. However, we would like to emphasize that the Kriging method is clearly powerful to search the most stable structure.

To confirm the applicability of this method to more complex structures, we applied it to TiO$_2$ and CeO$_2$ grain boundaries. The most stable structure of the rutile-TiO$_2$ Σ5[001]/(210) grain boundary is shown in Fig. 8.10c. To maintain charge neutrality, the termination planes were set to Ti and O on their respective sides. A total of 21,630 structures and energies need to be computed in order to find the most stable structure using conventional brute force calculations, while the Kriging method can find the most stable structure by performing only 42 calculations, achieving a more than 500-fold efficiency. In addition, the obtained structure was compared with the previously reported one (Fig. 8.10c) [32], clearly showing that the structure determined by the present strategy was correct.

Next, the present method was applied to the fluorite-CeO_2 $\Sigma3[110](111)$ grain boundary. In contrast to the other three grain boundaries, this one possesses a [110] rotation axis. Notably, the Kriging method can determine the most stable structure (which is in agreement with the one reported previously [33]) using only 12 calculations. These results indicate that the Kriging method is applicable to complex oxides and can potentially achieve efficiency improvements by factors of $\sim10^3$–10^4 over the conventional all-candidate calculation method.

Finally, the reasons behind the broad applicability of the Kriging method are discussed. As mentioned above, the Kriging method searches for stable structures in a three-dimensional data set, as shown in Fig. 8.5b, with extrapolation of this data

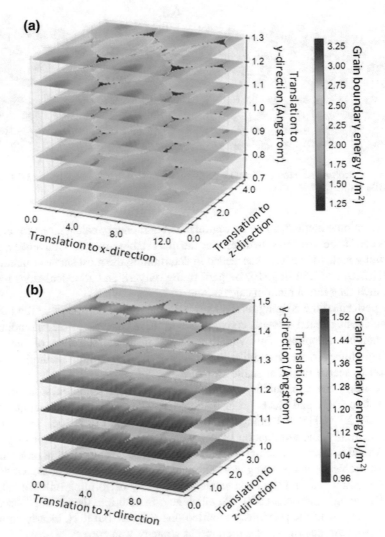

Fig. 8.11 Data space for structure searching. **a** Grain boundary energy plots in the data space for $\Sigma5[001](310)$ GB of MgO and **b** $\Sigma5[001](310)$ GB of Cu [15, 16]

space performed using the Gaussian process. The success of the Kriging method indicates the suitability of this extrapolation method for the present data space, with data spaces for MgO $\Sigma5[001](310)$ and Cu $\Sigma5[001](310)$ compared in Fig. 8.11a, b. The corresponding Cu data space was obtained in previous studies [12, 15]. Although these data spaces appear to be different from each other, their energy profiles are similar. Namely, the grain boundary energy gradually and continuously changes with changing rigid body translation, with no discrete large energy changes present in the data space. This fact indicates that the Kriging method is applicable for grain boundaries possessing a continuous energy surface.

8.3 Microscopic Approach for Interfaces

8.3.1 Scanning Transmission Electron Microscopy (STEM)

STEM is one branch of TEM techniques that has been extensively used for characterizing interface structures in many materials and devices. In recent years, STEM combined with aberration correction technology has enabled direct atom-by-atom imaging via annular dark-field (ADF) imaging. ADF imaging uses a doughnut-shaped annular detector to selectively collect high-angle scattered electrons, building up images from the variation in this signal with probe position in a raster scan. Since the integrated intensity of high-angle scattered electrons strongly scales with the atomic number of the atoms under the probe, this imaging (so-called Z-contrast imaging) can sensitively visualize heavy element atoms [34]. However, ADF can seldom reliably visualize light elements due to their weak power to scatter electrons at higher angles. While ADF imaging mainly uses electrons scattered at high-angles to form atom images, there are many other possible detector geometries for collecting electron signals and forming images. One is known as annular bright-field (ABF) imaging, which involves the selective collection of electrons inside the bright-field disk via a small annular detector [35]. It has been shown that ABF imaging can directly visualize light atoms and can even directly visualize H atom columns inside compound materials [36, 37]. Since ADF and ABF images can be obtained simultaneously from the same sample positions, both heavy and light element atomic structures can now be directly visualized by combining these two images. However, it is still very difficult to identify the atomic species using only the ADF and ABF image contrasts, especially at interface regions where structure and chemistry are drastically changing. Since STEM uses an atomic-scale electron probe, STEM-based analytical techniques such as energy-dispersive X-ray spectroscopy (EDS) and electron energy loss spectroscopy (EELS) can also achieve atomic resolution. In particular, by utilizing ultrasensitive silicon drift detectors (SDDs) with much higher count rates, atomic-scale EDS mapping is now becoming possible [38]. This capability should be very powerful to directly characterize dopant/impurity segregation behaviors in grain boundaries and heterointerfaces.

Thus, atomic-resolution STEM combined with spectroscopy will become an indispensable technique for characterizing atomic-scale structures and chemistry of interfaces.

8.3.2 Interface Structures Using Aberration-Corrected STEM

Two interface characterization studies using aberration-corrected STEM and EDS are highlighted in this section. One is on the solute segregation in a GB of ceramics [39] and the other is on the impurity segregation in a metal/ceramic heterointerface [40]. These studies demonstrate that atomic-resolution STEM is a powerful tool for directly understanding very complex segregation phenomena in materials.

8.3.2.1 Solute Segregation Behavior of a $\sum 3$ Grain Boundary in Yttria Stabilized Zirconia [39]

ZrO_2 doped with Y_2O_3 (YSZ) is one of the most important materials for use as an electrolyte in solid oxide fuel cells, where the overall ionic conductivity is strongly affected by the presence of GBs; such an effect may potentially originate from GB chemical inhomogeneity. Previous studies have shown that the Y solute atoms segregate to the GBs, and the amount of segregation is strongly dependent on the GB characteristics [41]. However, the atomic-scale mechanism of how Y solute actually segregate to GBs is still not well understood. In this study, we show an atomic-scale EDS mapping of a $\Sigma3[110]/\{111\}$ model GB in YSZ.

Figure 8.12 shows an ADF STEM image of the $\Sigma3[110]/\{111\}$ grain boundary of YSZ, where the GB atomic arrangement is in good agreement with previous high-resolution TEM studies [41]. However, it is almost impossible to distinguish Zr and Y atoms from the STEM image alone. Then, atomic-resolution STEM-EDS mapping was carried out in order to distinguish the two atomic components. Figure 8.13a, b show atomic-resolution EDS maps around the GB, where the Y and Zr maps clearly reveal the formation of characteristic ordered segregation structures. The intensity variation is further highlighted in the corresponding intensity profile shown in Fig. 8.13c, d.

It is noteworthy that in some atomic column layers, Y atoms are obviously depleted. This indicates that the Y atoms are not simply substituting in all of the cation sites around the GB to form segregation structures. Thus, we experimentally found that Y solute segregation formed atomically ordered extended structures across the GB within a range of approximately 3 nm. These experimental results are in good agreement with large-scale Monte Carlo simulations [39]. The simulation suggested that such processes can be driven by both the site-dependent segregation of Y due to strain and $Y\text{-}V_O$ interactions. Thus, recent advanced microscopy

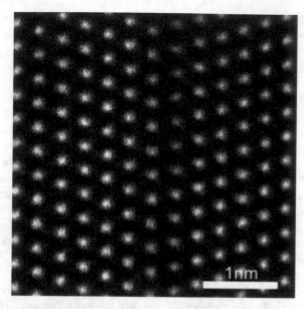

Fig. 8.12 ADF STEM image of the Σ3[110]/{111} grain boundary of YSZ (adopted from Ref: [39])

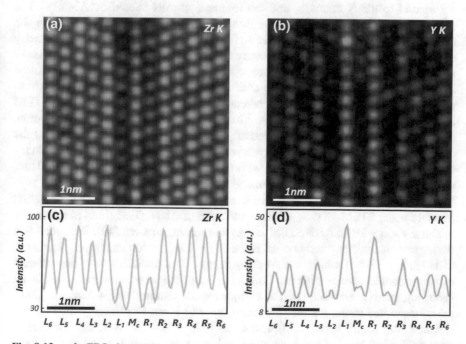

Fig. 8.13 a, b, EDS elemental maps for **a** Zr K map and **b** Y K map. **c, d,** normalized intensity profiles derived by summing the X-ray counts in the maps in the direction parallel to the GB for **c** Zr K and **d** Y K (adopted from Ref: [39])

combined with theory can shed new light on the fundamental mechanism of solute segregation behaviors in GBs.

8.3.2.2 Dopant Segregation Behavior in a Metal/Ceramic Interface [40]

Heterostructures between metals and ceramics have been widely used for power electronic devices requiring both high thermal performance and reliability in harsh environments. Since interfaces play a critical role in many properties, a fundamental understanding of the interface structure and formation mechanism is vitally important. One important possibility for obtaining heterointerfaces with better properties is to control dopant/impurity segregation behaviors. However, it has been very challenging to directly observe segregation structures at atomic dimensions in heterointerfaces. Here, atomic-resolution STEM-EDS mapping is shown to be a powerful tool for directly determining segregation structures in metal/ceramic heterointerfaces.

Figure 8.14 shows simultaneous cross-sectional (a) ADF and (b) ABF STEM images of an Al alloy (containing Si and Mg as major dopants) /AlN interface formed using a liquid phase bonding technique [40]. In the AlN bulk region, compared to the ADF image, the columns with weaker intensities in the ABF image correspond to the N columns, and the interface of AlN should be Al-polar. The atomic structures of the Al alloy region were not clearly resolved because the present viewing direction is not well-aligned along the certain high symmetry crystallographic axis. To clearly identify the interface atomic structure, we show noise-filtered images of the interface core region in Fig. 8.14c, d. From these images, we can divide the interface core region into three different layered structures labeled as the 1st, 2nd, and 3rd layers. Considering both ADF and ABF STEM images, the 1st, 2nd and 3rd layer interface structures are anion-cation-anion, cation-anion, and anion layers, respectively. However, it is difficult to determine the detailed atomic structure of the three layers from the STEM image contrast since dopant elements such as Al and Mg with close atomic numbers may coexist. Thus, we performed atomic-resolution chemical mapping using STEM-EDS.

Figure 8.15 shows atomically resolved chemical maps of the Al alloy/AlN interface using STEM-EDS. The elemental maps of Al, N, O, Mg, and Si are shown in comparison with an ABF STEM image and structure model [40]. We found that the highest signal in Mg and O maps is located at the 1st interfacial layer, but that of Si is slightly shifted within the Al alloy region. This indicates that these dopant elements should occupy different atomic layers at the interface region. In the 1st layer, Mg atoms are concentrated to a single atomic column layer, whereas O atoms are concentrated to the top and bottom of the Mg layer. Considering the bonding distances and angles between Mg and O columns in the 1st layer, this structure is very similar to a MgO_6 octahedron with rock salt structure. In the 2nd layer, a local maximum of Al can be found at the cation columns. Thus, the cation columns can be identified as Al columns. O and N could not be separated in the 2nd layer,

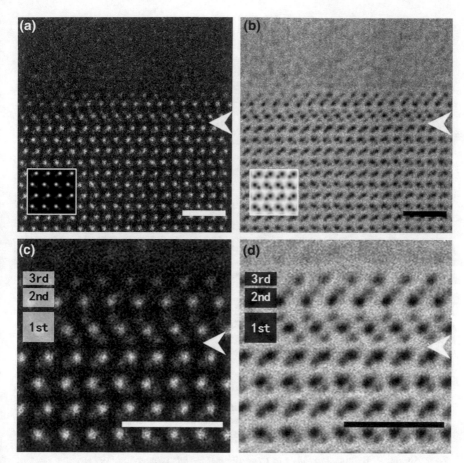

Fig. 8.14 Simultaneously obtained atomic-resolution ADF **a** and ABF **b** STEM images of an Al alloy/AlN interface (adopted from Ref: [40]). Calculated images of AlN bulk are superimposed in the lower left in **a** and **b**. The magnified views are shown in **c** and **d**

but the ABF image contrast of the 2nd layer is similar to the Al-N contrast in the AlN bulk structure, although with inverted polarity (N-polar). Thus, we consider that the main structure of the 2nd layer is an AlN_4 tetrahedral monolayer. The polarity of this layer is inverted from the AlN substrate due to the presence of the MgO interlayer. Theoretical simulations suggested that the interface between Al metal and N-polar AlN is much more energetically stable than that between Al metal and Al-polar AlN.

Thus, Al alloy /AlN heterointerfaces should be stabilized by the formation of self-organized atomic-scale layered structures with Mg dopant segregation. Atomic-resolution STEM-EDS is a powerful tool for directly determining heterointerface structures with dopant/impurity segregation.

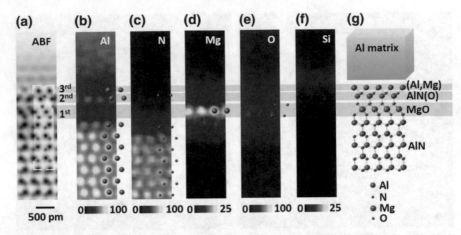

Fig. 8.15 a The averaged ABF STEM image and the corresponding elemental maps of **b** Al, **c** N, **d** Mg, **e** O, and **f** Si, respectively (adopted from Ref: [40]). The structure model of the heterointerface **g** determined from the experimental results is shown

Acknowledgements This work was supported by a Grant-in-Aid for Scientific Research on Innovative Areas "Nano Informatics" (Grant No. JP25106003) from the Japan Society for the Promotion of Science (JSPS).

References

1. Y. Ikuhara, Grain boundary atomic structures and light-element visualization in ceramics: combination of Cs-corrected scanning transmission electron microscopy and first-principles calculations. J. Electron. Microsc. (Tokyo) **60**, 173–188 (2011). https://doi.org/10.1093/jmicro/dfr049
2. A. Sutton, R. Balluffi, *Interfaces in Crystalline Materials* (Clarendon, 1995)
3. J.P. Buban, K. Matsunaga, J. Chen, N. Shibata, W.Y. Ching, T. Yamamoto, Y. Ikuhara, Grain boundary strengthening in alumina by rare earth impurities. Science **311**(2006) 212–215. https://doi.org/10.1126/science.1119839
4. K. Matsunaga, H. Nishimura, H. Muto, T. Yamamoto, Y. Ikuhara, Direct measurements of grain boundary sliding in yttrium-doped alumina bicrystals. Appl. Phys. Lett. **82**, 1179–1181 (2003). https://doi.org/10.1063/1.1555690
5. T. Mizoguchi, H. Ohta, H.-S. Lee, N. Takahashi, Y. Ikuhara, Controlling interface intermixing and properties of SrTiO3-based superlattices. Adv. Funct. Mater. **21**, 2258–2263 (2011). https://doi.org/10.1002/adfm.201100230
6. H. Ohta, S. Kim, Y. Mune, T. Mizoguchi, K. Nomura, S. Ohta, T. Nomura, Y. Nakanishi, Y. Ikuhara, M. Hirano, H. Hosono, K. Koumoto, Giant thermoelectric Seebeck coefficient of a two-dimensional electron gas in SrTiO3. Nat. Mater. **6**, 129–134 (2007). https://doi.org/10.1038/nmat1821
7. H. Sawada, N. Shimura, F. Hosokawa, N. Shibata, Y. Ikuhara, Resolving 45-pm-separated Si-Si atomic columns with an aberration-corrected STEM. Microscopy **64**, 213–217 (2015). https://doi.org/10.1093/jmicro/dfv014
8. S.J. Pennycook, P.D. Nellist, *Scanning Transmission Electron Microscopy* (Springer New York, New York, 2011). https://doi.org/10.1007/978-1-4419-7200-2
9. N. Tanaka, Scanning transmission electron microscopy of nanomaterials, in ed. by N. Tanaka (Imperial College Press, London, 2014), pp. i–xliv. https://doi.org/10.1142/9781848167902_fmatter

10. S. Ranganathan, On the geometry of coincidence-site lattices. Acta Crystallogr. **21**, 197–199 (1966). https://doi.org/10.1107/S0365110X66002615
11. D. Brandon, The structure of high-angle grain boundaries. Acta Metall. **14**, 1479–1484 (1966). https://doi.org/10.1016/0001-6160(66)90168-4
12. S. Kiyohara, H. Oda, T. Miyata, T. Mizoguchi, Prediction of interface structures and energies via virtual screening. Sci. Adv. **2**, e1600746–e1600746 (2016). https://doi.org/10.1126/sciadv.1600746
13. A.L.-S. Chua, N.A. Benedek, L. Chen, M.W. Finnis, A.P. Sutton, A genetic algorithm for predicting the structures of interfaces in multicomponent systems. Nat. Mater. **9**, 418–422 (2010). https://doi.org/10.1038/nmat2712
14. G. Schusteritsch, C.J. Pickard, Predicting interface structures: from SrTiO3 to graphene. Phys. Rev. B. **90**, 35424 (2014). https://doi.org/10.1103/PhysRevB.90.035424
15. S. Kiyohara, H. Oda, K. Tsuda, T. Mizoguchi, Acceleration of stable interface structure searching using a kriging approach. Jpn. J. Appl. Phys. **55**, 2–6 (2016). https://doi.org/10.7567/JJAP.55.045502
16. S. Kikuchi, H. Oda, S. Kiyohara, T. Mizoguchi, Bayesian optimization for efficient determination of metal oxide grain boundary structures. Phys. B Condens. Matter. (2017) (in press). https://doi.org/10.1016/j.physb.2017.03.006
17. T. Ueno, T.D. Rhone, Z. Hou, T. Mizoguchi, K. Tsuda, COMBO: an efficient Bayesian optimization library for materials science, Mater. Discov. 10–13, (2016). https://doi.org/10.1016/j.md.2016.04.001
18. R. Gómez-Bombarelli, J. Aguilera-Iparraguirre, T.D. Hirzel, D. Duvenaud, D. Maclaurin, M. A. Blood-Forsythe, H.S. Chae, M. Einzinger, D.-G. Ha, T. Wu, G. Markopoulos, S. Jeon, H. Kang, H. Miyazaki, M. Numata, S. Kim, W. Huang, S.I. Hong, M. Baldo, R.P. Adams, A. Aspuru-Guzik, Design of efficient molecular organic light-emitting diodes by a high-throughput virtual screening and experimental approach. Nat. Mater. **15**, 1120–1127 (2016). https://doi.org/10.1038/nmat4717
19. D. Wolf, Structure-energy correlation for grain boundaries in F.C.C. metals—III. Symmetrical tilt boundaries. Acta Metall. Mater. **38**, 781–790 (1990). https://doi.org/10.1016/0956-7151(90)90030-K
20. G. Hasson, J.-Y. Boos, I. Herbeuval, M. Biscondi, C. Goux, Theoretical and experimental determinations of grain boundary structures and energies: correlation with various experimental results. Surf. Sci. **31**, 115–137 (1972). https://doi.org/10.1016/0039-6028(72)90256-7
21. J.D. Gale, GULP: a computer program for the symmetry-adapted simulation of solids. J. Chem. Soc. Faraday Trans. **93**, 629–637 (1997)
22. F. Cleri, V. Rosato, Tight-binding potentials for transition metals and alloys. Comput. Simul. Mater. Sci. **205**, 233–253 (1991). https://doi.org/10.1007/978-94-011-3546-7_11
23. P. Ballo, V. Slugeň, Grain boundary sliding and migration in copper: the effect of vacancies. Comput. Mater. Sci. **33**, 491–498 (2005). https://doi.org/10.1016/j.commatsci.2004.09.049
24. M.A. Tschopp, D.L. Mcdowell, Asymmetric tilt grain boundary structure and energy in copper and aluminium. Philos. Mag. **87**, 3871–3892 (2007). https://doi.org/10.1080/14786430701455321
25. H. Nakashima, M. Takeuchi, Grain boundary energy and structure of α-Fe < 110 > symmetric tilt boundary, Tetsu-to-Hagane. **86**(2000) 357–362. https://doi.org/10.2355/tetsutohagane1955.86.5_357
26. R. Kurtz, H. Heinisch, The effects of grain boundary structure on binding of He in Fe. J. Nucl. Mater. **329–333**, 1199–1203 (2004). https://doi.org/10.1016/j.jnucmat.2004.04.262
27. A. Catlow, *Chapter 10 Interionic Potentials in Ionic Solids*, (n.d.)
28. A.V. Bandura, J.D. Kubicki, Derivation of force field parameters for TiO2–H2O systems from ab initio calculations (2003). https://doi.org/10.1021/JP034093T
29. L. Minervini, M.O. Zacate, R.W. Grimes, Defect cluster formation in M2O3-doped CeO2. Solid State Ionics **116**, 339–349 (1999). https://doi.org/10.1016/S0167-2738(98)00359-2

30. B.P. Uberuaga, X.-M. Bai, P.P. Dholabhai, N. Moore, D.M. Duffy, Point defect–grain boundary interactions in MgO: an atomistic study. J. Phys. Condens. Matter **25**, 355001 (2013). https://doi.org/10.1088/0953-8984/25/35/355001

31. Y. Yan, M.F. Chisholm, G. Duscher, A. Maiti, S.J. Pennycook, S.T. Pantelides, Impurity-induced structural transformation of a MgO grain boundary, Phys. Rev. Lett. **81** (1998), 3675–3678. https://doi.org/10.1103/PhysRevLett.81.3675

32. S.B. Sinnott, R.F. Wood, S.J. Pennycook, Ab initio calculations of rigid-body displacements at the $\Sigma 5$ (210) tilt grain boundary in TiO2. Phys. Rev. B. **61**, 15645–15648 (2000). https://doi.org/10.1103/PhysRevB.61.15645

33. B. Feng, H. Hojo, T. Mizoguchi, H. Ohta, S.D. Findlay, Y. Sato, N. Shibata, T. Yamamoto, Y. Ikuhara, Atomic structure of a $\Sigma 3$ [110]/(111) grain boundary in CeO2. Appl. Phys. Lett. **100**, 73109 (2012). https://doi.org/10.1063/1.3682310

34. S.J. Pennycook, L.A. Boatner, Chemically sensitive structure-imaging with a scanning transmission electron microscope. Nature **336**, 565–567 (1988). https://doi.org/10.1038/336565a0

35. S.D. Findlay, N. Shibata, H. Sawada, E. Okunishi, Y. Kondo, T. Yamamoto, Y. Ikuhara, Robust atomic resolution imaging of light elements using scanning transmission electron microscopy. Appl. Phys. Lett. **95**, 2–4 (2009). https://doi.org/10.1063/1.3265946

36. S.D. Findlay, T. Saito, N. Shibata, Y. Sato, J. Matsuda, K. Asano, E. Akiba, T. Hirayama, Y. Ikuhara, Direct imaging of hydrogen within a crystalline environment. Appl. Phys. Express **3**, 6–9 (2010). https://doi.org/10.1143/APEX.3.116603

37. R. Ishikawa, E. Okunishi, H. Sawada, Y. Kondo, F. Hosokawa, E. Abe, Direct imaging of hydrogen-atom columns in a crystal by annular bright-field electron microscopy. Nat. Mater. **10**, 278–281 (2011). https://doi.org/10.1038/nmat2957

38. M.W. Chu, S.C. Liou, C.P. Chang, F.S. Choa, C.H. Chen, Emergent chemical mapping at atomic-column resolution by energy-dispersive X-ray spectroscopy in an aberration-corrected electron microscope, Phys. Rev. Lett. **104**(2010), 1–4. https://doi.org/10.1103/PhysRevLett.104.196101

39. B. Feng, T. Yokoi, A. Kumamoto, M. Yoshiya, Y. Ikuhara, N. Shibata, Atomically ordered solute segregation behaviour in an oxide grain boundary. Nat. Commun. **7**, 11079 (2016). https://doi.org/10.1038/ncomms11079

40. A. Kumamoto, N. Shibata, K. Nayuki, T. Tohei, N. Terasaki, Y. Nagatomo, T. Nagase, K. Akiyama, Y. Kuromitsu, Y. Ikuhara, Atomic structures of a liquid-phase bonded metal/nitride heterointerface. Sci. Rep. **6**, 22936 (2016). https://doi.org/10.1038/srep22936

41. N. Shibata, F. Oba, T. Yamamoto, Y. Ikuhara §, Structure, energy and solute segregation behaviour of [110] symmetric tilt grain boundaries in yttria-stabilized cubic zirconia. Philos. Mag. **84**, 2381–2415 (2004). https://doi.org/10.1080/14786430410001693463

Chapter 9
High Spatial Resolution Hyperspectral Imaging with Machine-Learning Techniques

Motoki Shiga and Shunsuke Muto

Abstract Recent advances in scanning transmission electron microscopy (STEM) techniques have enabled us to obtain spectroscopic datasets such as those generated by electron energy-loss (EELS)/energy-dispersive X-ray (EDX) spectroscopy measurements in a PC-controlled way from a specified region of interest (ROI) even at atomic scale resolution, also known as hyperspectral imaging (HSI). Instead of conventional analytical procedures, in which the potential constituent chemical components are manually identified and the chemical state of each spectral component is successively determined, a statistical machine-learning approach, which is known to be more effective and efficient for the automatic resolution and extraction of the underlying chemical components stored in a huge three-dimensional array of an observed HSI dataset, is used. Among the statistical approaches suitable for processing HSI datasets, methods based on matrix factorization such as principal component analysis (PCA), multivariate curve resolution (MCR), and nonnegative matrix factorization (NMF) are useful to find an essential low-dimensional data subspace hidden in the HSI dataset. This chapter describes our developed NMF method, which has two additional terms in the objective function, and which is particularly effective for analyzing STEM-EELS/EDX HSI datasets: (i) a soft orthogonal penalty, which clearly resolves partially overlapped spectral components in their spatial distributions and (ii) an automatic relevance determination (ARD) prior, which optimizes the number of components involved in the observed

M. Shiga
Department of Electrical, Electronic and Computer Engineering,
Gifu University, 1-1, Yanagido, Gifu 501-1193, Japan
e-mail: shiga_m@gifu-u.ac.jp

M. Shiga
Precursory Research for Embryonic Science and Technology, Japan Science
and Technology Agency, 4-1-8, Honcho, Kawaguchi, Saitama 332-0012, Japan

S. Muto (✉)
Electron Nanoscopy Division, Advanced Measurement Technology Center,
Institute of Materials and Systems for Sustainability, Nagoya University,
Nagoya 464-8603, Japan
e-mail: smuto@imass.nagoya-u.ac.jp

© The Author(s) 2018
I. Tanaka (ed.), *Nanoinformatics*, https://doi.org/10.1007/978-981-10-7617-6_9

179

data. Our analysis of real STEM-EELS/EDX HSI datasets demonstrates that the soft orthogonal penalty is effective to obtain the correct decomposition and that the ARD prior successfully identifies the correct number of physically meaningful components.

Keywords Non-negative matrix factorization · Scanning transmission electron microscopy · Hyperspectral image analysis · Electron energy-loss spectroscopy Energy-dispersive X-ray spectroscopy

9.1 Introduction

Current scientific analytical instruments are mostly computer-controlled and based on digital circuits. This facilitates automated measurements because the experimental procedures can be specified by using program code. For instance, recent advances in scanning transmission electron microscopy (STEM) techniques, including the development of brighter electron sources, digitally controlled operation, detectors with higher sensitivity, and sophisticated online signal processing, have enabled us to obtain comprehensive information not only on the local structures but also on the chemistry of solids by concurrently applying spectroscopic methods such as electron energy-loss (EELS) and energy-dispersive X-ray (EDX) spectroscopy to a specified region of interest (ROI). The spectrometers collect a set of spectra, each from the subnanometer area of the sample, using subnanometric incident electron probe scanning over the two-dimensional ROI with a subnanometric step width. This method is known as hyperspectral imaging (HSI). The typical data acquisition time is now reduced to several minutes for an entire EELS dataset with 2,000 energy channels over $10^4 = 100 \times 100$ pixels (sampling points). Accordingly, the associated volume of data to be analyzed has been drastically increasing. In this context, statistical analysis methods could be more effective to thoroughly extract information embedded in massive amounts of data without any preconception, rather than relying on conventional spectral analysis of sampling points detected manually based on the insight of experts.

Among the various statistical approaches, principal component analysis (PCA) [1–3] is one of the most fundamental and popular methods. PCA successively casts mutually orthogonal eigenvectors (basis vectors) and associated score images (spatial intensity distributions of the corresponding basis vectors) in the order of significance, that is, in the order of the magnitude of eigenvalues, by way of the singular value decomposition of the HSI data matrix consisting of the experimental spectra as its row vectors. Trebbia and Bonnet [2] and Bosman et al. [3] applied PCA to EELS-HSI datasets, and not only detected exotic chemical bonding states in the samples, but also effectively filtered statistical noise from the HSI data matrix by reconstructing this matrix with a few essential basis spectra and their spatial intensity distributions. Parish and Brewer [4] studied the validity of PCA in a quantitative composition analysis of the constituent phases in their

EDX-HSI data. Note that, in their treatment, the phase overlapping areas were masked for exclusion from the quantification process; otherwise, the derived phase compositions could be biased with respect to the actual ones. These reports on PCA assumed that each pixel contains a linear combination of principal components with the orthogonality condition intrinsic to PCA. Using simulated atomic resolution EELS-HSI data, Lichteret and Verbeeck [5] pointed out that, when the noise level exceeds the intensity of the signal of interest, the signal intensities are distributed over a number of principal components, and are thus usually considered as noise. This behavior seems statistically natural, but we would not notice this phenomenon in actual experimental data. On the other hand, Spiegelberg and Rusz recently reexamined the applicability of PCA to noisy EELS data [6]. In order to estimate the amount of bias present in each principal component, Lichtert and Verbeeck [5] proposed evaluation criteria which, however, do not exhibit the correct asymptotic behavior considering the size of the dataset. Spiegelberg and Rusz [6] proposed alternative evaluation criteria, taking the size of the dataset into account.

Dobigeon and Brun [7] compared the results obtained by applying PCA, independent component analysis (ICA) [8], vertex component analysis (VCA) [9], and Bayesian linear unmixing (BLU) [10] to experimental EELS-HSI data. They eventually found that BLU provided the most plausible spatial distributions for the constituent spectral components, presumably because of its more realistic modeling of the EELS-HSI data. Spiegelberg et al. [11] also discussed a set of such data decomposition methods. In particular, they established randomized VCA (RVCA), an extension of VCA for application to noisy data, and compared its efficiency with that of minimum volume simplex analysis (MVSA) and BLU.

Over a decade, our research group has been developing an alternative method to nonnegative matrix factorization (NMF), or multivariate curve resolution (MCR) for the analysis of EELS-HSI [12, 13]. We consider this approach to be successful because NMF naturally restricts both the spatial intensities and basis spectra to nonnegative values. Contrary to NMF, the methods mentioned above such as PCA allow the spatial intensities and spectra to have negative values, which hampers the direct physical interpretation of the resolved spectral profiles. We adopted the modified alternating least-square (MALS) fitting algorism of NMF [14] to map the different phases in the degradation of Li battery cathodes [15–19] and the chemical states of nitrogen in nitrogen-doped TiO_2 [20, 21]. We also successfully applied NMF to a series of EELS datasets for the extraction of atom site-specific core-loss spectra, where the relative excitation probabilities of the spectra varied with the diffraction condition because of the electron channeling effects [22–25]. In these applied data analyses, the nonnegative constraint of the elements of extracted basis spectra and spatial intensity distributions were effective, and the resulting spectra extracted by NMF were consistent with the computational results obtained by first principles calculations [15–19, 22–25].

In general, approaches such as PCA and NMF are known as matrix factorization because these methods factorize a HSI data matrix into the product of two thin matrices, i.e. matrices of the spatial intensity distribution and basis spectra, with some suitable constraints resulting from the designed model. The next section first briefly formalizes the problem setting of matrix factorization with HSI data [26, 27]. We then present our proposed NMF [26], which presents two advantages with respect to HSI analysis against conventional NMFs: (i) spatially clear decomposition of overlapping intensity distributions achieved by introducing a spatially orthogonal penalty term and (ii) automatic selection of a number of essential chemical components by introducing a penalty term of an automatic relevance determination (ARD) prior distribution. Our analysis of real STEM-EDX/EELS HSI datasets demonstrates that the spatial orthogonal penalty is effective to obtain the correct decomposition and the ARD prior can successfully select the correct number of physically meaningful components.

9.2 Methodology

9.2.1 Mathematical Formulation of HSI Data

The observed HSI data are stored in a three-dimensional array termed a data cube $D(x, y, E)$, which is a function of the two-dimensional spatial position (x, y) on the specimen and the absorption/emission energy E. For the convenience of mathematical manipulation, the data cube is often transformed to a two-dimensional $N_{xy} \times N_{ch}$ matrix X, where $N_{xy} = N_x \times N_y$ is the number of pixels, i.e. the product of the number of scanning steps N_x and N_y along the spatial x- and y-axis, respectively, and N_{ch} is the number of detector channels. After the transformation, the observed spectrum at position (x, y) is stored in a row of matrix X. A basic statistical method to extract a few essential basis spectra and their spatial intensity distribution assumes that the spectral intensity at each sample pixel is represented by a linear combination of the basis spectra associated with the underlying chemical components (states or phases). Assuming that the number of essential chemical components in the observed spatial region is K, which is much smaller than the size of matrix X, this analysis can be formulated by matrix factorization, which factorizes HSI data matrix X into low rank (or thin) matrices of the spatial intensity distribution C and basis spectra S:

$$X \approx CS^{\mathrm{T}}, \tag{9.1}$$

where the size of C is $N_{xy} \times K$ and the size of S is $N_{ch} \times K$, and superscript T denotes a matrix transpose. Each column vector of S (referred to as loading in multivariate analysis) is a basis spectrum of a chemical component. On the other hand, each column vector of C (referred to as score) is a spatial intensity distribution over the ROI positions. Hence, each row vector of C is the intensities of K chemical components at a spatial position. Using the i-th column of matrix C, a two-dimensional spatial distribution of the i-th chemical component can be reconstructed by rearranging the elements such that they are returned to the original two-dimensional position.

The matrix factorization can identify both spatial intensity matrix C and spectral matrix S by minimizing the reconstruction error, which is the distance between observation X and the reconstruction, i.e. CS^T. This identification is possible because matrix X, which consists of a huge number of elements with a relatively much smaller K, is highly redundant. Thus, the identification problem is equal to that intended to find the essential subspace where the original X occurs. This approach can identify plausible C and S with much higher signal-to-noise ratios (SNRs) than those manually selected from the small number of representative observed spatial points, i.e. point-to-point analysis.

Matrix factorization needs to assume a suitable restriction of C and S because the optimization problem results in many local minima. Principal component analysis (PCA) identifies C and S by minimizing the squared error $\|X - CS^T\|^2$ with the orthogonal constraints in both C and S. Owing to the orthogonal constraint, PCA can easily find the global solution using a singular value decomposition (SVD) algorithm. However, PCA can generate unnatural C and S, in which the element can include negative spatial intensities and spectral values. Moreover, the strong orthogonal constraint cannot allow overlaps to exist among the chemical components in both spatial and spectral space. These problems require the outputs by PCA to be adjusted to obtain physically meaningful insights. We overcame these problems by using an approach involving non-negative matrix factorization, in which the elements of C and S are not allowed to be negative.

9.2.2   Non-negative Matrix Factorization with a Gaussian Noise Model

This section presents a formal mathematical description of our model and algorithms to provide the concept of our developed NMF. Let $X \in R_+^{N_{xy} \times N_{ch}}$ be an HSI data matrix, where R_+ is the set of all nonnegative real numbers. NMF factorizes X into two thin matrices $C \in R_+^{N_{xy} \times K}$ and $S \in R_+^{N_{ch} \times K}$, where K is much smaller than both N_{xy} and N_{ch}. Hence, the factorization model is given by

$$X = CS^T + \varepsilon \qquad (9.2)$$

where $\varepsilon \in R^{N_{xy} \times N_{ch}}$ is a noise matrix of which the elements are generated statistically independent of each other. In our problem setting, only X is observed, whereas C and S are not observed. The goal of NMF is to identify the optimal C and S under a suitable noise model ε. One of the most common models is a Gaussian noise model, in which an element of a noise matrix is generated from a Gaussian distribution:

$$p\left(\varepsilon_{ij}|0,\sigma^2\right) = \frac{1}{\sqrt{2\pi\sigma^2}}\exp\left\{-\frac{\varepsilon_{ij}^2}{2\sigma^2}\right\}, \tag{9.3}$$

where σ^2 is the noise variance. Since the statistically independent assumption of $\varepsilon = X - \overline{X}$, where $\overline{X} = CS^T$ is the noiseless data matrix, the log-likelihood function of matrix X is given by

$$\log p\left(X|\overline{X},\sigma^2\right) = -\frac{1}{2\sigma^2}\sum_{i=1}^{N_{xy}}\sum_{j=1}^{N_{ch}}\left(X_{ij}-\overline{X}_{ij}\right)^2 - \frac{N_{xy}N_{ch}}{2}\log 2\pi\sigma^2 \tag{9.4}$$

Taking a common statistical estimation approach of the maximum likelihood estimation, i.e. maximizing $\log p(X|\overline{X},\sigma^2)$, C and S can be optimized using only data matrix X. By taking the negative value of the log-likelihood, i.e. $-\log p(X|\overline{X},\sigma^2)$, and neglecting σ^2 in the first term and the second term, the optimization problem is transformed into the minimization of the squared error function between observation X and reconstruction $\overline{X}$:

$$D_{EU}\left(X|\overline{X}\right) = \frac{1}{2}\sum_{i=1}^{N_{xy}}\sum_{j=1}^{N_{ch}}\left(X_{ij}-\overline{X}_{ij}\right)^2. \tag{9.5}$$

Contrary to PCA, the minimization problem of Eq. (9.5) over both C and S is non-convex, and contains a number of local minima. The optimization algorithm for an NMF does not always converge to the global optimum of C and S. Hence, it is necessary to run the optimization algorithm multiple times from different initializations, resulting in considerable computational cost. The computational efficiency has been improved by developing fast optimization algorithms such as matrix multiplication (MM) [28], alternating least-squares (ALS) [29], and hierarchical alternating least-squares (HALS) [30]. In general, MM is sensitive to the initial configuration, whereas the other algorithms are not. Among these approaches, HALS offers the best convergence to local minima. Hence, we adopted the HALS framework for the optimization of our new NMF model. Another problem presented by NMFs is that the number of chemical components needs to be manually selected in advance, which inevitably introduces a problem similar to that of PCA if the noise level is larger than the signal intensities. As the number of components increases, the reconstruction error naturally decreases. However, this decrease is not essential to identify C and S because it results in overfitting to observed data when the number of components is excessively large. Thus, relying on the reconstruction error only cannot be useful to identify the essential number of physically meaningful components.

To overcome the above difficulties in STEM-EELS/EDX HSI data analysis, we developed a new NMF model that imposes the following penalty terms on the spatial intensity matrix C: (i) a spatial orthogonal penalty [31] and (ii) a sparse penalty to optimize the number of components, termed an automatic relevance determination (ARD) prior [32]. For the optimization of low-rank matrices C and S, we further developed an algorithm based on hierarchical alternating least-squares (HALS) [30], which is more efficient than the matrix multiplication (MM) [28] used before [32]. The following section describes these extended models and their optimization algorithms.

9.2.3 Optimization Algorithms with Soft Spatial Orthogonal Constraint

A goal of HSI data analysis is to identify the pure spectra and spatial intensity distributions of each chemical component from the spectra and distribution of a mixture of chemical components, i.e. observed matrix X. The basic NMF model, e.g. the minimization of Eq. (9.5), often generates unresolved spectra and spatial distributions that still contain spatially overlapped or unnaturally unresolved spectra because the basic NMF induces sparse decomposition of C and S. However, the EELS spectrum of a pure chemical component is not sparse, meaning that the intensities of an EELS spectrum are more than zero over all energy bands, whereas the intensities of an EDX spectrum are almost zero except for the peak positions. Hence, poor resolution is more problematic in STEM-EELS analysis than in EDX analysis.

Our approach to solve the above problem entails introducing the spatial orthogonal constraint [31]. This constraint ensures that spectral matrix S is relatively more relaxed than C and then S can be non-sparse. Because the exact orthogonal constraint is too strict, we used weight parameter w to relax this constraint, which is known as a soft spatial orthogonal constraint. Then our objective function of $C_{\cdot k}$ to be minimized is formulated as follows:

$$\frac{1}{2}\sum_{i=1}^{N_{xy}}\sum_{j=1}^{N_{ch}}\left(\left[X^{(k)}\right]_{ij} - \left[C_{\cdot k}S_{\cdot k}^{\mathrm{T}}\right]_{ij}\right)^2 + w \cdot \xi_k C_{\cdot k}^{\mathrm{T}} c^{(k)} \quad \text{s.t.} \quad \|C_{\cdot k}\|_2 = 1, \qquad (9.6)$$

where

$$X^{(k)} = X - CS + C_{\cdot k}S_{\cdot k}^{\mathrm{T}}, \quad k = 1, \ldots, K, \qquad (9.7)$$

$$c^{(k)} = \sum_{m \neq k} C_{\cdot m}, \quad k = 1, \ldots, K. \qquad (9.8)$$

Parameter $w, 0 \leq w \leq 1$, is important to adjust the orthogonal penalty and ξ_k is the Lagrange multiplier for the exact orthogonal constraint of C. When $w = 1$ the optimized C is an exact orthogonal matrix in which any chemical components do NOT overlap. When $w = 0$, among all the components, the optimized components in C may extensively overlap. The optimal value of w depends on the situation, such as the spatial resolution of the data (step width of STEM-HSI) and localization of chemical components. Thus, the optimal value of w must be chosen according to the measurement level.

Applying some algebra to Eq. (9.6) enables us to obtain an analytical solution in terms of the Lagrange multiplier ξ_k. Substituting the obtained ξ_k, the update rule of $C_{.k}$ is given by

$$C_{.k} = \left[X^{(k)} S_{.k} - w \frac{c^{(k)\mathrm{T}} X^{(k)} S_{.k}}{c^{(k)\mathrm{T}} c^{(k)}} C_{.k} \right]_+ , \quad k = 1, \ldots, K \qquad (9.9)$$

where the operator $[A]_+$ replaces all negative values in matrix A with zeros. Hence, it can be calculated by $[A]_+ = \{A + \mathrm{abs}(A)\}/2$, where function abs outputs a matrix consisting of the absolute value of the elements in A. The second term weighted by w is due to the orthogonal penalty term. After applying Eq. (9.9), each column of C should be normalized by

$$C_{.k} \leftarrow C_{.k}/\|C_{.k}\|_2, \quad k = 1, \ldots, K. \qquad (9.10)$$

Thus, we omit the normalization of S, and the update is given by

$$S_{.k} = \left[\left(X^{(k)} \right)^{\mathrm{T}} C_{.k} \right]_+ , \quad k = 1, \ldots, K. \qquad (9.11)$$

Figure 9.1 provides the pseudo-code of this NMF, which we named SO-NMF.

9.2.4 Probabilistic View of a NMF Model with an Automatic Relevance Determination Prior

Optimizing the number of components using only the observed HSI data is practically important. Maximum likelihood estimation (or an estimation based on minimizing errors) cannot be effective for the optimization because it causes overfitting of the HSI data when the number of components is large. This overfitting problem is avoided by using a Bayes estimation (or a maximum a posteriori (MAP) estimation) with a prior distribution of scale parameters (relevance weights) [32]. The process of choosing only the important components is known as automatic relevance determination (ARD).

To perform ARD in NMF, we assume a prior distribution for C using an exponential distribution with a scale parameter λ_k for the probability density of column k of C, i.e. $C_{.k}$:

	Input: Data matrix X, the weight of orthogonal constraint w, the number of components K, the maximum number of iterations T_{max}, the number of initializations R_{max} **Output:** Spatial intensities of components C and basis spectra S
1:	For r from 1 to R_{max}:
2:	$t = 0$
3:	While $t < T_{max}$ and $L^{(t,r)} \neq L^{(t-1,r)}$ (not converged)
4:	$t = t + 1$
5:	For k from 1 to K:
6:	Update $C_{\cdot k}$ by Eq. (9.9)
7:	Normalize $C_{\cdot k}$ by Eq. (9.10)
8:	end
9:	For k from 1 to K:
10:	Update $S_{\cdot k}$ by Eq. (9.11)
11:	end
12:	Compute $L^{(t,r)} = D_{EU}(X|CS^T)$ by Eq. (9.5)
13:	end
14:	end
15:	Choose the best optimization results by $r_{best} = \operatorname{argmin}_r L^{(T_{max},r)}$

Fig. 9.1 Pseudocode of our NMF with the soft orthogonal constraint (SO-NMF)

$$p(C_{nk}|\lambda_k) = \frac{1}{\lambda_k}\exp\left(-\frac{C_{nk}}{\lambda_k}\right), \quad n = 1, \ldots, N_{xy}, \quad k = 1, \ldots, K. \quad (9.12)$$

The above density distribution generates nonnegative random values with a large probability density around zero, resulting in a sparse matrix of C. For the prior distribution of λ_k, we assume an inverse-Gamma distribution:

$$p(\lambda_k|a, b) = \frac{b^a}{\Gamma(a)}\lambda_k^{-(a+1)}\exp\left(-\frac{b}{\lambda_k}\right), \quad k = 1, \ldots, K, \quad (9.13)$$

where a and b are hyper-parameters to adjust the sparseness of λ_k. On the other hand, the probability density distribution of column k of S, i.e. $p(S_{\cdot k})$, is assumed to be uniformly distributed on the unit hyper-sphere in $\mathbf{R}_+^{N_{ch}}$. Using Eqs. (9.4), (9.12), and (9.13), the negative log-likelihood function of an NMF model with ARD priors is given by

$$L(C, S, \lambda, \sigma^2) = -\log p(X|\overline{X}, \sigma^2) - \sum_{i=1}^{N_{xy}} \sum_{k=1}^{K} \log p(C_{ik}|\lambda_k)$$

$$- \sum_{k=1}^{K} \log p(S_{\cdot k}) - \sum_{k=1}^{K} \log p(\lambda_k|a, b)$$

$$= \frac{N_{xy} N_{ch}}{2} \log 2\pi\sigma^2 + \frac{1}{2\sigma^2} \sum_{i=1}^{N_{xy}} \sum_{j=1}^{N_{ch}} (X_{ij} - \overline{X}_{ij})^2 \qquad (9.14)$$

$$+ \sum_{k=1}^{K} \frac{1}{\lambda_k} \left(b + \sum_{i=1}^{N_{xy}} C_{ik} \right) + (N_{xy} + a + 1) \sum_{k=1}^{K} \log \lambda_k$$

$$+ K(a\log b - \log \Gamma(a)),$$

With regard to the optimization of C, $L(C, S, \lambda, \sigma^2)$ is a penalized likelihood function with the L_1 norm of C, resulting in a sparse matrix C. The NMF minimizing $L(C, S, \lambda, \sigma^2)$ is referred to as ARD–NMF.

Because the simultaneous optimization of $L(C, S, \lambda, \sigma^2)$ over all C, S, λ, and σ^2 is non-convex, multiple optimizations from different initial configurations are required. To update C and S, we use HALS [30], which updates each column $C_{\cdot k}$ and $S_{\cdot k}$ alternately. Applying some algebra to the minimization of $L(C, S, \lambda, \sigma^2)$, we obtain the following update rule for $C_{\cdot k}$:

$$C_{\cdot k} = \left[X^{(k)} S_{\cdot k} - \frac{\sigma^2}{\lambda_k} \right]_+, \qquad k = 1, \dots, K \qquad (9.15)$$

The second term in Eq. (9.15) is attributable to the ARD prior, which induces the sparse matrix of C. The update rule for $S_{\cdot k}$ by HALS is given by

$$S_{\cdot k} = \frac{\widetilde{s}_k}{\|\widetilde{s}_k\|_2}, \qquad k = 1, \dots, K \qquad (9.16)$$

where $\|x\|_2$ is the L_2 norm of vector x and

$$\widetilde{s}_k = \left[\left(X^{(k)} \right)^{\mathrm{T}} C_{\cdot k} \right]_+, \qquad k = 1, \dots, K \qquad (9.17)$$

Similarly, the update rules for the relevance weight λ and σ^2 to minimize $L(C, S, \lambda, \sigma^2)$ with all other quantities fixed are given by

$$\lambda_k = \frac{b + \sum_{i=1}^{N_{xy}} C_{ik}}{N_{xy} + a + 1}, \qquad k = 1, \dots, K \qquad (9.18)$$

$$\sigma^2 = \frac{1}{N_{xy}N_{ch}} \sum_{i=1}^{N_{xy}} \sum_{j=1}^{N_{ch}} \left(X_{ij} - \overline{X}_{ij}\right)^2 \tag{9.19}$$

The hyper-parameter b can be set using an approximate empirical estimator [32] as follows:

$$b = \frac{(a-1)\sqrt{N_{ch}}}{K} \frac{1}{N_{xy}N_{ch}} \sum_{i=1}^{N_{xy}} \sum_{j=1}^{N_{ch}} X_{ij}, \tag{9.20}$$

In our experiments, the hyper-parameter a was set to $a = 1 + \delta$, where $\delta = 10^{-16}$, to choose the minimum number of components with the minimum $L(C, S, \lambda, \sigma^2)$. After the optimization of ARD-NMF, the relevance (or importance) values of components are given by $\lambda_k, k = 1, \ldots, K$. Because the values of redundant components cannot be exactly zero, we empirically set a threshold value to remove such components.

9.2.5 Optimization Algorithm for C with Both ARD and Spatial Orthogonal Constraint

When we simply combine the soft orthogonal constraint and the ARD effect using both penalty terms, then the update rule of $C_{.k}$ can be obtained as follows:

$$C_{.k} = \left[X^{(k)}S_{.k} - \frac{\sigma^2}{\lambda_k} - w \frac{c^{(k)T}\left(X^{(k)}S_{.k} - \sigma^2/\lambda_k\right)}{c^{(k)T}c^{(k)}} C_{.k} \right]_+, \quad k = 1, \ldots, K \tag{9.21}$$

In this update, $C_{.k}$ should not be renormalized to reduce the effect of the orthogonal constraint for irrelevant components. We propose Eq. (9.21) as an update rule for $C_{.k}$ when the orthogonal constraint is necessary. Figure 9.2 shows the pseudo-code of our proposed NMF algorithm, which we named ARD-SO-NMF. In the special case without the orthogonal constraint, i.e. $w = 0$, the algorithm is simply ARD-NMF. Line 12–20 has the purpose of merging the components when the spectra are similar. In this procedure, the similarity is evaluated by using the cosine similarity and the spectra are considered to be the same when the value exceeds 0.99. This operation is necessary to choose the correct number of components because the orthogonality condition with $w > 0$ enforces splitting of the components even when the spectra are exactly the same. Our MATLAB and Python codes are available at https://github.com/MotokiShiga.

Input: Data matrix X, the weight of orthogonal constraint w, the maximum number of components K, the maximum number of iterations T_{max}, the number of initializations R_{max}

Output: Spatial intensities of components C, basis spectra S and relativeness λ

1:   Set hyper-parameter $a = 1 + \delta$ and compute b by Eq. (9.20)
2:   For r from 1 to R_{max}:
3:     $t = 0$
4:     While $t < T_{max}$ and $L^{(t,r)} \neq L^{(t-1,r)}$ (not converged)
5:       $t = t + 1$
6:       For k from 1 to K:
7:         Update $C_{\cdot k}$ by Eq. (9.21)
8:       end
9:       For k from 1 to K:
10:       Update $S_{\cdot k}$ by Eq. (9.16)
11:       end
12:       For k from 1 to K:
13:         For m from 1 to $(k-1)$:
14:           If $S_{\cdot k}^{T} S_{\cdot m} > 0.99$:
15:            $S_{\cdot m} = 1/\sqrt{N_{ch}}$
16:            $C_{\cdot k} = C_{\cdot k} + C_{\cdot m}$
17:            $C_{\cdot m} = 0$
18:           end
19:         end
20:       end
21:       Update λ by Eq. (9.18)
22:       Update σ^2 by Eq. (9.19)
23:       Compute $L^{(t,r)} = L(C, S, \lambda, \sigma^2)$ by Eq. (9.14)
24:     end
25:   end
26:   Choose the best optimization results by $r_{best} = \text{argmin}_r \, L^{(T_{max}, r)}$

Fig. 9.2 Pseudocode of our NMF with ARD and the soft orthogonal constraint (ARD-SO-NMF)

9.3 Application

9.3.1 Experimental Procedures

A real dataset was acquired from a cross-sectional TEM (XTEM) sample of a Si diode, prepared by a focused ion beam (FIB) technique. We recorded the HSI data for Si-$L_{2,3}$ including zero-loss peak (ZLP) using a JEOL JEM-1000 K RS ultra-high voltage S/TEM of Nagoya University, operated at 1000 kV, with a Gatan Quantum equivalent EEL spectrometer of which the energy dispersion was set to 0.2 eV/channel.

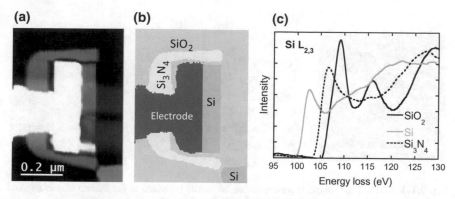

Fig. 9.3 **a** Cross-sectional ADF-STEM image, **b** spatial distribution of three components, and **c** their reference Si-$L_{2,3}$ spectra of silicon diode test sample

The sample thickness of the measured area was estimated at 0.1 μm from the low-loss spectrum. The energy drift of the spectra during the acquisition was corrected by ZLP alignment and calibration. After the energy calibration, the pre-edge background modeled by a power law was subtracted to isolate the Si-$L_{2,3}$ spectrum. Figure 9.3 shows an annular dark-field STEM (ADF-STEM) image of the ROI of the Si diode, a manually validated component map and spectra.

Another experimental STEM-EELS HSI dataset was prepared by measuring the atomic resolution EELS-HSI of Mn_3O_4. Polycrystalline Mn_3O_4 with a spinel crystal structure was obtained, and a TEM sample was prepared by conventional ion milling as previously described [24]. We measured the Mn-$L_{2,3}$ HSI using a JEOL ARM-200F aberration-corrected STEM, operated at 200 kV, with the Gatan Quantum EELS having an energy dispersion of 0.5 eV/channel. The average full width at half maximum (FWHM) of ZLP collected simultaneously (Dual EELS mode) with Mn $L_{2,3}$ was approximately 2 eV. The thickness of the measured area was approximately 40 nm, estimated from the low-loss spectra. Prior to applying NMF to the data, the energy drift of the spectra during the acquisition was corrected using the dual EELS mode synchronized with the ZLP alignment and calibration. After the energy calibration, the pre-edge background intensities were subtracted by modeling them with a power law.

Figure 9.4a–c show the ADF-STEM image, schematic projected structure of the Mn_{Tet} (divalent Mn occupying the tetrahedral site, Mn_{Oct} (trivalent Mn at the octahedral site) and O (oxygen) columns along the present incident beam direction and the extracted site-specific spectra, respectively. In (a) the heavier element (Mn) alone appears bright. These data are more difficult to analyze, because the inner shell excitation is delocalized by a certain distance and the neighboring atomic columns simultaneously contribute to the spectrum intensity at a sampling point due to electron channeling effects [25] and orbital hybridization between the elements.

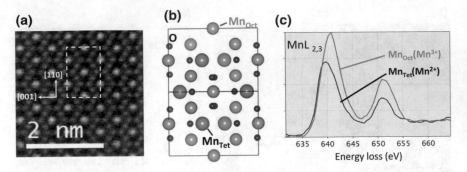

Fig. 9.4 **a** ADF-STEM image, **b** corresponding atom-site positions in the framed area of (**a**), and **c** Mn-$L_{2,3}$ reference spectra for STEM-EELS-HSI data from Mn_3O_4

For all datasets, even after the above pre-processing, a few elements in X had small negative values due to background removal. Thus, we replaced these values with zeros. To normalize the scale of X, all elements were divided by the average of the elements in X.

An STEM-EDX-HSI dataset was acquired from a sintered ceramic composite of Y-doped ZrO_2–$LaSrMnO_3$ (supplied by courtesy of Dr. T. Mori of the National Institute of Materials Science), which exhibits a distinct composition variation across the electron transparent sample area. A thin film was prepared for TEM by using an FIB technique. We measured the EDX-HSI using a JEOL 2100F S/TEM

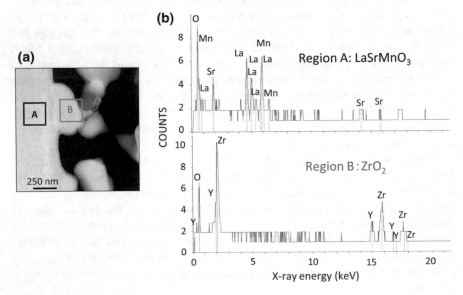

Fig. 9.5 ADF-STEM image of $LaSrMnO_3$-Y doped ZrO_2 ceramic composite sample (**a**) and typical EDX counts per pixel from framed areas (**b**)

operated at 200 kV, equipped with a JEOL EDX silicon drift detector, Dry SD60GV. Figure 9.5 shows the ADF-STEM image and typical counts (spectra) in representative points, corresponding to the two different phases, where the maximum net peak counts per pixel do not exceed 10 counts, and have a typical sparse feature that is suitable for testing the relevance of the proposed method.

9.3.2   Spatial Orthogonal Constraint on STEM-EELS Data

We evaluated the effect of the orthogonal constraint by changing the value of w with a fixed number of components, i.e. SO-NMF. We used the two STEM-EELS-HSI datasets described in Sect. 9.3.1. Because neither the spatial distribution maps nor the spectra in the datasets are sparse, the conventional NMF optimization has multiple local minima. Thus, reaching the global minima (or a good local minimum) is difficult. Our aim in this experiment was to verify that the orthogonal constraint reduces the search space and that SO-NMF generates a reasonable decomposition of NMF.

9.3.2.1   XSTEM-EELS Data from a Silicon Device

The method was first applied to the dataset from the Si diode sample, as shown in Fig. 9.3 in Sect. 9.3.1. The number of components in SO-NMF was set to $K = 3$, which is the number of reference components. In the result with $w = 0$ (no orthogonal constraint: first row in Fig. 9.6), the third spectral components exhibit unnatural intensity decreases at 110 eV, where a sharp peak from the first spectral component is overlaid. This can happen in EELS-HSI under certain conditions [17]. This sudden lowering in intensity disappears when spatial orthogonality ($w \geq 0.01$) is included, as seen in Fig. 9.6. Slight cross-talk between the second and third components remains in both the spatial distribution maps and spectra for $w = 0.01$. The resolved spectral profiles and their spatial distributions are almost the same for $w \geq 0.05$, which effectively reproduces the spectra and expected spatial distributions, although the spatial phase separation seems (unnaturally) overly emphasized for $w = 1.0$.

9.3.2.2   Atomic Resolution STEM-EELS of Mn₃O₄

Next, we validated the method using the atomic resolution $Mn-L_{2,3}$ SI data from the Mn_3O_4 spinel sample (cf. Fig. 9.4 in Sect. 9.3.1). The number of components in SO-NMF was set to $K = 3$, which is the number of components determined by ARD-SO-NMF in Sect. 9.3.3.3. The SO-NMF results for $0 \leq w \leq 1$ are shown in Fig. 9.7, with the score images in the first, second, and third columns and the resolved spectral profiles in the fourth column. In the case without spatial

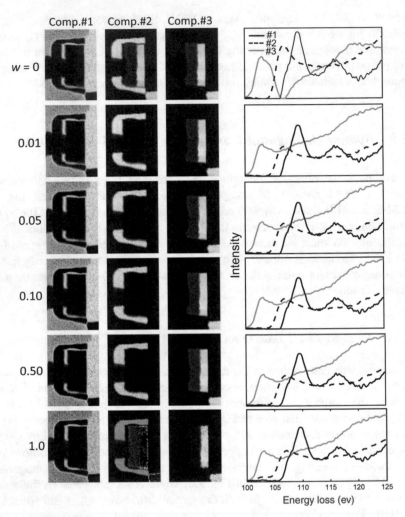

Fig. 9.6 Results of SO-NMF with various weights of spatial orthogonality constraint for Si-$L_{2,3}$ STEM-EELS-HSI data

orthogonality ($w = 0$), the resolved spectral profiles are inconsistent with the expected reference profiles (Fig. 9.4c), the peak at around 640 eV of component 2 shifted to the left. Further, component 3 exhibits a physically unnatural intensity drop at the distinct peak positions of component 1, as for the case of no orthogonality (Fig. 9.7, top-right figure). With a small spatial orthogonality ($w = 0.01$) included, the spectral shapes converged to those consistent with the reference spectra and the additional component localized on the oxygen columns. It can be seen that there is an optimum value of w for reproducing good spectral profiles and plausible spatial distributions. As w increases, the spatial distributions become more orthogonal to each other, whereas the resolved spectral shapes converge to one

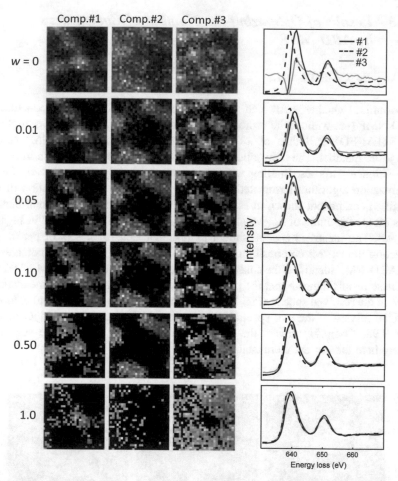

Fig. 9.7 Results of SO-NMF with various weights of the spatial orthogonality constraint for Mn-$L_{2,3}$ STEM-EELS-HSI data

form. For $w \geq 0.5$ the spatial distributions are far from the actual projected structures, even though the resolved spectral shapes are essentially identical.

We subsequently focus on the additional third component, the spatial distribution of which was found to be localized on the projected oxygen atom positions. This localization was attributed to the electron channeling effect [25], which is responsible for propagating the incident electron wave function along the neighboring Mn columns for a sample exceeding a certain thickness when the electron probe is placed on the oxygen column. The resolved spectrum of the third component actually exhibits a spectral profile characteristic of the weighted average of the other two components, because each oxygen atom is coordinated with trivalent Mn_{Oct} and divalent Mn_{Tet} atoms.

9.3.3 Results of Optimizing the Number of Components by ARD-NMF

9.3.3.1 STEM-EDX Data

To examine whether ARD can select the correct number of components, our ARD-NMF (i.e. without the orthogonal constraint imposed, $w = 0$) was applied to the STEM-EDX-HSI data of a Y-doped $ZrO_2(YSZ)$–$LaSrMnO_3(LSM)$ ceramic composite material. The conventional elemental distributions are shown in Fig. 9.8 for reference purposes. Starting with 10 components, only two survived after the optimization algorithm terminated, as shown in Fig. 9.9. The distribution of each identified component shown in Fig. 9.9a is consistent with the elemental distributions of: (1) the union of La, Mn, and Sr and (2) the union of Zr and Y in Figs. 9.5 and 9.8. The identified spectra shown in Fig. 9.9c consist of sets of peaks, each reflecting the correct composition of YSZ or LSM in Fig. 9.5. This indicates that our ARD-NMF identified the constituent phases correctly for this STEM-EDX data.

These results indicate that the present method effectively removed the statistical noise in the resolved spectra (Fig. 9.9c), and the score images (Fig. 9.9a) exhibit no artificial mixing of the two spectral components. Note that a 10-nm layer of LSM (Fig. 9.9a: Comp.#1) covers the YSZ substrate surface; this can be seen more clearly here than in the elemental maps.

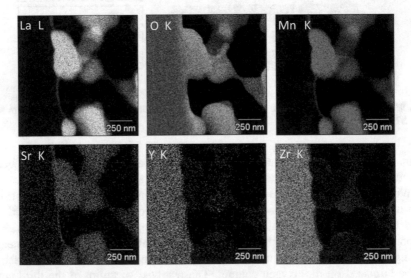

Fig. 9.8 EDX elemental maps of $LaSrMnO_3$-Y-doped ZrO_2 ceramic composite sample

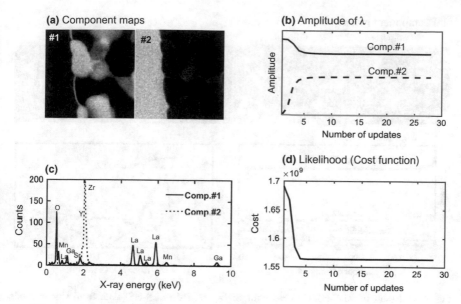

Fig. 9.9 Result of ARD-NMF for STEM-EDX-HSI data

9.3.3.2   XSTEM-EELS Data from a Silicon Device

We then applied both ARD-NMF and ARD-SO-NMF with $K = 10$ to the EELS-HSI data of Si-$L_{2,3}$ energy-loss near edge structure (ELNES) obtained from a cross-sectional Si diode sample. The reference component spectra (Fig. 9.3c) are not sparse, that is, nonzero values range over the energy-loss axis. We compared the results of ARD-NMF with those from ARD-SO-NMF to verify that the orthogonal constraint produces a clearer decomposition for non-sparse data. The results are shown in Figs. 9.10 and 9.11.

Figure 9.10b shows that ARD-NMF selected four components whereas the reference contains three. In Fig. 9.10a, the generated component distribution of components 1 and 3 exhibit extensive overlap, whereas the actual spectra of these components are not overlapped, as also seen in the case of $w = 0$ in Fig. 9.6. This result was attributed to a property of basic NMF, which induces a sparse decomposition on both spatial and spectra matrices.

The ARD-SO-NMF results (with $w = 0.1$) are shown in Fig. 9.11b. Figure 9.11c shows that the identified spectra are consistent with the reference spectra shown in Fig. 9.3c. The spatial distributions of components obtained by ARD-SO-NMF (Fig. 9.11a) are clearly separated, whereas those resulting from ARD-NMF (Fig. 9.10a) overlap extensively. This difference demonstrates the effect of the orthogonal constraint. In this case, ARD-SO-NMF selected three components and their spectra are consistent with their reference spectra, whereas the spectra by ARD-NMF ($w = 0$) display unnatural reductions in intensity. Thus, these results

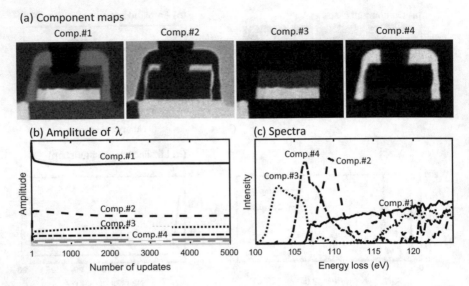

Fig. 9.10 Result of ARF-NMF ($w = 0$) for Si-$L_{2,3}$ STEM-EELS-HSI data from silicon diode sample

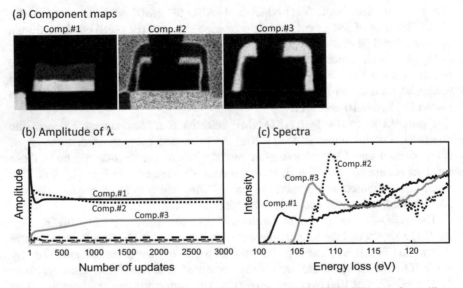

Fig. 9.11 Result of ARF-SO-NMF ($w = 0.01$) for Si-$L_{2,3}$ STEM-EELS-HSI data from silicon diode sample

suggest that the method can effectively detect subtle spectral differences by introducing the orthogonal constraint.

9.3.3.3 Atomic Resolution STEM-EELS of Mn_3O_4

The ARD-NMF technique was also applied to the experimental atomic resolution
Mn-$L_{2,3}$ SI data from the Mn_3O_4 spinel sample. Figures 9.12a–c show the
ARD-NMF results with $K = 10$. As shown in Fig. 9.12b, ARD-NMF selected three
components and eliminated the other seven components during the optimization.
Thus, ARD-NMF detected an additional component other than those related to the
two Mn sites, as discussed in Sect. 9.3.2.2. The spatial distributions of the resolved
components shown in Fig. 9.12a and c are basically consistent with the projected
Mn_{Oct} and a Mn_{Tet} atom positions in Figs. 9.4b and 9.4c, respectively, the relative
chemical shifts of which are also consistent with their valence states. However, the
boundary between the components is less clear because of the delocalization of the
chemical bonding states, and the resolved spectral profiles (Fig. 9.12c) are inconsistent with the expected theoretical profiles, with component 2 exhibiting physically unnatural intensity decreases at the distinct peak positions of component 1.
Because an ARD prior induces sparseness, this problem often occurs when
ARD-NMF is applied to EELS in which different spectral profiles largely overlap.

We overcame this problem by applying ARD-SO-NMF with $w = 0.01$.
Figure 9.13a demonstrates that, because of the orthogonal condition, the

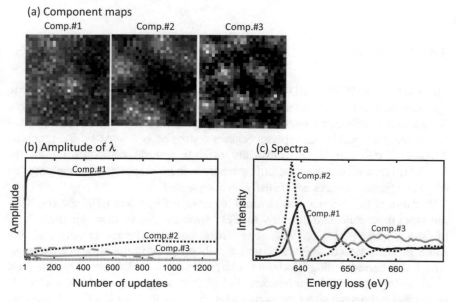

Fig. 9.12 Result of ARF-NMF ($w = 0$) for Mn-$L_{2,3}$ STEM-EELS-HSI data from Mn_3O_4

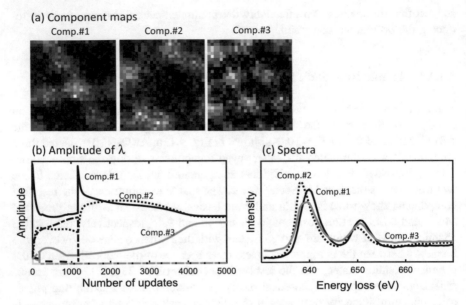

Fig. 9.13 Result of ARF-SO $=$ NMF ($w = 0.01$) for Mn-$L_{2,3}$ STEM-EELS-HSI data from Mn$_3$O$_4$

components are separated more clearly. Especially, overlaps between the first component and the others are resolved with greater clarity, as shown in Fig. 9.13a, and detection of the third component was improved.

9.4 Discussion

The ARD-SO-NMF and ARD-NMF algorithms proposed in this study were able to optimize the number of spectral components for both the EDX and EELS datasets. An additional orthogonal constraint was required when neither the spatial distribution nor the spectra were sparse. Such a constraint is offered by the proposed ARD-SO-NMF. Our NMF realistically extracts spectral components from the EELS-HSI data when the spatial orthogonality penalty is appropriate, implying that different spectral features are spatially well separated.

Because of the differing complexities intrinsic to EELS and EDX spectra, NMF processes these datasets differently. An EDX spectrum can be characterized by a set of Gaussian-like peaks, generally separated in energy, whereas an elemental core EELS includes various spectral components overlapped in the same energy range, where the corresponding electronic energy levels in solids are approximately continuously distributed. Furthermore, the spectral components of EDX are mostly sparse and orthogonal along the energy axis, contrary to those in EELS. NMF with only an error function as the objective function prefers the orthogonal basis spectra in the energy axis because of their completeness. On the other hand, NMF with spatial orthogonality models the practical situation, in which an EELS-HSI dataset is assumed to be more orthogonal in space than in energy, more accurately.

There are several local minima in the likelihood functions. The type of NMF algorithm appears to eventually achieve a more appropriate minimum, although it is not mathematically possible to prove the dependence. Moreover, because of the computational cost, it is difficult to obtain all of the local minima, even when we apply the spatial orthogonality constraint and sparse priors for the ARD effect, i.e. by using a small value of K, in the data matrix. The present NMF method is capable of minimizing and extracting the objective function of particular solutions by systematically varying the weight of the spatial orthogonality in the object function. In both of the EELS examples presented herein, an increase in w caused the resolved components to be distributed more widely over the sample space and their spectral shapes to become less sparse (or orthogonal). This change in spectral shape, which is prone to be sparse under the basic NMF, is clearly resolved and exhibits the composition more accurately when the orthogonal constraint is applied. Hence, the proposed NMF can identify chemical states from the resolved spectra more accurately than existing methods that do not use spatial orthogonality. In the case of the atomic resolution HSI of Mn_3O_4, the method resulted in physically meaningless solutions when we overestimated the spatial orthogonality. In general, we can reach solutions that are physically more realistic/interpretable, comparable to the theoretical spectra predicted by first principles calculations or reference experimental spectra, by changing the value of w systematically and understanding the extent to which the spectral shapes and spatial distributions of the resolved components vary. This scheme seems much more effective and pragmatic than estimating the solution bounds by repeating the decomposition routines with many different initial random numbers in the loading or score matrices.

In this respect the proposed SO constraint may fail when the spatial distributions of the component states strongly overlap with each other. Spiegelberg et al. proposed an alternative scheme to extract nonnegative source signals of strongly mixed data [33], instead of imposing the present SO constraint. By randomly drawing samples from the space of positive spectra in the signal subspace spanned by the prominent principal components, a sampled dataset of which the spectral components can be conveniently extracted using, e.g., VCA or NMF, is obtained with a large probability. These components typically correspond well to the pure spectra of the original data assuming that the spectral components are orthogonal to each other in at least one channel.

Existing processing schemes can produce controversial results should the spectral background be subtracted in advance before the statistical processing, and this probably depends on the type of spectral data being processed. Although background subtraction is generally considered to lose important spectral information, we believe background subtraction to be necessary in the present framework because our NMF assumes that no background structure is incorporated. As demonstrated in the supplementary material in a previous paper [26], our proposed NMF was unable to provide the expected correct results for STEM-EELS SI without background subtraction, which thus presents further work for the future, i.e. incorporating a background structure in our model.

9.5 Summary

We proposed a new multivariate curve resolution method based on NMF with two penalty terms: (i) a soft orthogonal constraint to effectively resolve overlapping spectra, and (ii) an ARD prior to optimize the number of components. Validations using experimental STEM-EDX/EELS SI data demonstrated that the ARD prior successfully resolved the correct number of physically interpretable spectral components. The soft orthogonal constraint was effective for STEM-EELS HSI data that were neither sparse in the spatial nor the spectral regions. The proposed SO-NMF and ARD-SO-NMF schemes can successfully resolve physically meaningful components by reducing the search space for low-rank matrices, even in cases where conventional NMF is unable to correctly resolve the components. These advantages reduce the costs of HSI data analysis and of extracting hidden spectral information from experimental data using objective and statistical measures rather than empirical knowledge. The proposed method is applicable to any type of HSI dataset, such as that generated by Raman spectroscopy, infrared absorption, and time-of-flight mass spectroscopy. Future prospects would include investigating the ability of the present ARD-NMF scheme to correctly detect small amounts of significant phases.

Acknowledgements This work was in part supported by Grants-in-Aid for Scientific Research on Innovative Areas "Nano Informatics" (Grant No. 25106004, 26106510 and 16H00736), KIBAN-KENKYU A (Grant No. 26249096) and KIBAN-KENKYU B (Grant No. JP16H02866) from the Japan Society for the Promotion of Science (JSPS) and by Precursory Research for Embryonic Science and Technology (Grant No. JPMJPR16N6) from Japan Science and Technology Agency (JST).

References

1. N. Bonnet, N. Brun, C. Colliex, Ultramicroscopy **77**, 97 (1999)
2. P. Trebbia, N. Bonnet, Ultramicroscopy **34**, 165 (1990)
3. M. Bosman, M. Watanabe, D.T.L. Alexander, V.J. Keast, Ultramicroscopy **106**, 1024 (2006)
4. C.M. Parish, L.N. Brewer, Ultramicroscopy **110**, 134 (2010)
5. S. Lichtert, J. Verbeeck, Ultramicroscopy **125**, 35 (2013)
6. J. Spiegelberg, J. Rusz, Ultramicroscopy **175**, 40 (2017)
7. N. Dobigeon, N. Brun, Ultramicrosc. **120**, 25 (2012)
8. N. Bonnet, D. Nuzillard, Ultramicroscopy **102**, 327 (2005)
9. J. Nascimento, J. Bioucas-Dias, I.E.E.E. Trans, Geosci. Remote Sens. **43**, 898 (2005)
10. N. Dobigeon, S. Moussaoui, M. Coulon, J.-Y. Tourneret, A.O. Hero, I.E.E.E. Trans, Signal Process. **57**, 4355 (2009)
11. J. Spiegelberg, J. Rusz, T. Thersleff, K. Pelckmans, Ultramicroscopy **174**, 14 (2017)

12. S. Muto, T. Yoshida, K. Tatsumi, Mater. Trans. **50**, 964 (2009)
13. P.G. Kotula, M.R. Keenan, J.R. Michael, Microsc. Microanal. **9**, 1 (2003)
14. J.H. Wang, P.K. Hopke, T.M. Hancewicz, S.L. Zang, Anal. Chim. Acta **476**, 93 (2003)
15. S. Muto, Y. Sasano, K. Tatsumi, T. Sasaki, K. Horibuchi, Y. Takeuchi, Y. Ukyo, J. Electrochem. Soc. **156**, A371 (2009)
16. S. Muto, K. Tatsumi, T. Sasaki, H. Kondo, T. Ohsuna, K. Horibuchi, Y. Takeuchi, Electrochem. Solid State Lett. **13**, A115 (2010)
17. Y. Kojima, S. Muto, K. Tatsumi, H. Oka, H. Kondo, K. Horibuchi, Y. Ukyo, J. Power Sources **196**, 7721 (2011)
18. S. Muto, K. Tatsumi, Y. Kojima, H. Oka, H. Kondo, K. Horibuchi, Y. Ukyo, J. Power Sources **205**, 449 (2012)
19. Y. Honda, S. Muto, K. Tatsumi, H. Kondo, K. Horibuchi, T. Sasaki, J. Power Sources **291**, 85 (2015)
20. T. Yoshida, S. Muto, J. Wakabayashi, Mater. Trans. **48**, 2580 (2007)
21. J. Senga, K. Tatsumi, S. Muto, Y. Yoshida, J. Appl. Phys. **118**, 115702 (2015)
22. K. Tatsumi, S. Muto, J. Phys. Condens. Matter **21**, 104213 (2009)
23. Y. Yamamoto, K. Tatsumi, S. Muto, Mater. Trans. **48**, 2590 (2007)
24. K. Tatsumi, S. Muto, Y. Yamamoto, H. Ikeno, S. Yoshioka, I. Tanaka, Ultramicroscopy **106**, 1019 (2006)
25. N.R. Lugg, G. Kothleitner, N. Shibata, Y. Ikuhara, Ultramicroscopy **151**, 150 (2015)
26. M. Shiga, K. Tatsumi, S. Muto, K. Tsuda, Y. Yamamoto, T. Mori, T. Tanji, Ultramicroscopy **170**, 43 (2016)
27. M. Shiga, S. Muto, K. Tatsumi, T. Tsuda, Trans. Mat. Res. Soc. Jpn. **41**, 333 (2016)
28. D.D. Lee, H.S. Seung, Adv. Neural Inform. Process. Syst. **13**, 556 (2001)
29. M.W. Berry, M. Browne, A.N. Langville, V.P. Pauca, R.J. Plemmons, Comput. Stat. Data Anal. **52**, 155 (2007)
30. A. Cichocki, IEICE Trans. Fund. Electron. Commun. Comput. **92**, 708 (2009)
31. K. Kimura, Y. Tanaka, M. Kudo, In *Proceedings of the 6th Asian Conference on Machine Learning,* Nha Trang, Vietnam, 26–28 Nov. 2014
32. V.Y.F. Tan, C. Fevotte, I.E.E.E. Trans, Pat. Anal. Mach. Intell. **35**, 1592 (2013)
33. J. Spiegelberg, S. Muto, M. Ohtsuka, K. Pelckmans, J. Rusz, Ultramicrosc. **182**, 205 (2017)

Part III
Materials Developments

Chapter 10
Fabrication, Characterization, and Modulation of Functional Nanolayers

Hiromichi Ohta and Hidenori Hiramatsu

Abstract Regions of a few nanometers at the surface or interface of a material exhibit various functional properties, which differ from those of the bulk because the electrons and/or ions receive different potentials due to the incoherent atomic arrangement. High-quality epitaxial films of functional materials called "nanolayers" are important to utilize such functional properties. However, fabrication of high-quality nanolayers of complex materials with complicated crystal structures is usually challenging due to the difference in the thermochemical properties of the constituents. In this chapter, epitaxial growth techniques, especially "reactive solid-phase epitaxy" of functional oxides and chalcogenides, are reviewed based on the authors' efforts. Additionally, this chapter reviews several modulation methods of optical, electrical, and magnetic properties of functional oxide nanolayers.

Keywords Nanolayers · Epitaxial growth method · Functional oxides and chalcogenides · Modulation methods

H. Ohta (✉)
Research Institute for Electronic Science, Hokkaido University,
N20W10, Kita-ku, 001-0020 Sapporo, Japan
e-mail: hiromichi.ohta@es.hokudai.ac.jp

H. Hiramatsu (✉)
Laboratory for Materials and Structures,
Institute of Innovative Research, Tokyo Institute of Technology,
4259 Nagatsuta-cho, R3-1, 226-8503 Midori-ku, Yokohama, Japan
e-mail: h-hirama@mces.titech.ac.jp

H. Hiramatsu
Materials Research Center for Element Strategy, Tokyo Institute of Technology,
4259 Nagatsuta-cho, SE-6, 226-8503 Midori-ku, Yokohama, Japan

© The Author(s) 2018
I. Tanaka (ed.), *Nanoinformatics*, https://doi.org/10.1007/978-981-10-7617-6_10

10.1 Epitaxial Growth and Characterization of Functional Nanolayers

Regions of a few nanometers at the surface or interface of a material often exhibit various functional properties, which differ from those of the bulk due to the fact that the electrons and/or ions receive different potentials due to the incoherent atomic arrangement. High-quality epitaxial films of functional materials called "nanolayers" are important to utilize such functional properties.

In this chapter, epitaxial growth techniques, especially "reactive solid-phase epitaxy" of functional oxides and chalcogenides, are reviewed based on the authors' efforts. Additionally, this chapter reviews several modulation methods of optical, electrical, and magnetic properties of functional oxide nanolayers.

10.2 Pulsed Laser Deposition

Pulsed laser deposition (PLD) is a physical vapor deposition technique [1]. By irradiating focused laser pulses of an excimer laser or a higher (3rd or 4th) harmonic of a Nd:YAG laser onto the target material (single crystals or ceramics or powder), which is located in an ultrahigh vacuum chamber, films can be deposited on the substrate as a result of vaporization of the target materials occur during laser irradiation (Fig. 10.1). PLD is one of the most powerful techniques for epitaxial

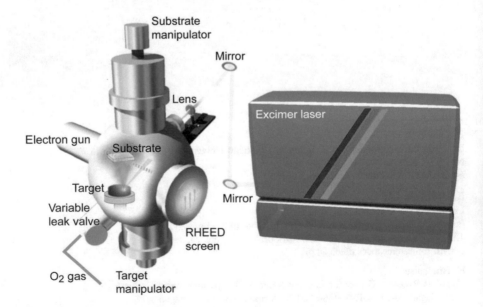

Fig. 10.1 Schematic illustration of a PLD system

film growth of inorganic solids, especially oxides. It has several advantages compared to the other deposition techniques, such as sputtering. In the case of PLD, the chemical composition of the resultant film is almost same as that of the target material, although generally it differs from the target because the chemical species show different sputtering yields in the case of sputtering. Moreover, the atmosphere in the PLD chamber can be widely controlled from an ultrahigh vacuum to $\sim 10^2$ Pa, allowing a thermodynamically nonequilibrium crystalline phase of a material to be fabricated.

As an example, PLD growth and characterization of the $SrTiO_3$-$SrNbO_3$ solid solution system are explained. $SrTiO_3$ has attracted increasing attention as the next generation of *oxide electronics* [2]. Doping with the appropriate substituent, such as Nb^{5+} (Ti^{4+} site) or La^{3+} (Sr^{2+} site), easily varies the charge carrier concentration of $SrTiO_3$ from insulating to metallic ($n_{3D} \sim 10^{21}$ cm^{-3}). Electron-doped $SrTiO_3$ is one of the most extensively studied materials for thermoelectric applications [3, 4]. In 2001, Okuda et al. [5] synthesized $Sr_{1-x}La_xTiO_3$ ($0 \leq x \leq 0.1$) single crystals by the floating-zone method. They reported that the crystals exhibit a large power factor ($S^2 \cdot \sigma$) of 2.8–3.6 mW m^{-1} K^{-2} at room temperature. Later, Ohta et al. reported the carrier transport properties of Nb- and La-doped $SrTiO_3$ single crystals (carrier concentration, $n \sim 10^{20}$ cm^{-3}) at high temperatures (~ 1000 K) to clarify the intrinsic thermoelectric properties of these materials [6].

The experimental discovery of unusually large thermopower outputs from superlattices and two-dimensional electron gases in $SrTiO_3$ [7, 8] spurred substantial research efforts into $SrTiO_3$ superlattices [9, 10] and heterostructures [11–13] for thermoelectric applications. For example, a superlattice composed of one unit cell (uc) of $SrTi_{0.8}Nb_{0.2}O_3$ and 10 uc of $SrTiO_3$ exhibits a giant thermopower, most likely due to an electron confinement effect. Although electron confinement is strongly correlated with the electronic structure [14, 15], a full understanding of the fundamental electronic phase behavior of the $SrTi_{1-x}Nb_xO_3$ solid solution system has yet to be developed.

Although high-quality single crystals of $SrTi_{1-x}Nb_xO_3$ species with $x > 0.1$ are not available due to the low solubility limit of Nb in the lattice [16], epitaxial films with these material compositions can be fabricated by PLD [17]. As summarized in Fig. 10.2, pure $SrTiO_3$ (space group $Pm\bar{3}m$, cubic perovskite structure, $a = 3.905$ Å) is an insulator with a bandgap of 3.2 eV. The bottom of the conduction band is composed of triply degenerate, empty Ti $3d$–t_{2g} orbitals, while the top of the valence band is composed of fully occupied O $2p$ orbitals [18]. The valence state of Ti ions in crystalline $SrTiO_3$ is 4 + (Ti $3d^0$). On the other hand, pure $SrNbO_3$ (space group $Pm\bar{3}m$, cubic perovskite structure, $a = 4.023$ Å) is a metallic conductor [19–21]. The valence state of the Nb ion is 4 + (Nb $4d^1$). In between $SrTiO_3$ and $SrNbO_3$ in the $SrTi_{1-x}Nb_xO_3$ ss, there are two possible types of valence state changes in the Ti and Nb ions, as shown in Fig. 10.2b and c. In the case of isovalent substitution (Fig. 10.2b), the mole fraction of Ti^{4+} proportionally decreases with increasing Nb^{4+} (x). On the other hand, heterovalent substitution, in which two Ti^{4+} or Nb^{4+} ions are substituted by adjacent (Ti^{3+}/Nb^{5+}) ions, can occur, as shown in

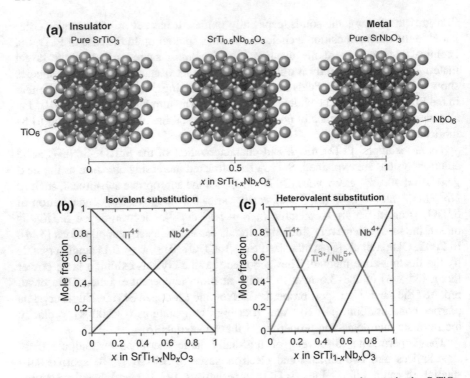

Fig. 10.2 Schematic of the crystal structure and possible valence state changes in the SrTiO$_3$-SrNbO$_3$ solid solution system. **a** Schematic of the crystal structure. Pure SrTiO$_3$ is an insulator with a bandgap of 3.2 eV, in which the valence state of the Ti ions (blue, TiO$_6$) is 4 + (Ti 3d^0). In contrast, pure SrNbO$_3$ is a metal, in which the valence state of the Nb ions (Red, NbO$_6$) is 4 + (Nb 4d^1). **b, c** Possible valence state changes of the Ti and Nb ions in the SrTiO$_3$-SrNbO$_3$ solid solution system: **b** isovalent substitution, where Ti^{4+} is substituted by Nb^{4+} and **c** heterovalent substitution, where two Ti^{4+}/Nb^{4+} ions are substituted by adjacent Ti^{3+}/Nb^{5+} ions. Reprinted with permission from [22]. © 2017 AIP

Fig. 10.2c. Based on these considerations, we focused on the valence state changes of Ti and Nb ions in the SrTi$_{1-x}$Nb$_x$O$_3$ ss.

Zhang et al. [22] fabricated approximately 100 nm-thick SrTi$_{1-x}$Nb$_x$O$_3$ ($x = 0.05$, 0.1, 0.2, 0.3, 0.4, 0.5, 0.55, 0.6, 0.7, 0.8, 0.9, and 1.0) epitaxial films by PLD using dense ceramic disks of a SrTiO$_3$-SrNbO$_3$ mixture. Insulating (001) LaAlO$_3$ (pseudo-cubic perovskite, $a = 3.79$ Å) was used as the substrate. The growth conditions were precisely controlled with a substrate temperature of 850 °C, an oxygen pressure of $\sim10^{-4}$ Pa, and a laser fluence of 0.5–1 J cm^{-2} pulse^{-1}, yielding a growth rate of 0.3 pm pulse^{-1}.

Figure 10.3a summarizes the Xray reciprocal space mappings (RSMs) around the ($\bar{1}$03) diffraction spot of LaAlO$_3$ (overlaid). Intense diffraction spots from ($\bar{1}$03) SrTi$_{1-x}$Nb$_x$O$_3$ are seen together with those from the LaAlO$_3$ substrate, indicating that incoherent heteroepitaxial growth of the target materials occurs for all x compositions. The peak positions of the diffraction spots from each composition

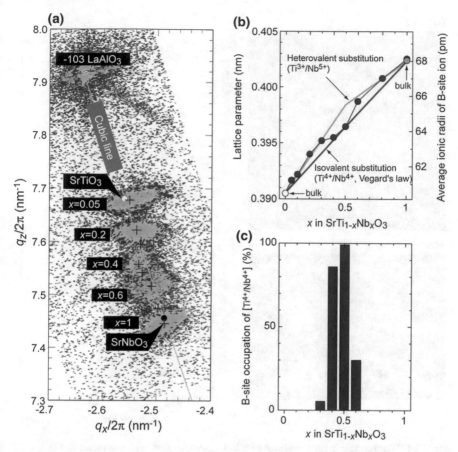

Fig. 10.3 Crystallographic characterization of the $SrTi_{1-x}Nb_xO_3$ epitaxial films on a (001) $LaAlO_3$ single-crystal substrate. **a** Xray reciprocal space mappings around the ($\bar{1}03$) diffraction spot of the $SrTi_{1-x}Nb_xO_3$ epitaxial films. The location of the $LaAlO_{-3}$ diffraction spot, $(q_x/2\pi, q_z/2\pi) = (-2.64, 7.92)$, corresponds to the pseudo-cubic lattice parameter of $LaAlO_3$ ($a = 0.379$ nm). Red symbols (+) indicate the peak positions of the $SrTi_{1-x}Nb_xO_3$ epitaxial films. **b** Changes in the lattice parameters of the $SrTi_{1-x}Nb_xO_3$ films (circles, left axis) superimposed with isovalent/heterovalent substitution lines (black line: isovalent substitution, gray line: heterovalent substitution, right axis), calculated using Shannon's ionic radii [23]. **c** Changes in the B-site occupation by [Ti^{4+}/Nb^{4+}] derived from the data in (**b**). Reprinted with permission from [22]. © 2017 AIP

correspond well with the cubic line ($q_z/q_x = -3$), suggesting that no epitaxial strain is induced in the films. It should be noted that a slight tetragonal distortion is observed in the $x = 0.4$ ($c/a = 1.0057$) and 0.5 ($c/a = 1.0050$) samples.

From the RSMs of the $SrTi_{1-x}Nb_xO_3$ films, we extracted the lattice parameters using the formula $a = (2\pi/q_x \cdot 2\pi/q_x \cdot 6\pi/q_z)^{1/3}$. Figure 10.3b plots the lattice parameters of the $SrTi_{1-x}Nb_xO_3$ film as a function of x. We observed an M-shaped trend along with a general increase in the lattice parameter with increasing x. In order to analyze the changes in the lattice parameter, we calculated the average

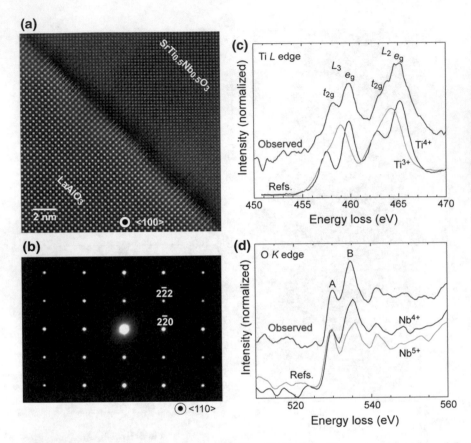

Fig. 10.4 Electron microscopy analyses of a $SrTi_{1-x}Nb_xO_3$ film with a composition of $x = 0.5$. **a** HAADF-STEM image acquired with the electron beam incident along the <100> direction. Periodic misfit dislocations (~8.5 nm interval) at the heterointerface are indicated by red lines. **b** Selected-area electron diffraction pattern acquired with the electron beam incident along the <110> direction. **c, d** EELS spectra acquired around the **c** Ti L edge and **d** O K edge. EELS spectra for Ti^{3+}/Ti^{4+} [27] and Nb^{4+}/Nb^{5+} [28] from previous studies are also plotted for comparison. Reprinted with permission from [22]. © 2017 AIP

ionic radii in the crystal structure and used Shannon's ionic radii as a comparison [23]: Ti^{4+} (60.5 pm), Ti^{3+} (67.0 pm), Nb^{4+} (68.0 pm), and Nb^{5+} (64.0 pm). In the ranges of $0.05 \leq x \leq 0.3$ and $x \geq 0.6$, the observed lattice parameters closely follow the heterovalent substitution line, suggesting that two Ti^{4+} or Nb^{4+} ions are substituted by adjacent (Ti^{3+}/Nb^{5+}) ions [24]. On the other hand, at $x = 0.4$ and 0.5, the observed lattice parameter correspond well with the isovalent substitution line. Moreover, at $x = 0.5$, the B-site occupation of $[Ti^{4+}/Nb^{4+}]$ is almost 100%, as shown in Fig. 10.3c.

Figure 10.4a shows a cross-sectional HAADF-STEM image of the $SrTi_{0.5}Nb_{0.5}O_3$ film. Periodical mismatch dislocations with intervals of ~8.5 nm are seen at the heterointerfaces. If the strain in the thin film is fully relaxed by such misfit dislocations, it is possible to calculate the spacing between dislocations (d) from $d = \mathbf{b}/\delta$, where $\mathbf{b}$ is the Burgers vector and δ is the lattice mismatch between thin film and substrate [25]. Using the lattice parameters obtained from XRD [$\delta = (q_{x\,sub} - q_{x\,film})/q_{x\,film} = +0.0435$], the estimated dislocation spacing is 8.7 nm, suggesting that the dislocations fully relax the strain in the film. Although superspots originating from the (111) diffraction are often observed in $AB_{0.5}B'_{0.5}O_3$ compositions that crystallize in B-site-ordered double perovskite structures [26], they are not observed in the $SrTi_{0.5}Nb_{0.5}O_3$ film (Fig. 10.4b). This is most likely due to the slight tetragonal distortion of the crystal structure. Figure 10.4 shows the EELS spectra acquired around the Ti L (c) and O K edges (d). The reported EELS spectra of Ti^{3+}/Ti^{4+} [27] and Nb^{4+}/Nb^{5+} [28] are plotted for comparison. In the Ti L edge spectrum (c), t_{2g} and $e_{.g.}$ peak splitting is clearly observed for Ti L_3, indicating that the dominant valence state of Ti is 4+. In the O K edge spectra (d), two intense peaks (assigned as A and B) are clearly observed where peak B has a higher intensity than peak A. A previous study noted that this is a characteristic feature of Nb^{4+} [28]. The peak intensity ratio A/B is calculated to be 0.66, which roughly corresponds with the Nb^{4+} spectrum (0.66).

By using abovementioned films, Zhang and Ohta et al. clarified the thermo-electric phase diagram for the $SrTi_{1-x}Nb_xO_3$ ($0.05 \leq x \leq 1$) solid solution system (Fig. 10.5). They observed two thermoelectric phase boundaries in the system, which originate from the step-like decrease in the carrier effective mass at $x \sim 0.3$ and from the local minimum in the carrier relaxation time at $x \sim 0.5$. The origins of these phase boundaries are related to the isovalent/heterovalent B-site substitution. The parabolic Ti 3d orbitals dominate the electron conduction for compositions with $x < 0.3$, whereas the Nb 4d orbital dominates when $x > 0.3$. At $x \sim 0.5$, a tetragonal distortion of the lattice, in which the B-site is composed of Ti^{4+} and Nb^{4+} ions, leads to the formation of tail-like impurity bands, which maximize electron scattering. These results provide a foundation for further research to improve the thermoelectric performance of $SrTi_{1-x}Nb_xO_3$.

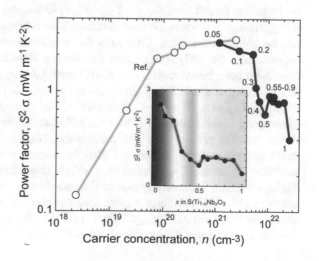

Fig. 10.5 Thermoelectric phase diagram for the $SrTiO_3$-$SrNbO_3$ solid solution system. The thermoelectric power factor ($S^2 \cdot \sigma$) of the $SrTiO_3$-$SrNbO_3$ solid solution system is plotted along with the previously reported values [7]. The x dependence of $S^2 \cdot \sigma$ is shown in the inset. The system's thermoelectric phase boundaries are clearly seen at $x \sim 0.3$ and ~ 0.5. Reprinted with permission from [22]. © 2017 AIP

10.3 Reactive Solid-Phase Epitaxy

As explained above, PLD is a powerful technique for epitaxial film growth of metal oxides. Although there are many reports on epitaxial oxide film growth by the PLD method, it is still difficult to fabricate epitaxial films of complex oxides composed of different vapor pressure elements, especially alkali metals. Since complex oxides have high melting points (>1500 °C), substrates must be heated at high temperature (>800 °C : ~60% of the melting point) in a vacuum during PLD. If the multiplex elements have different vapor pressures at the substrate temperature, the chemical composition of the resultant film completely differs from that of the target because re-vaporization of the high vapor pressure element occurs during deposition.

To overcome this issue, Ohta et al. developed the "Reactive Solid-Phase Epitaxy (R-SPE)" method in 2003 (Fig. 10.6) [29]. In this method, an epitaxial film of a monoxide, which is a component of a complex oxide, is fabricated by PLD. Then the film is heated at high temperatures with another member of the target oxide (thin film or powder). During the heat treatment, a solid-solid reaction between the monoxide film and the other member elements occurs while maintaining the crystallographic orientation. Using the R-SPE method, epitaxial films of $InMO_3(ZnO)_m$ (M = Ga and In, m = integer) [29], $ZnRh_2O_4$ [30], $LaCuOCh$ (Ch = S or Se) [31], and Na_xCoO_2 ($x \sim 0.8$) [32] have been fabricated. Thus, the R-SPE method effectively fabricates epitaxial films of complex oxides composed of high vapor pressure elements. In this section, recent progress of "reactive solid-phase epitaxy" of functional oxides and chalcogenides is reviewed.

10.3.1 $Na_{\approx 2/3}MnO_2$ Epitaxial Film

Layered alkali ion-containing metal oxides (LAMO), A_xMO_2 (A: alkali metal and M: transition metal) have received considerable interest as candidate materials for energy storage and conversion applications. This is because their chemical potential can be readily controlled. Changing the concentration of A^+ in the interspace between adjacent MO_2 layers tunes the valence state of the M ion. In particular, there have been many studies on cobalt-based A_xCoO_2 because the A^+ concentration is easily controlled and these oxides possess a two-dimensional electronic structure. Li_xCoO_2 ($0 \leq x \leq 1$) is one of the best cathode active materials in commercial Li-ion batteries because the Li^+ concentration can be controlled by an electrochemical process [33, 34]. Meanwhile, Na_xCoO_2 ($x \sim 0.8$) is a promising thermoelectric material, which can directly convert a temperature difference into electricity. Additionally, it exhibits a rather large thermopower even though it displays metallic conductivity due to the two-dimensional nature of the electronic structure [35, 36]. Furthermore, the two-dimensional electronic structure of the bilayer hydrated crystal $Na_{\approx 0.3}CoO_2 \cdot 1.3H_2O$ allows it to exhibit superconductivity at a critical temperature T_c of ~4 K [37, 38].

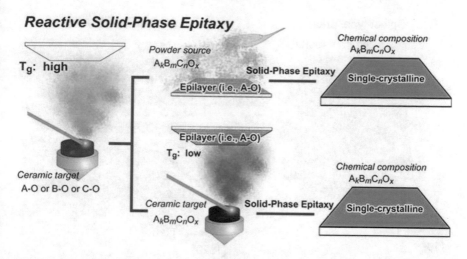

Fig. 10.6 Schematic diagram of the "Reactive Solid-Phase Epitaxy" method. A bilayer laminate composed of a thin epitaxial layer of simple oxide (A–O or B–O or C–O) or metal grown on a substrate and a polycrystalline layer or powder source of target $A_kB_mC_nO_x$ is thermally annealed at high temperatures (~1000 °C). The solid-state reaction at high temperatures leads to the formation of a thin single-crystalline layer on the substrate, which may act as "an epitaxial template" for successive homoepitaxial SPE growth of the film

Unlike A_xCoO_2 systems, the physical properties of A_xMnO_2 have yet to be clarified, although Na_xMnO_2 has recently been proposed as a new candidate for the cathode active material in Na-ion batteries [39–41] because Mn (Clarke number: 0.09) is more abundant than Co (Clarke number: 0.004). At least two crystallographic phases of $NaMnO_2$ are known; low-temperature α-$NaMnO_2$ [39, 42] has an O3 layered structure with monoclinic symmetry and high-temperature β-$NaMnO_2$ [40] has a *Pmnm* structure with orthorhombic symmetry. It should be noted that the crystal symmetry of Na_xMnO_2 strongly depends on x. The low-temperature phase $Na_{0.67}MnO_2$ [41, 43] has a P2 layered structure with hexagonal symmetry. Recently, Billaud et al. [41] reported that $Na_{0.67}MnO_2$ with a P2 structure exhibits a high capacity of 175 mA h g^{-1} with a good capacity retention.

However, there are a few studies on the electrical conductivity of Na_xMnO_2 [44]. The most rational reason is the lack of large single crystals. It is difficult to measure the intrinsic electrical property using powder compacts due to severe electron scattering. High-quality epitaxial films may be a solution to clarify the electrical conductivity of Na_xMnO_2.

In 2017, Katayama et al. fabricated Na_xMnO_2 epitaxial films by the R-SPE method using the following procedure (Fig. 10.7) [45]. First, a 70 nm-thick MnO_y thin film was heteroepitaxially grown on a (0001) α-Al_2O_3 substrate (10 × 10 × 0.5 mm) by PLD using a KrF excimer laser ($\lambda = 248$ nm, 20 ns, 10 Hz, ~1.5 J cm^{-2} $pulse^{-1}$) to ablate the Mn_2O_3 ceramic disk. During the deposition, the substrate temperature and oxygen pressure were kept at 700 °C and ~10^{-2} Pa, respectively. After deposition,

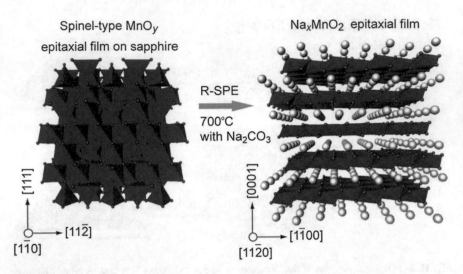

Fig. 10.7 Schematic of the crystal structure change from MnO$_y$ to Na$_x$MnO$_2$ during R-SPE [Gray: Na, blue: Mn, red: O]. First, a spinel-type MnO$_y$ film is heteroepitaxially grown on a sapphire substrate. The MnO$_y$ epitaxial film, covered with Na$_2$CO$_3$ powder, is heated at 700 °C in air. As a result, Na$^+$ ions are supplied into the MnO$_y$ film together with O^{2-} ions during the heating, forming the Na$_x$MnO$_2$ epitaxial film with a layered crystal structure. Reprinted with permission from [45]. © 2017 ACS

the film was cooled to RT in the PLD chamber. The film's top surface was completely covered with another sapphire plate, and the sandwiched specimen was subsequently preserved in Na$_2$CO$_3$ powder. Then the film was heated at 700 °C for 30 min in air to supply Na$^+$ and O^{2-} into the MnO$_y$ film. During heat treatment, the film color changed from light brown to dark brown.

Fig. 10.8 HAADF-STEM image of the Na$_{\approx 2/3}$MnO$_2$/ α-Al$_2$O$_3$ substrate interface. Incident direction of the electron beam is [1$\bar{1}$00] Na$_{\approx 2/3}$MnO$_2$. The atomically sharp heterointerface is clearly seen. Reprinted with permission from [45]. © 2017 ACS

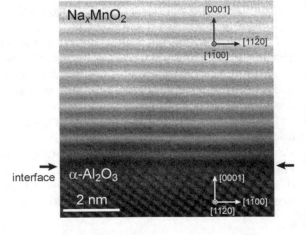

Fig. 10.9 Electrical
conductivity. Temperature
dependence of the electrical
conductivity for the
Na$_{\approx 2/3}$MnO$_2$ and the
Na$_{\approx 0.61}$MnO$_2$ ≈ 0.42H$_2$O
films measured by the AC
impedance method in air. The
resultant films show electrical
conductivities of ~1 mS cm^{-1}
for Na$_{\approx 2/3}$MnO$_2$ film and
~0.1 mS cm^{-1} for
Na$_{\approx 0.61}$MnO$_2$ ≈ 0.42H$_2$O
film at RT. Inset shows the
σT–$1/T$ plots for both films;
the activation energy (E_a) for
electron hopping conduction
is 0.47 eV. Reprinted with
permission from [45]. © 2017
ACS

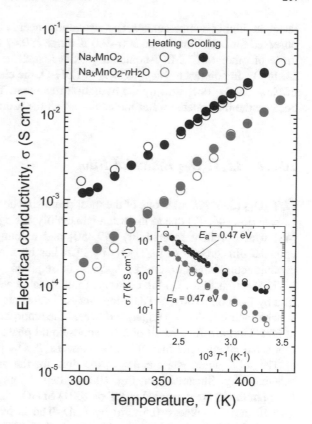

Figure 10.8 shows a cross-sectional HAADF-STEM image of the R-SPE grown
Na$_{\approx 2/3}$MnO$_2$ film around the interface observed from the direction of
Na$_{\approx 2/3}$MnO$_2$‖α-Al$_2$O$_3$. The stripe patterns correspond to the layered structure of
Na$_{\approx 2/3}$MnO$_2$. It should be noted that an interfacial layer is not observed, confirming
that the present sample has an atomically sharp interface between the film and
substrate, contrary to the previously reported observation for a Na$_{\approx 0.8}$CoO$_2$ film.

Figure 10.9 summarizes the temperature (T) dependence of the electrical con-
ductivity (σ) for the Na$_{\approx 2/3}$MnO$_2$ and hydrated Na$_{\approx 0.61}$MnO$_2$ ≈ 0.42H$_2$O epitaxial
films. It should be noted that the $\sigma - T$ curves for both films do not show a
remarkable hysteresis in the heating–cooling cycles ranging from RT to 400 K,
suggesting that the absorbed water does not significantly contribute to σ because the
surface-adsorbed water should be released at 100–150 °C, although a slight deviation from a straight line at ~100 °C is observed in hydrated Na$_{\approx 0.61}$MnO$_2$ ≈ 0.42
H$_2$O film. At RT, σ of the Na$_{\approx 2/3}$MnO$_2$ epitaxial film is ~1 mS cm^{-1}, which is two
orders of magnitude larger than that of an α-Na$_{0.70}$MnO$_{2.25}$ single crystal (~0.5 μS
cm^{-1}). In contrast, σ of the Na$_{\approx 0.61}$MnO$_2$ ≈ 0.42H$_2$O film is ~0.1 mS cm^{-1}, which
is comparable to that of the Na$_x$MnO$_2$ · nH$_2$O ceramic (~0.05 mS cm^{-1}). In both
cases, σ increases exponentially with temperature because electron hopping becomes

faster at higher temperatures. The activation energy for electron hopping (E_a) observed for both films in the 300–400 K range is 0.47 eV, which is comparable to those of other Mn^{3+}/Mn^{4+}-containing oxides, such as α-$MnO_{2-\delta}$, Li_xMnO_2, and $LiMn_2O_4$. In contrast to $Na_{\approx 0.35}CoO_2 \cdot 1.3H_2O$, the electron hopping conductivity of $Na_{\approx 2/3}MnO_2$ decreases by the hydration treatment. This decrease is most likely because the intercalated water molecules affect the ratio of Mn^{3+}/Mn^{4+}.

10.3.2 $Li_4Ti_5O_{12}$ Epitaxial Film

$Li_4Ti_5O_{12}$ (S.G. $Fd\bar{3}m$) is one of the most promising anode active materials of solid Li-batteries, [46, 47] due to its structural stability during charge/discharge reactions [48] with excellent reversibility [49, 50] and a long cycle life [51]. Although epitaxial film growth of $Li_4Ti_5O_{12}$ by PLD has been reported, [52, 53] the target ceramic containing excess Li species is required to fabricate stoichiometric $Li_4Ti_5O_{12}$ thin films. On the contrary, Li et al. fabricated an amorphous $Li_4Ti_5O_{12}$ film by PLD on a (001) $SrTiO_3$ single-crystal substrate at room temperature with a stoichiometric $Li_4Ti_5O_{12}$ target, and heated the amorphous film with molten $LiNO_3$ at 600 °C in air. As a result of this "solid–liquid phase epitaxy", they successfully fabricated an epitaxial film of single-phase $Li_4Ti_5O_{12}$ [54].

The solid–liquid phase epitaxy procedure for the growth of $Li_4Ti_5O_{12}$ films is schematically illustrated in Fig. 10.10. Step 1(a): Amorphous Li-Ti-O films (100 nm thick) are deposited at RT on (001) $SrTiO_3$ single-crystal substrates (area: 10×10 mm^2, thickness: 0.5 mm) by PLD. The in-plane lattice mismatch is too large for $Li_4Ti_5O_{12}$ to coherently grow on (001) $SrTiO_3$ substrate, where the lattice mismatch between cubic $Li_4Ti_5O_{12}$ (the half of a-axis lattice parameter, $a/2 = 0.4176$ nm) and $SrTiO_3$ ($a = 0.3905$ nm) is estimated to be -6.9%. A KrF excimer laser with an energy fluence of $\sim$2 J cm^{-2} pulse^{-1} and a repetition rate of 10 Hz is used to ablate a ceramic target of stoichiometric $Li_4Ti_5O_{12}$. The oxygen pressure during film deposition is kept at a low P_{O2} of 1.0×10^{-3} Pa and the deposition rate is 3.3 nm min^{-1}. Step 2(b): The resultant film is covered with $LiNO_3$ powder. Then it is heated at 600 °C for 30 min in air at temperature increasing rate of 40 °C/min in an Al_2O_3 crucible using an electric furnace. During the heating process, the $LiNO_3$ powder melts due to its low melting point of 261 °C and entirely covers the Li-Ti-O film at 600 °C. The film is naturally cooled to RT in the furnace. Step 3(c): The resultant film is washed by distilled water since the film surface is covered with the remaining $LiNO_3$ film. The resultant film surface looks very clean, indicating that $LiNO_3$ is successfully removed.

Figure 10.11a–c shows the out-of-plane XRD patterns for $Li_4Ti_5O_{12}$ films [(a) as-deposited, (b) heated at 600 °C without $LiNO_3$, and (c) heated at 600 °C with $LiNO_3$]. Only the intense diffraction peaks of 00 l $SrTiO_3$ are observed in the as-deposited $Li_4Ti_5O_{12}$ film, indicating that the film is amorphous (Fig. 10.11a).

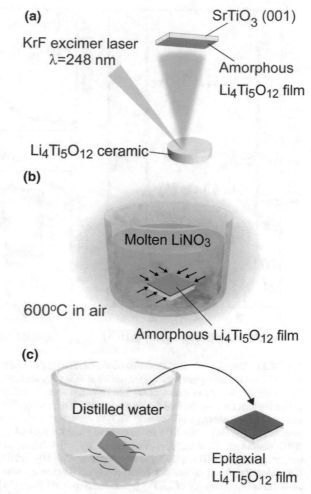

Fig. 10.10 Solid–liquid phase epitaxy of the $Li_4Ti_5O_{12}$ film. **a** Step 1: An amorphous $Li_4Ti_5O_{12}$ film is deposited at RT on a (001) $SrTiO_3$ single-crystal substrate by PLD using a dense $Li_4Ti_5O_{12}$ ceramic as the target. **b** Step 2: The resultant film is heated at 600 °C for 30 min in air with a $LiNO_3$ powder, which melts during the heating process, in an Al_2O_3 crucible using electric furnace. Then the film is naturally cooled down to RT in the furnace. **c** Step 3: The $Li_4Ti_5O_{12}$ epitaxial film is obtained after the resultant film is washed with distilled water since the film surface is covered with remaining $LiNO_3$ film. Reprinted from [54]. © 2016 The Japan Society of Applied Physics

After heating the amorphous $Li_4Ti_5O_{12}$ film without $LiNO_3$ powder at 600 °C in air, the 004 anatase-TiO_2 diffraction peak is observed, but the $Li_4Ti_5O_{12}$ diffraction peak is not seen in the out-of-plane XRD pattern (Fig. 10.11b). The c-axis lattice parameter (0.954 nm) for the TiO_2 phase is almost the same as 0.951 nm for the pure anatase-TiO_2 bulk [55].

In contrast, the intense diffraction peak of 004 $Li_4Ti_5O_{12}$ is observed after the film is heated with molten $LiNO_3$ at 600 °C (Fig. 10.11c). The full width at half maximum (FWHM) value of the out-of-plane rocking curve ($\Delta\omega$) for 004 $Li_4Ti_5O_{12}$ diffraction is ~0.8°, indicating that the $Li_4Ti_5O_{12}$ film is preferentially oriented perpendicular to the substrate surface (Fig. 10.11d). The chemical composition of the $Li_4Ti_5O_{12}$ film could not be accurately estimated when the Li/Ti ratio is an extremely large value of ~5.0, presumably due to the residual adhesive $LiNO_3$

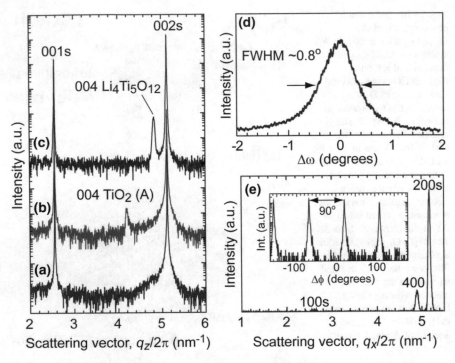

Fig. 10.11 Out-of-plane XRD patterns of the $Li_4Ti_5O_{12}$ films [**a** as-deposited, **b** heated at 600 °C without $LiNO_3$ (solid-phase epitaxy), **c** heated at 600 °C with $LiNO_3$ (solid–liquid phase epitaxy)]. Only intense diffraction peaks of 00 l $SrTiO_3$ are seen in the as-deposited $Li_4Ti_5O_{12}$ film (**a**), indicating that the as-deposited film is amorphous. The intense diffraction peak of 004 $Li_4Ti_5O_{12}$ is observed in (**c**), although 004 anatase TiO_2 is crystallized when the film is heated without $LiNO_3$ powder (**b**). FWHM of the Xray rocking curve ($\Delta\omega$) for the 004 $Li_4Ti_5O_{12}$ is ~0.8° (**d**). **e** In-plane XRD pattern of the $Li_4Ti_5O_{12}$ film grown by solid–liquid phase epitaxy. Only the intense diffraction peak of 400 $Li_4Ti_5O_{12}$ is seen together with $h00$ $SrTiO_3$. The ϕ scan of 400 $Li_4Ti_5O_{12}$ diffraction [(**e**) inset] shows a fourfold rotational symmetry with every 90° rotation originating from the cubic symmetry of $Li_4Ti_5O_{12}$ lattice. Reprinted from [54]. © 2016 The Japan Society of Applied Physics

and/or Li species incorporated into $SrTiO_3$ substrate. However, the film density characterized by the Xray reflectivity measurements for the $Li_4Ti_5O_{12}$ film is 3.5 g cm^{-3}, which is consistent with the 3.48 g cm^{-3} of the $Li_4Ti_5O_{12}$ phase. In contrast, that of the $Li_4Ti_5O_{12}$ film heated without $LiNO_3$ is 4.1 g cm^{-3}, approaching 3.90 g cm^{-3} of the anatase TiO_2 due to the decrease of in Li content in the film. At this stage, the actual vaporization temperature of Li in the $Li_4Ti_5O_{12}$ film has yet to be examined, but these results suggest that molten $LiNO_3$ plays an essential role in suppressing the vaporization of Li species and enables crystallization of the $Li_4Ti_5O_{12}$ phase at relatively low temperature (600 °C) by solid–liquid phase epitaxy.

10.3.3 KFe₂As₂ Epitaxial Film

A novel attractive property has recently been theoretically predicted for KFe_2As_2. Pandey et al. [56] reported that KFe_2As_2, which is the end member of the 122-type iron-based superconductors $(Ba_{1-x}K_x) Fe_2As_2$ (i.e., $x = 1$) with a critical temperature ≈ 3 K [57], may exhibit a large spin Hall conductivity (SHC), which is comparable to that of Pt [58]. That is, it exhibits 10^4 times larger SHC (2×10^4 Ω^{-1} m^{-1}) than that of a semiconductor (0.5 Ω^{-1} m^{-1}) [59]. Such a high SHC originates from the strong spin-orbit coupling of the Fe 3d states with Dirac cones below the Fermi level of heavily hole-doped KFe_2As_2. Indeed, high-resolution angle-resolved photoemission spectroscopy experiments show that the electron pocket at the M point of $(Ba_{1-x}K_x)Fe_2As_2$ completely disappear for KFe_2As_2 due to its heavily self-hole-doped nature ($Ba^{2+} \leftrightarrow K^+ + hole$) [60].

Because KFe_2As_2 contains alkali metal K as its main constituent, it is very air sensitive. Therefore, the thin-film growth of KFe_2As_2 is difficult due to two intrinsic properties: its extremely hygroscopic nature and the high vapor pressure of potassium. Thin-film growth of KFe_2As_2 and electrical measurements with device patterning are challenging issues. These issues were solved by combining room-temperature pulsed laser deposition using K-rich KFe_2As_2 bulk targets with thermal crystallization in a KFe_2As_2 powder after encapsulation in an evacuated silica-glass tube. All of the setup processes must be conducted in a vacuum chamber and a dry Ar atmosphere in a glove box (Fig. 10.12) [61]. Optimized KFe_2As_2 films on (La, Sr)(Al, Ta)O_3 single-crystal substrates are obtained by crystallization at 700 °C. These films are strongly c-axis oriented. Electrical measurements were performed with thin films protected by grease passivation to block reaction with the atmosphere. The KFe_2As_2 films exhibit a superconductivity transition at 3.7 K, which is the same as that of bulk KFe_2As_2. This result is the first demonstration of a superconducting KFe_2As_2 thin film.

The obtained KFe_2As_2 films are, however, not epitaxial films, but c-axis orientated ones without an in-plane orientation. This is attributed to the maximum thermal annealing temperature up to 700 °C in the conventional annealing method sealed in an evacuated silica-glass tube. When we raised the annealing temperature to >700 °C, the films decompose into Fe_2As, FeAs, and Fe. It indicates that the gas-tightness of this synthesis condition is poor >700 °C for K and the alkali metal component K does not remain in the films.

Therefore, an improved solid-phase epitaxy technique using a custom-made alumina vessel, which realizes a high annealing temperature of 1000 °C without vaporization of K from the films, was developed, and high-quality heteroepitaxial KFe_2As_2 thin films on MgO single crystals were successfully obtained (Fig. 10.13) [62]. This result demonstrates that this solid-phase epitaxy technique is a powerful method for the complex compounds with extremely high vapor pressures, such as K.

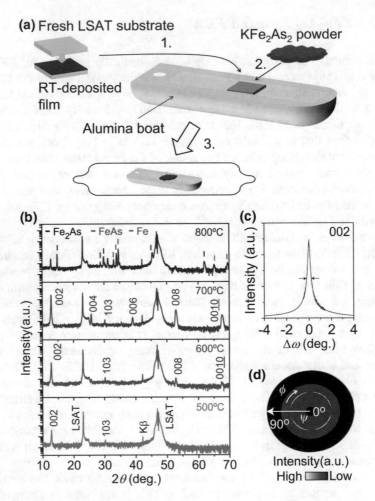

Fig. 10.12 Solid-phase epitaxy for KFe_2As_2 film growth. **a** Set up before thermal annealing. **b** XRD patterns of the films annealed at $T_a = 500–800$ °C buried in KFe_2As_2 powder. **b** Out-of-plane rocking curve of the 002 diffraction of the KFe_2As_2 film annealed at $T_a = 700$ °C. **c** Pole figure of the 103 diffraction of the KFe_2As_2 film annealed at $T_a = 700$ °C. Reprinted from ref. [61]. Copyright © 2014 American Chemical Society

10.3.4 (Sn, Pb)Se Epitaxial Film

SnSe is usually a p-type semiconductor and has the orthorhombic GeS-type layered crystal structure composed of an alternating stack of $(Sn^{2+}Se^{2-})_2$ layers along the a-axis. In contrast, a simple binary selenide PbSe has a cubic rock-salt (RS-) type structure. Different from SnSe, the RS-type structure is thermodynamically stable for PbSe at room temperature. Comparing these crystal structures, one expects that a smaller hole effective mass and a higher hole mobility would be realized if the

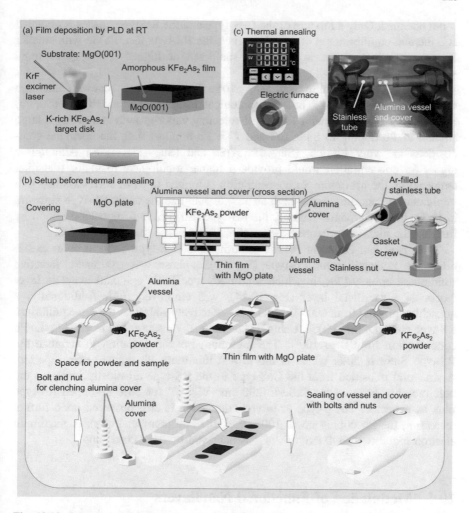

Fig. 10.13 Improved solid-phase epitaxy for KFe$_2$As$_2$ film growth. Reprinted from ref. [62]. Copyright © 2016 The Japan Society of Applied Physics

crystal structure of SnSe is changed from the thermal equilibrium GeS-type one to the RS-type one because the three-dimensional network of high-coordination number polyhedra [sixfold (PbSe$_6$) in the RS-type one] forms larger band dispersions than two-dimensional layered structures [threefold (SnSe$_3$) in the GeS-type one]. As found in the SnSe–PbSe phase diagram, isovalent Pb^{2+} ions can substitute for part of the Sn^{2+} sites in the orthorhombic GeS-type SnSe at thermal equilibrium [63]. For example, at 400 K, ~20% Pb^{2+} can occupy the Sn^{2+} sites in the orthorhombic GeS-type structure, while more than 60% Pb^{2+} substitution is necessary to stabilize the RS-type structure in (Sn, Pb)Se. On the other hand, in the intermediate Pb concentration region between 20 and 60%, single-phase (Sn, Pb)Se is not obtained. Only a mixture of the GeS-type and the RS-type phases is obtained at a thermal equilibrium at 400 K. However, the RS-type (Sn, Pb)Se with smaller Pb concentrations down to ~40% is stabilized at higher temperatures (e.g., ~1100 K).

Recently, RS-type SnSe and (Sn, Pb)Se have gathered renewed attention because they are expected to be new topological insulators. So far, RS-type (Sn, Pb) Se single crystals are grown using a self-selecting vapor-growth method by a large amount of Pb doping to SnSe (63 and 77% Pb doping) [i.e., the chemical compositions are very Pb-rich, $(Sn_{0.37}Pb_{0.63})Se$ and $(Sn_{0.23}Pb_{0.77})Se$.] Although a higher Sn concentration would provide a higher topological insulator transition temperature, the maximum Sn concentration is limited to 37%, corresponding to a minimum Pb concentration as high as 63% to stabilize the cubic RS-type structure in SnSe. From the phase diagram, the RS-type (Sn, Pb)Se composition region by freezing the high-temperature RS-type (Sn, Pb)Se phase may be extended.

Thus, isovalent Pb doping to the orthorhombic GeS-type SnSe in order to stabilize the nonequilibrium RS-type (Sn, Pb)Se phase was examined. Reactive solid-phase epitaxy [29], in which a thin RS-type PbSe epitaxial template layer works as a sacrificial layer (Fig. 10.14), was employed [64]. Additionally, a quenching process from 600 °C to RT also effectively stabilizes the nonequilibrium RS-type epitaxial (Sn, Pb)Se. Using this technique, we succeeded in varying the Pb concentration from 0 to 100%. The minimum Pb concentration to stabilize the RS-type SnSe is 50%, which is the lowest minimum Pb content ever reported. A structural transition from the GeS-type to the RS-type drastically increases the hole mobility from 60 for SnSe to 290 cm^2 V^{-1} s^{-1} for 58% Pb-doped RS-type film, as expected. The p-type to n-type conversion is also observed upon further increasing the Pb doping up to 100% (i.e., the end member PbSe). A maximum electron mobility of 340 cm^2 V^{-1} s^{-1} is achieved by 61% Pb doping.

10.4 Modulation of Functional Nanolayers

State-of-the-art information storage devices such as USB flash drives are electronic data storage devices, which store digital information by electrical resistivity changes of semiconducting silicon using the electric field effect to process information into "words" consisting of various combinations of the numbers "0" and "1". Since miniaturization technology has already reached its limit, epoch-making technology is strongly required to further improve the storage capacity.

We have demonstrated multi-information memory devices, which are composed of metal oxides showing both an electrical resistivity change and magnetism/color change simultaneously by a redox reaction of the metal oxides. Three-terminal thin-film transistor (TFT) structures were fabricated on a functional metal oxide using an insulating oxide, water-infiltrated calcium aluminate (C12A7) with

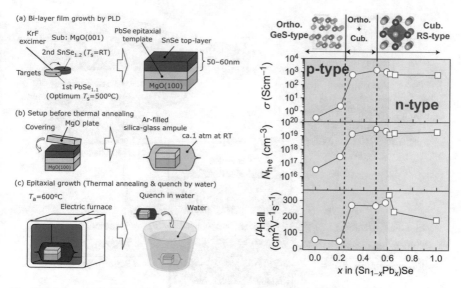

Fig. 10.14 Reactive solid-phase epitaxy for the $(Sn_{1-x}Pb_x)S$ epitaxial films and their carrier transport properties at room temperature as a function of x. σ, $N_{h,e}$, and μ_{Hall} show the electrical conductivity, carrier concentration of hole or electron, and Hall mobility, respectively. Reprinted from ref. [64]. Copyright © 2016 American Chemical Society

mesoporous structure, as the gate insulator. We utilized H^+/OH^- ions in the water to change the valence state of the metal oxides by applying a gate voltage since H^+/OH^- ions are strong reducing/oxidizing agents for metal oxides. Upon changing the valence of the transition metal ion, the metal oxide changes from an insulator to a metal as well as from nonmagnetic to magnetic or from colorless transparent (invisible) to visible, as schematically illustrated in Fig. 10.15. Although the present device requires a relatively long storage time (a few seconds) because it utilizes mobile ion diffusion in the functional metal oxide, it has great merits. For example, it has nonvolatile operations, which mean no standby power is required after storing information. The present multi-information storage device should be useful for Internet of Things (IoT) technologies.

As IoT technologies become more ubiquitous, the information gathered annually is rapidly increasing as various machines, as well as personal computers, are connected to the internet. State-of-the-art information storage devices such as USB flash drives are electronic data storage devices. They store digital information using an electrical resistivity change of semiconducting silicon and an electric field effect that process information into "words" consisting of various combinations of the numbers "0" and "1". Although the storage capacity of such devices increased annually due to miniaturization techniques, the limit of such miniaturization techniques has already been reached. Consequently, complicated multi-levelization techniques such as "0", "1", "2", and "3" are utilized to improve the storage

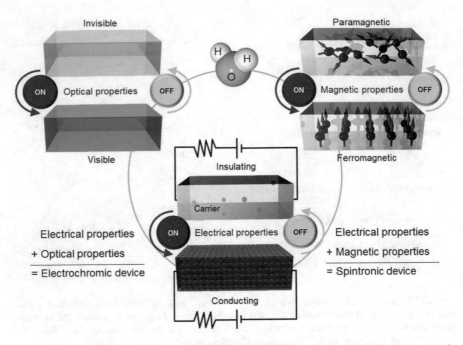

Fig. 10.15 Schematic concept of a reversible conversion of optical-, electrical-, and magnetic properties of functional metal oxides using an electrochemical redox reaction of metal oxides with H^+ and OH^- of liquid water. For example, by combining the electrical properties and the optical properties, a novel electrochromic device, which can store A/B in addition to 0/1 for storing information, can be developed

capacity. To further improve the storage density, epoch-making technology is strongly required.

To overcome this obstacle, we proposed the following idea: utilizing functional materials whose optical transmittance or magnetism can be dramatically changed together with a change in the electrical resistivity. Such features would be very useful to improve the storage capacity. For example, multiple storing/reading of information becomes possible when vision and electrical signals are combined with displays. Such devices are appropriate for future IoT technologies. However, it is impossible to use optical transmittance or magnetic properties in case of semi-conductor Si. In addition, it is impossible to use the electrical resistivity change in case of a magnetic metal.

Our research has focused on metal oxides because some metal oxides exhibit changes in their optical property or magnetic property together with the electrical property via an oxidation/reduction (redox) reaction. Generally, such redox reactions occur at a high temperature (several hundred degrees) heat treatment in an oxidizing or reducing atmosphere. However, this method is inappropriate for device operations. On the other hand, redox reactions using electrochemistry such chemical battery cells occur at room temperature. The latter technique is appropriate for practical applications, but the device must be sealed to prevent electrolyte leakage.

Unexpectedly, in 2010, Ohta et al. found that water, which is automatically absorbed into the mesoporous structure of an insulating oxide due to capillary action, can be a good electrolyte for this purpose [65]. Three-terminal thin-film transistor (TFT) structures were fabricated on a functional metal oxide using an insulating oxide, calcium aluminate ($12CaO \cdot 7Al_2O_3$, C12A7) with mesoporous structure (namely CAN, calcium aluminate with nanopores) as the gate insulator (Fig. 10.16). C12A7 can be prepared by PLD at room temperature under a relatively high oxygen atmosphere of ~5 Pa. CAN films contain many mesopores (~10 nm in diameter) whose volume fraction is ~30% [65]. Temperature desorption spectra of the CAN film reveal that the mesopores are fully occupied with molecular water. The AC conductivity of the CAN film is ~10^{-9} S cm^{-1} at room temperature, [66] which is comparable to that of ultrapure water. Thus, water moisture in air is automatically absorbed in the mesopores of CAN film by capillary action.

In 2016, Katase and Ohta et al. utilized H^+/OH^- ions in a water-infiltrated CAN film to change the valence of the metal oxides by applying a gate voltage because H^+/OH^- ions are strong reducing/oxidizing agents for metal oxides. As the valence of the transition metal ion changes, the metal oxide changes from an insulator to a metal, a nonmagnetic to a magnetic material, or from transparent to a black color. Although the present device requires a relatively long storage time (a few seconds) since it utilizes mobile ion diffusion in the functional metal oxide, it has great merits. For example, it has nonvolatile operations, which mean no standby power is required after storing information. The present multi-information storage devices would be useful for IoT technologies.

The authors have developed two multi-information memory devices, which use a magnetic [67] or optical [68] signal along with an electronic signal to double the storage capacity in these "multiplex writing/reading" devices. In addition to the binary 0/1 method of storing information in a state-of-the-art memory device, the present devices can also store A/B for the information. More details for each type of memory device are provided below.

10.4.1 Utilizing Antiferromagnetic Insulator/Ferromagnetic Metal Conversion in SrCoO$_{2.5+\delta}$ [67]

To realize "multiplex writing/reading" devices, material selection is the most information factor. Katase and Ohta et al. choose strontium cobaltite, SrCoO$_{2.5+\delta}$, for this purpose because SrCoO$_{2.5}$ is an antiferromagnetic insulator and SrCoO$_3$ is a ferromagnetic metal [69, 70]. The valence state of the cobalt ion in SrCoO$_{2.5+\delta}$ can be controlled from 3 + (SrCoO$_{2.5}$) to 4 + (SrCoO$_3$) by changing the excess oxygen

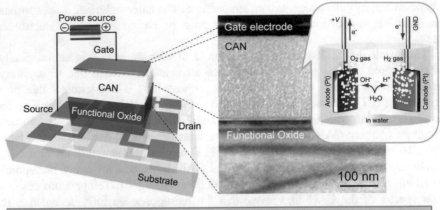

CAN: Calcium Aluminate with Nanopore

Matrix	Nanopores
Amorphous 12CaO·7Al$_2$O$_3$ (a-C12A7)	Fully occupied with ultra pure water
ε_r ~12	Average diameter ~10 nm
	Volume fraction ~30 vol.%
	Percolation AC conductivity ~10^{-9} S cm^{-1}

Fig. 10.16 Schematic illustration of CAN (calcium aluminate with nanopore) gated functional oxide thin-film transistor with a three-terminal electrodes geometry. Since 30 volume percent of the CAN film is occupied with liquid water, H$^+$ and OH$^-$ ions in the CAN film move with a gate voltage application. The percolation AC conductivity is ~10^{-9} S cm^{-1}, which is comparable to that of ultrapure water. Changing the valence of the transition metal ion by a gate voltage application changes the functional oxide from an insulator to a metal as well as from a nonmagnetic to a magnetic material or from colorless and transparent to a black color

content (δ) from 0 to 0.5. Since the crystal structures of SrCoO$_{2.5}$ (brownmillerite) and SrCoO$_3$ (perovskite) are similar, the authors expected that the topotactic redox reaction between SrCoO$_{2.5}$ and SrCoO$_3$ can be controlled electrochemically. By utilizing this phenomenon for three-terminal thin-film transistors with water containing a mesoporous glass gate insulator, we developed a multi-information memory device. This device can be utilized not only to change the electrical resistivity (0/1) but also to change of magnetic property (A/B), as schematically shown in Fig. 10.17.

The three-terminal TFT device, which is composed of an epitaxial SrCoO$_{2.5}$ film (30 nm, active channel material), an amorphous Na-Ta-O film with a mesoporous structure (300 nm, gate insulator), and an amorphous WO$_3$ film (20 nm, proton absorber), was prepared by PLD on (001) SrTiO$_3$ single-crystal substrate. It should be noted that we recently developed an amorphous Na-Ta-O film with a mesoporous structure, which can be used as an alkaline solution. When a negative gate voltage (−3 V) is applied between the gate and source electrodes, OH$^-$ ions, which are contained in the mesoporous glass, penetrate into SrCoO$_{2.5+\delta}$. Finally, SrCoO$_3$

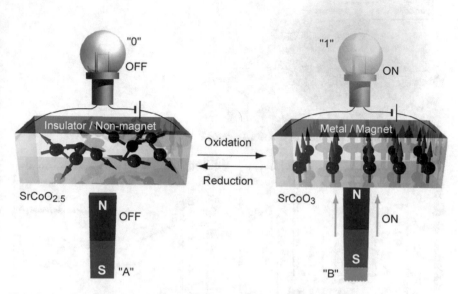

Fig. 10.17 Principle of a multi-memory device using antiferromagnetic insulator/ferromagnetic metal conversion in $SrCoO_{2.5+\delta}$. This device would store both A/B and 0/1 information. Reprinted with permission from [67]. © 2016 John Wiley and Sons

is formed in 3 s. On the contrary, a positive gate voltage (+3 V) application to the gate–source electrodes reduces $SrCoO_3$ into $SrCoO_{2.5}$ in 3 s [67].

Figure 10.18a shows a schematic of the device structure, which is similar to conventional three-terminal thin-film transistors. The channel (source–drain) length and width are 800 μm and 400 μm, respectively. The electrodes E1–E4 are used to measure the sheet resistance (R_s). Figure 10.18b shows the changes in R_s of the device. Before the device operation (state A), R_s increases with decreasing temperature, indicating an insulating behavior. When a negative gate voltage (−3 V) is applied for 3 s (state B), R_s decreases by three orders of magnitude and shows a metallic temperature dependence. After that, the device returns to the original state when a positive gate voltage (+3 V) is applied for 3 s. The device is reversibly operable (Fig. 10.18b, inset).

Figure 10.18c shows the changes in the magnetic state of the device at states A, B, and C. At states A and C, the magnetic moment is zero, indicating that $SrCoO_{2.5+\delta}$ ($\delta = 0$) is an antiferromagnetic state. At state B, the device shows a ferromagnetic behavior with a Curie temperature of 275 K, indicating that $SrCoO_{2.5+\delta}$ ($\delta \sim 0.5$) is a ferromagnetic state. These results clearly demonstrate that both electrical resistivity and magnetism changes can be used in the present device.

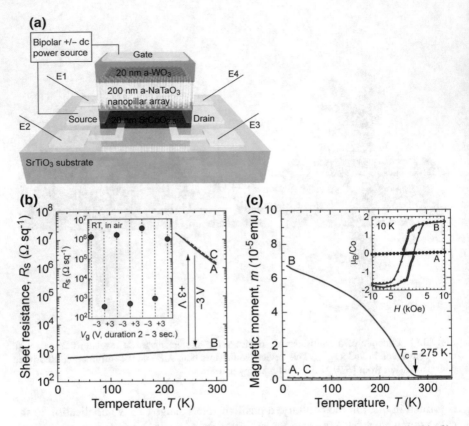

Fig. 10.18 a Schematic device structure similar to conventional three-terminal thin-film transistors. **b** Temperature dependence of the sheet resistance of the device. **a** Virgin state, **b** after applying a negative V_g of -3 V, and **c** subsequent application of $+3$ V (dotted line). The inset shows the cyclability at RT in air. **c** $m-T$ curves of the SrCoOx layer at states A–C in (**b**) measured under $H = 20$ Oe applied parallel to the in-plane direction. The inset shows a magnetic hysteresis loop at 10 K at states A and B. Reprinted with permission from [67]. © 2016 John Wiley and Sons

10.4.2 Utilizing a Colorless Transparent Insulator/Dark Blue Metal Conversion in H_xWO_3 [68]

Katase and Ohta et al. [68] have also developed a new information display/storage device using a three-terminal thin-film transistor structure on an electrochromic material, which has been attracted attention as an "electric curtain". The device shows a color change (colorless transparent/dark blue) together with an electrical conductivity change (insulator/metal) by applying a gate voltage. Since the device can be fabricated at room temperature, low cost fabrication is possible. Thus, larger area devices are easily fabricated. For example, the present device is applicable as an information display/storage on a window glass.

Protonation/deprotonation of tungsten trioxide (WO_3), known as an electrochromic material, is converted reversibly from a colorless transparent insulator to a dark blue metal [71]. By utilizing this phenomenon in a three-terminal thin-film transistor with a water-containing mesoporous glass gate insulator, a multi-information memory device has been realized. This device can be utilized using not only a change in the electrical resistivity (0/1) but also a change in optical transmittance (A/B), as schematically shown in Fig. 10.19.

The three-terminal TFT device composed of an amorphous WO_3 film (100 nm, active channel material), a mesoporous CAN film (300 nm, gate insulator), a polycrystalline NiO film (50 nm, oxygen absorber), and amorphous ITO films (20 nm, gate, source, and drain electrodes) was prepared by PLD at room temperature on a glass substrate. When a positive gate voltage (a few volts) is applied between the gate and the source electrodes, H^+ and OH^- ions, which are contained in the mesoporous glass, diffuse to the WO_3 and NiO sides, respectively, forming HWO_3 and NiOOH. Since the resultant dark blue colored HWO_3 shows a metallic electrical conductivity, the channel (drain–source) becomes electrically conductive. On the contrary, a negative gate voltage (a few volts) application to the gate–source electrodes results in HWO_3 and NiOOH returning to WO_3 and NiO, respectively. This conversion can be reversibly operated and the degree of change can be controlled by the applied gate voltage.

Figure 10.20a schematically depicts the device structure composed of a-WO_3 (80 nm), CAN (300 nm), and NiO (20 nm)/ITO (20 nm) layers. Transparent ITO thin films are used for all the electrodes. All the films are deposited by PLD at room temperature. The device was fabricated on a transparent glass substrate (1 cm × 1 cm) as shown in Fig. 10.20b. The channel (source–drain) length and

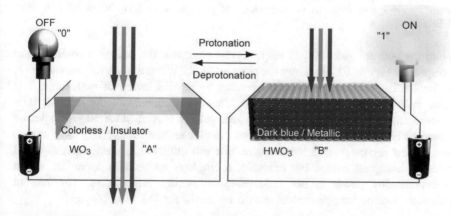

Fig. 10.19 Principle of a multi-memory device using a colorless transparent insulator (amorphous WO_3)/dark blue metal (amorphous HWO_3) conversion in H_xWO_3. This device would store both A/B and 0/1 information. This device should be suitable as a smart window or a smart mirror, which can display or store information. Furthermore, the present device can be used as an "electronic curtain" as the whole surface of window glass can be switched reversibly from colorless and transparent to dark blue. Reprinted from [68]. © 2016 NPG

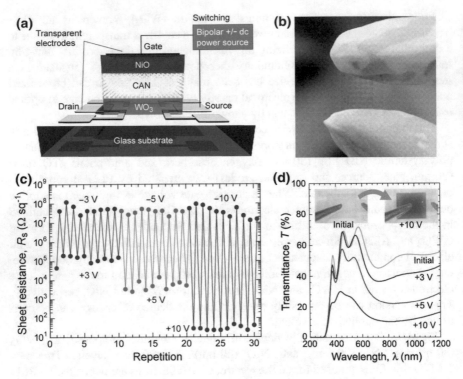

Fig. 10.20 a Schematic device structure. **b** Transparent device on a glass substrate. **c** Repeatable switching of the sheet resistance. **d** Optical transmission spectra. Before the operation, the device is an insulator ($R_s \sim 10^8 \ \Omega \ sq^{-1}$) and fully transparent in the visible light region. When a positive gate voltage is applied to the device for 10 s, R_s decreases several orders of magnitude and the device becomes dark blue due to electrochemical protonation of WO_3. Reprinted from [68]. © 2016 NPG

width are 800 µm and 400 µm, respectively. Note that the device is colorless and transparent. Figure 10.20c shows the changes in R_s. When a positive gate voltage is applied to the device for 10 s, R_s decreases by several orders of magnitude. The color becomes dark blue (Fig. 10.20d). The device reverts to the original state (transparent, insulator) when a negative gate voltage is applied for 10 s. The device is reversibly operable. Although the present device requires a relatively long storage time (a few seconds) since it utilizes mobile ion diffusion in the functional metal oxide, it has great merits. For example, it employs nonvolatile operation, which means standby power is not required after storing information. The present multi-information storage devices would be useful for IoT technologies.

Acknowledgements The authors would like to thank Prof. T. Katase (Tokyo Tech.), Dr. N. Li, Dr. S. Katayama, Mr. Y. Zhang, Prof. T. Kamiya (Tokyo Tech.), and Prof. H. Hosono (Tokyo Tech.) for the valuable discussions and experimental assistance. This work was supported by a

Grant-in-Aid for Scientific Research on Innovative Areas (25106007). H. Ohta was also supported by the Asahi Glass Foundation. H. Hiramatsu was also supported by Support for TokyoTech Advanced Research (STAR).

References

1. R. Eason, *Pulsed Laser Deposition of Thin Films Applications-Led Growth of Functional Materials.* (Wiley, Inc., UK, 2007)
2. R. Ramesh, D.G. Schlom, MRS Bull. **33**, 1006 (2008)
3. H. Ohta, K. Sugiura, K. Koumoto, Inorg. Chem. **47**, 8429 (2008)
4. J.W. Fergus, J. Eur. Ceram. Soc. **32**, 525 (2012)
5. T. Okuda, K. Nakanishi, S. Miyasaka, Y. Tokura, Phys. Rev. B **63**, 113104 (2001)
6. S. Ohta, T. Nomura, H. Ohta, K. Koumoto, J. Appl. Phys. **97**, 034106 (2005)
7. H. Ohta, S. Kim, Y. Mune, T. Mizoguchi, K. Nomura, S. Ohta, T. Nomura, Y. Nakanishi, Y. Ikuhara, M. Hirano, H. Hosono, K. Koumoto, Nature Mater. **6**, 129 (2007)
8. H. Ohta, T. Mizuno, S.J. Zheng, T. Kato, Y. Ikuhara, K. Abe, H. Kumomi, K. Nomura, H. Hosono, Adv. Mater. **24**, 740 (2012)
9. W.S. Choi, H. Ohta, H.N. Lee, Adv. Mater. **26**, 6701 (2014)
10. P. Delugas, A. Filippetti, M.J. Verstraete, I. Pallecchi, D. Marre, V. Fiorentini, Phys. Rev. B **88**, 045310 (2013)
11. T.A. Cain, S. Lee, P. Moetakef, L. Balents, S. Stemmer, S.J. Allen, Appl. Phys. Lett. **100**, 161601 (2012)
12. I. Pallecchi, F. Telesio, D.F. Li, A. Fete, S. Gariglio, J.M. Triscone, A. Filippetti, P. Delugas, V. Fiorentini, D. Marre, Nature Commun. **6**, 6678 (2015)
13. S. Shimizu, S. Ono, T. Hatano, Y. Iwasa, Y. Tokura, Phys. Rev. B **92**, 165304 (2015)
14. L.D. Hicks, M.S. Dresselhaus, Phys. Rev. B **47**, 12727 (1993)
15. N.T. Hung, E.H. Hasdeo, A.R.T. Nugraha, M.S. Dresselhaus, R. Saito, Phys. Rev. Lett. **117**, 036602 (2016)
16. C. Rodenbucher, M. Luysberg, A. Schwedt, V. Havel, F. Gunkel, J. Mayer, R. Waser, Sci Rep-Uk **6**, 32250 (2016)
17. T. Tomio, H. Miki, H. Tabata, T. Kawai, S. Kawai, J. Appl. Phys. **76**, 5886 (1994)
18. L.F. Mattheiss, Phys. Rev. B **6**, 4718 (1972)
19. S.A. Turzhevsky, D.L. Novikov, V.A. Gubanov, A.J. Freeman, Phys. Rev. B **50**, 3200 (1994)
20. D. Oka, Y. Hirose, S. Nakao, T. Fukumura, T. Hasegawa, Phys. Rev. B **92**, 205102 (2015)
21. X.X. Xu, C. Randorn, P. Efstathiou, J.T.S. Irvine, Nature Mater. **11**, 595 (2012)
22. Y. Zhang, B. Feng, H. Hayashi, T. Tohei, I. Tanaka, Y. Ikuhara, H. Ohta, J. Appl. Phys. **121**, 185102 (2017)
23. R.D. Shannon, Acta Crystallogr. A **32**, 751 (1976)
24. Y. Ishida, R. Eguchi, M. Matsunami, K. Horiba, M. Taguchi, A. Chainani, Y. Senba, H. Ohashi, H. Ohta, S. Shin, Phys. Rev. Lett. **100**, 056401 (2008)
25. Y. Ikuhara, P. Pirouz, Microsc. Res. Tech. **40**, 206 (1998)
26. P.K. Davies, J.Z. Tong, T. Negas, J. Am. Ceram. Soc. **80**, 1727 (1997)
27. D.A. Muller, N. Nakagawa, A. Ohtomo, J.L. Grazul, H.Y. Hwang, Nature **430**, 657 (2004)
28. C.L. Chen, Z.C. Wang, F. Lichtenberg, Y. Ikuhara, J.G. Bednorz, Nano Lett. **15**, 6469 (2015)
29. H. Ohta, K. Nomura, M. Orita, M. Hirano, K. Ueda, T. Suzuki, Y. Ikuhara, H. Hosono, Adv. Funct. Mater. **13**, 139 (2003)
30. H. Ohta, H. Mizoguchi, M. Hirano, S. Narushima, T. Kamiya, H. Hosono, Appl. Phys. Lett. **82**, 823 (2003)
31. H. Hiramatsu, K. Ueda, H. Ohta, M. Orita, M. Hirano, H. Hosono, Appl. Phys. Lett. **81**, 598 (2002)

32. H. Ohta, S.W. Kim, S. Ohta, K. Koumoto, M. Hirano, H. Hosono, Cryst. Growth Des. **5**, 25 (2005)
33. K. Mizushima, P.C. Jones, P.J. Wiseman, J.B. Goodenough, Mater. Res. Bull. **15**, 783 (1980)
34. T. Ohzuku, A. Ueda, J. Electrochem. Soc. **141**, 2972 (1994)
35. I. Terasaki, Y. Sasago, K. Uchinokura, Phys. Rev. B **56**, 12685 (1997)
36. M. Lee, L. Viciu, L. Li, Y.Y. Wang, M.L. Foo, S. Watauchi, R.A. Pascal, R.J. Cava, N. P. Ong, Nature Mater. **5**, 537 (2006)
37. K. Takada, H. Sakurai, E. Takayama-Muromachi, F. Izumi, R.A. Dilanian, T. Sasaki, Nature **422**, 53 (2003)
38. R.E. Schaak, T. Klimczuk, M.L. Foo, R.J. Cava, Nature **424**, 527 (2003)
39. X.H. Ma, H.L. Chen, G. Ceder, J. Electrochem. Soc. **158**, A1307 (2011)
40. J. Billaud, R.J. Clement, A.R. Armstrong, J. Canales-Vazquez, P. Rozier, C.P. Grey, P.G. Bruce, J. Am. Chem. Soc. **136**, 17243 (2014)
41. J. Billaud, G. Singh, A.R. Armstrong, E. Gonzalo, V. Roddatis, M. Armand, T. Rojob, P.G. Bruce, Energy Environ. Sci. **7**, 1387 (2014)
42. R.O. Jean-Paul Parant, M. Devalette, C. Fouassier, P. Hagenmuller, J. Solid State Chem. **3**, 1 (1971)
43. A. Caballero, L. Hernan, J. Morales, L. Sanchez, J.S. Pena, M.A.G. Aranda, J. Mater. Chem. **12**, 1142 (2002)
44. S. Hirano, R. Narita, S. Naka, J. Cryst. Growth **54**, 595 (1981)
45. S. Katayama, T. Katase, T. Tohei, B. Feng, Y. Ikuhara, H. Ohta, Cryst. Growth Des. **17**, 1849 (2017)
46. K. Zaghib, M. Simoneau, M. Armand, M. Gauthier, J. Pow. Sources **81**, 300 (1999)
47. G.X. Wang, D.H. Bradhurst, S.X. Dou, H.K. Liu, J. Pow. Sources **83**, 156 (1999)
48. F. Ronci, P. Reale, B. Scrosati, S. Panero, V.R. Albertini, P. Perfetti, M. Di Michiel, J.M. Merino, J. Phys. Chem. B **106**, 3082 (2002)
49. S. Scharner, W. Weppner, P. Schmid-Beurmann, J. Electrochem. Soc. **146**, 857 (1999)
50. M. Kitta, T. Akita, Y. Maeda, M. Kohyama, Langmuir **28**, 12384 (2012)
51. K.M. Colbow, J.R. Dahn, R.R. Haering, J. Pow. Sources **26**, 397 (1989)
52. M. Hirayama, K. Kim, T. Toujigamori, W. Cho, R. Kanno, Dalton T **40**, 2882 (2011)
53. A. Kumatani, T. Ohsawa, R. Shimizu, Y. Takagi, S. Shiraki, T. Hitosugi, Appl. Phys. Lett. **101** (2012)
54. N. Li, T. Katase, Y. Zhu, T. Matsumoto, T. Umemura, Y. Ikuhara, H. Ohta, Appl. Phys. Express **9** (2016)
55. J.K. Burdett, T. Hughbanks, G.J. Miller, J.W. Richardson, J.V. Smith, J. Am. Chem. Soc. **109**, 3639 (1987)
56. S. Pandey, H. Kontani, D.S. Hirashima, R. Arita, H. Aoki, Phys. Rev. B **86** (2012)
57. K. Sasmal, B. Lv, B. Lorenz, A.M. Guloy, F. Chen, Y.Y. Xue, C.W. Chu, Phys. Rev. Lett. **101** (2008)
58. T. Kimura, Y. Otani, T. Sato, S. Takahashi, S. Maekawa, Phys. Rev. Lett. **98** (2007)
59. N.P. Stern, S. Ghosh, G. Xiang, M. Zhu, N. Samarth, D.D. Awschalom, Phys. Rev. Lett. **97** (2006)
60. T. Sato, K. Nakayama, Y. Sekiba, P. Richard, Y. M. Xu, S. Souma, T. Takahashi, G.F. Chen, J.L. Luo, N.L. Wang, H. Ding, Phys. Rev. Lett. **103** (2009)
61. H. Hiramatsu, S. Matsuda, H. Sato, T. Kamiya, H. Hosono, Acs Appl. Mater. Inter. **6**, 14293 (2014)
62. T. Hatakeyama, H. Sato, H. Hiramatsu, T. Kamiya, H. Hosono, Appl. Phys. Express **9** (2016)
63. A.A. Volykhov, V.I. Shtanov, L.V. Yashina, Inorg. Mater. **44**, 345 (2008)
64. T. Inoue, H. Hiramatsu, H. Hosono, T. Kamiya, Chem. Mater. **28**, 2278 (2016)
65. H. Ohta, Y. Sato, T. Kato, S. Kim, K. Nomura, Y. Ikuhara, H. Hosono, Nature Commun. **1**, 118 (2010)
66. H. Ohta, J. Mater. Sci. **48**, 2797 (2013)
67. T. Katase, Y. Suzuki, H. Ohta, Adv. Electron. Mater. **2** (2016)
68. T. Katase, T. Onozato, M. Hirono, T. Mizuno, H. Ohta, Sci. Rep-UK **6** (2016)

69. H. Taguchi, M. Shimada, M. Koizumi, J. Solid State Chem. **29**, 221 (1979)
70. Y. Takeda, R. Kanno, T. Takada, O. Yamamoto, M. Takano, Y. Bando, Z. Anorg. Allg. Chem. **541**, 259 (1986)
71. C.G. Granqvist, Sol. Energy Mater. Sol. Cells **60**, 201 (2000)

Chapter 11
Grain Boundary Engineering of Alumina Ceramics

Satoshi Kitaoka, Tsuneaki Matsudaira, Takafumi Ogawa,
Naoya Shibata, Miyuki Takeuchi and Yuichi Ikuhara

Abstract Oxygen permeability through alumina wafers was evaluated at high temperatures up to 1923 K to elucidate the mass-transfer mechanisms of polycrystalline alumina and serve as a model for protective alumina film formed on heat-resistant alloys. Oxygen permeation proceeded via grain boundary (GB) diffusion of oxygen from the higher oxygen partial pressure (P_{O2}) surface side to the lower P_{O2} surface side, along with the simultaneous GB diffusion of aluminum in the opposite direction to maintain the Gibbs–Duhem relationship. Oxygen GB diffusion coefficients in the vicinity of the P_{O2}(hi) surface were lower than those of oxygen GB self-diffusion without an oxygen potential gradient ($d\mu_O$). When $d\mu_O$ was applied to the wafer, the oxygen and aluminum fluxes at the outflow side of the wafer were significantly larger than those at the inflow side. Ln (Y and Lu) and Hf segregation at the GBs selectively reduced the diffusivity of oxygen and aluminum, respectively. Thus, the mesoscopic arrangements of segregating dopants, which were selected by taking into consideration the behavior of the diffusion species and the role of dopants, enabled the alumina film to have enhanced oxygen shielding capability and structural stability at high temperatures. Furthermore, the GB diffusion data derived from the oxygen permeation experiments were compared to those for alumina scale formed by the so-called two-stage oxidation of alumina-forming alloys.

Keywords Alumina · Grain boundary · Oxygen permeation
Diffusion · High temperature

S. Kitaoka (✉) · T. Matsudaira · T. Ogawa · N. Shibata · Y. Ikuhara
Japan Fine Ceramics Center, Nagoya 456-8587, Japan
e-mail: kitaoka@jfcc.or.jp

N. Shibata · M. Takeuchi · Y. Ikuhara
The University of Tokyo, Tokyo 113-8656, Japan

© The Author(s) 2018
I. Tanaka (ed.), *Nanoinformatics*, https://doi.org/10.1007/978-981-10-7617-6_11

11.1   Introduction

Polycrystalline α-alumina scale can play a key role to enable heat-resistant alloys that include aluminum to be applied as hot section components of airplane engines, gas turbines, and heat treatment furnaces in combustion environments. The α-alumina scale acts as a protective film against further oxidation of the alloys at high temperatures. Growth of the alumina scale is determined by the solid-state diffusion of both oxygen and aluminum along the grain boundaries (GBs) in response to their respective chemical potentials. Thus, it is expected that the durability of hot section components would be determined by the mass transport of oxygen and aluminum through the scale.

For scale growth by inward oxygen GB diffusion, the annihilation and production of oxygen vacancies proceed at the scale-gas and scale-metal interfaces by reactions (11.1) and (11.2), respectively [1]:

$$O_2 + 2V_O^{\bullet\bullet} + 4e' \rightarrow 2O_O^{\times} \tag{11.1}$$

$$2Al_M \rightarrow 3V_O^{\bullet\bullet} + 2Al_{Al}^{\times} + 2V_M + 6e' \tag{11.2}$$

Scale growth also occurs by outward aluminum GB diffusion. Aluminum vacancies are produced at the scale-gas interface by reaction (11.3) and are annihilated at the scale-metal interface by reaction (11.4) [1]:

$$3O_2 \rightarrow 4V_{Al}''' + 6O_O^{\times} + 12h^{\bullet} \tag{11.3}$$

$$Al_M + V_{Al}''' + 3h^{\bullet} \rightarrow Al_{Al}^{\times} + V_M \tag{11.4}$$

Although these reactions are expressed with holes or electrons on opposite sides, the concentrations of electrons (n) and holes (p) are related by another equilibrium constant [2]:

$$K_i = n \times p \tag{11.5}$$

When the alloys are oxidized through alumina scale under high oxygen partial pressures (P_{O2}) (such as in air), i.e., when they are subjected to a steep oxygen potential gradient ($d\mu_O$), the outward GB diffusion of aluminum produces new alumina on the GB surface of the scale, which results in the formation of GB ridges [3]. However, such ridges do not form in a low-P_{O2} environment, such as in a purified argon flow, where oxidation of the alloys could proceed thermodynamically [3]. The mass-transfer mechanisms in the scale appear to be strongly dependent on the extent of $d\mu_O$ to which the scale is exposed.

There have been many studies on oxygen GB diffusion in polycrystalline alumina using either secondary ion mass spectroscopy (SIMS) [4–7] or nuclear reaction analysis (NRA) [8] to determine depth profiles of [18]O (oxygen tracer) after high temperature exchange with ^{18}O-enriched oxygen. The oxygen diffusion coefficients of single GBs were recently determined by a SIMS-^{18}O line profiling technique at each GB near the surface of an alumina cross section [1]. The activation energies reported for the oxygen GB diffusion in the scale tend to be larger than those for the corresponding self-diffusion data. Thus, the application of a dμ_O suggests there is some influence on the oxygen GB diffusivity. However, there has been only one report [1] of GB self-diffusion coefficients for aluminum in alumina in the absence of a dμ_O and no data with application of a dμ_O. One of the likely reasons for this is the lack of an appropriate radioactive tracer, such as ^{26}Al with a very low specific activity and an extremely long half-life of 7.2×10^5 years, which makes it very difficult to perform radiotracer diffusion experiments. Consequently, for the mutual GB diffusion of both oxygen and aluminum in alumina during application of a dμ_O, it has yet to be clarified whether or not these ions migrate with a synergistic effect.

Alumina-forming alloys typically contain small quantities of oxygen-reactive elements (REs) (e.g., Y, La, Zr, and Hf) to improve their oxidation resistance. The REs segregate to GBs during alumina scale growth by oxidation of the alloys [9]. The REs have been considered to primarily decrease the aluminum GB diffusivity with respect to the oxygen diffusivity, according to ^{18}O depth profiling in scale after two-stage oxidation experiments [10–13]. In addition, the REs are considered to inhibit scale growth by effectively blocking the GB diffusion of aluminum due to an ionic-size mismatch because the ionic sizes of the REs are larger than that of Al^{3+}. However, the GB segregated REs diffused toward the scale surface together with aluminum during high-temperature oxidation for long periods, which resulted in the precipitation of RE-rich particles on the surface [9]. The addition of 0.05 at% Hf to a Fe–Cr–Al alloy was more effective for a reduction of the scale growth rate during oxidation of the alloy at 1427 K than a similar amount of Y-dopant [14]. Thus, Hf^{4+} is more effective than Y^{3+}, although the ionic radius of Hf^{4+} is midway between those of Al^{3+} and Y^{3+}. Therefore, there is little correlation between the ionic radius and suppressed scale growth [14]. The localized changes in the bonding strength between oxygen and aluminum or oxygen coordination of these segregated cations [15] may be related to these phenomena.

Both oxygen and aluminum not only interdiffuse along the GBs in growing scale, but their migration is simultaneously affected by various factors, such as dμ_O, the REs, impurities, and the diffusion length. Therefore, it is extremely difficult to quantitatively determine the degree of influence for individual factors that influence the movement of each diffusion species. The oxygen permeability technique with polycrystalline α-alumina wafer, which served as a model scale, is thus expected to be very useful to accurately evaluate mass-transfer through the wafers because the dμ_O applied to the wafers and the diffusion length are constant [16–24].

In this study, the mass-transfer mechanisms along the GBs in α-alumina are investigated using the oxygen permeation technique with ^{18}O$_2$ at high temperatures. This is followed by further improvement of the oxygen shielding capability and structural stability of alumina on the basis of the flux distribution analysis. Finally, the mass-transfer through the actual scales is discussed by comparing the diffusion data determined from oxygen permeation trials with literature values for the scales.

11.2   Experimental Procedures

11.2.1   Oxygen Permeability Measurements

Polycrystalline alumina wafer specimens with or without REs such as Ln (Lu, Y) and Hf, which were cut from the sintered bodies and polished to a mirror-like finish, served as a model scale for the measurement of oxygen permeability constants using a technique described in detail elsewhere [16–24]. Ln doping was expected to effectively retard mass-transfer in alumina under application of a $d\mu_O$ because Ln can significantly improve high-temperature GB creep resistance in polycrystalline alumina [25–27]. For the single RE-doped samples, a portion of the dopant was segregated at the GBs, and the remaining dopant was precipitated mainly at GBs as crystalline phases containing the dopant, which were identified as $Al_5Ln_3O_{12}$ and monoclinic-HfO_2 (m-HfO_2). Furthermore, mass-transfer along single GBs in two types of non-doped alumina bicrystal wafers was also evaluated by the oxygen permeation technique to clarify the correlation between the mass-transfer along each GB and the GB structural characteristics [18].

Figure 11.1 shows a schematic diagram of the oxygen permeability apparatus [23]. Each wafer specimen was placed between two alumina tubes under an Ar gas flow in a furnace, with Pt gaskets to create a seal between the wafer and the tubes.

Fig. 11.1   Schematic diagram of the oxygen permeability apparatus [23]

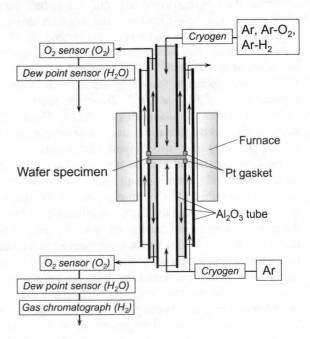

The P_{O2}, included as an impurity in the Ar gas, was monitored at the outlets of the upper and lower chambers that enclosed the wafer and the alumina tubes using a zirconia oxygen sensor at 973 K. The partial pressure of water vapor (P_{H2O}), another impurity in the Ar gas, was measured at room temperature using an optical dew point sensor. A gas-tight seal was achieved in both chambers by heating to 1893–1923 K, after which the wafer was kept at temperatures above 1773 K for 3 h in Ar at a flow rate of 1.67×10^{-6} m^3/s^1 for measurement of the oxygen permeability constants. Either Ar or Ar containing 1 vol% H$_2$ were subsequently introduced into both chambers at the same temperature.

Once the P_{O2} and P_{H2O} values were constant, an equilibrium state was reached, and these were taken as background levels. Other gases with different P_{O2}, such as pure O$_2$ and Ar gas containing either 0.01–10 vol% O$_2$ or 0.01–1 vol% H$_2$, were then introduced into one of the chambers, which caused the wafer to be subjected to a steep $d\mu_O$. The partial pressure of H$_2$ was measured at room temperature using gas chromatography. The oxygen permeation flux was considered to have reached a steady state when the monitored values of P_{O2}, P_{H2O}, and P_{H2} at the outlets became constant. The P_{O2} in each chamber at a high testing temperature, with the wafer subjected to a $d\mu_O$, was calculated thermodynamically from the P_{O2} measured at 973 K, or from the P_{H2O} and P_{H2} measured at room temperature. High purity polycrystalline alumina has excellent oxygen shielding properties; therefore, oxygen permeability measurements using a zirconia oxygen sensor must be conducted at high temperatures to accelerate the mass-transfer in the alumina wafers and aid in the detection of small amounts of oxygen molecules that permeate through the wafers. Oxygen permeation was detected for all polycrystalline wafers but not for a single-crystal wafer; therefore, permeation was considered to occur preferentially along the GBs with a strong dependence on the GB density S_{gb} (i.e., the grain size) of the wafers. Therefore, the oxygen permeability constant was calculated using:

$$\frac{PL}{S_{gb}} = \frac{C_p \cdot Q \cdot L}{V_{st} \cdot S \cdot S_{gb}}, \tag{11.6}$$

where P is the oxygen permeability, L is the wafer thickness, C_p is the concentration of permeated oxygen (P_{O2}/P_T, where P_T = total pressure), Q is the flow rate of the test gases, V_{st} is the standard molar volume of an ideal gas, and S is the permeation area of the wafer. S_{gb} values were determined by image analysis of the wafer surface microstructures after the oxygen permeation tests using scanning electron microscopy (SEM). The S_{gb} values of the bicrystal wafers were reduced by a factor of 10^5 compared with the polycrystalline alumina wafers; therefore, the amount of permeated oxygen could not be detected because it was below the lower detection limit of the oxygen sensor. The mass-transfer along each GB, especially aluminum diffusivity, was evaluated by measuring the surface profiles around the GB on both surfaces of the bicrystal wafer using atomic force microscopy (AFM) [18].

11.2.2 Determination of Oxygen GB Diffusion Coefficients for Each GB

The oxygen diffusion coefficients near the high-P_{O2} surface were determined using a SIMS-^{18}O line profiling technique at each GB [1, 24, 28, 29]. First, ^{18}O mapping of a wafer cross section was performed using SIMS with a beam diameter of 50 nm. The oxygen GB diffusion coefficient was then determined for individual GBs using Eq. (11.7) [30]:

$$D_{gb}\delta = 1.322\sqrt{\frac{D_L}{t}}\left(-\frac{\partial(ln(C_y - C_{bg}))}{\partial y^{6/5}}\right)^{-5/3}, \qquad (11.7)$$

where y is the penetration depth along each GB, t is the exposure time, D_L is the lattice diffusion coefficient for oxygen in sapphire, and C_y and C_{bg} are the respective fractions of ^{18}O at the penetration distance along each GB and the natural abundance (0.00204). D_L is also likely to depend on μ_O in the wafer, similar to the GB diffusion coefficient of Eq. (11.17). However, D_L was assumed to be constant at 5×10^{-20} m^2/s at 1873 K [6] because μ_O was almost constant in the immediate vicinity of the P_{O2}(hi) surface [21, 22]. The oxygen GB diffusion coefficients were determined from Eq. (11.7) within the range that corresponded to the normalized positions of the wafer, x/L. The β values (defined as $\delta(D_{gb}/D_L^{-1})/2(D_L t)^{1/2}$) for the oxygen GB diffusion coefficients must be sufficiently large ($\beta > 10$) to allow the use of Eq. (11.7). In the present work, all β values were larger than 100 and thus met the requirement.

11.3 Results and Discussion

11.3.1 Oxygen Permeation

Figure 11.2 shows the effect of the steady-state P_{O2} in the upper chamber on the oxygen permeability constants of non-doped and RE-doped samples [16, 17, 19, 23]. P_{O2} in the lower chamber was held constant at approximately 1 Pa. When a $d\mu_O$ is formed by the combination of P_{O2} less than 10^{-3} Pa and P_{O2} of ca. 1 Pa (low P_{O2} region), the oxygen permeability constants decreased with an increase in P_{O2} for all the samples. The oxygen permeability constants for the Hf-doped sample were comparable to those for the non-doped sample, whereas those for the Lu- and Y-doped samples were approximately one-third of those for the other samples. In the low P_{O2} region, all curve slopes corresponded to similar power constants of n = $-1/6$. For all the samples exposed to the low P_{O2} region, GB grooves were observed on both surfaces with a similar morphology to that formed by conventional thermal etching. The absence of GB ridges on the higher-P_{O2} (P_{O2}(hi)) surface suggests that aluminum migration played a small role in oxygen permeation. Therefore, the power

Fig. 11.2 Effect of equilibrium P_{O2} in the upper chamber on the oxygen permeability constants of polycrystalline alumina at 1923 K. The open symbols indicate the data for specimens exposed to a $d\mu_O$ that resulted from a P_{O2} of 1 Pa in the lower chamber and a P_{O2} in the upper chamber [23]

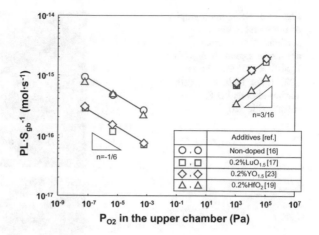

	Additives [ref.]
○ , O	Non-doped [16]
□ , □	0.2%LuO$_{1.5}$ [17]
◇ , ◇	0.2%YO$_{1.5}$ [23]
△ , △	0.2%HfO$_2$ [19]

P_{O2} in the upper chamber (Pa)

constant is applicable to the defect surface reaction given in Eq. (11.1) on the P_{O2}(hi) surface, and the reverse reaction proceeds on the opposite, lower-P_{O2} (P_{O2}(lo)) surface (P_{O2}(hi) $>>P_{O2}$(lo)).

In contrast, when a $d\mu_O$ was generated by a combination of P_{O2} above 10^3 Pa and a P_{O2} of ca. 1 Pa (high P_{O2} region), the oxygen permeability constants increased with P_{O2} for all the wafers. The oxygen permeability constants for the Hf-doped sample were about half of those of the non-doped, Lu-doped, and Yu-doped samples. All the slopes under high P_{O2} ($>10^3$ Pa) are comparable to each other and correspond to a power constant of $n = 3/16$, which suggests that the defect surface reaction given in Eq. (11.3) progresses on the P_{O2}(hi) surface side (formation of new alumina), while the reverse reaction occurs on the P_{O2}(lo) surface side (decomposition of alumina). In this case, GB ridges with heights of a few micrometers were observed on the P_{O2}(hi) surface, while deep crevices were formed at the GBs on the P_{O2}(lo) surface, as shown in Fig. 11.3. This result supports the participation of the defect surface reaction given by Eq. (11.3). In contrast, co-doping with both Lu and Hf increased the oxygen permeation for both P_{O2} regions and the corresponding power constants were maintained [19]. The formation of cubic-HfO$_2$ particles segregated at the GBs, which contain a large amount of oxygen vacancies due to a Lu solid solution, was considered to make it difficult to suppress oxygen permeation by co-doping.

Oxygen permeation is known to be controlled by the GB diffusion of oxygen and aluminum. According to the GB disconnection model, [1, 2, 31] oxygen vacancies are created by the reverse reaction of Eq. (11.1) at P_{O2}(lo) surface ledges and migrate by surface diffusion to the closest GBs, where they are annihilated at jogs on disconnections to form positively charged jogs. The oxygen GB disconnections, which carry some of the free space and all of the positive charge of the oxygen vacancies, migrate toward the P_{O2}(hi) surface. The charged jogs on the oxygen GB disconnections just below the P_{O2}(hi) surface then reform oxygen vacancies that migrate to surface ledges and are annihilated according to the reaction in Eq. (11.1).

Fig. 11.3 SEM micrographs
of the surfaces and cross
sections of non-doped
alumina exposed to $P_{O2}(hi)/$
$P_{O2}(lo) = 10^5$ Pa/1 Pa at
1923 K for 10 h: **a** $P_{O2}(hi)$
surface side and **b** $P_{O2}(lo)$
surface side [20]

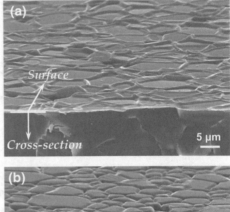

In contrast, aluminum vacancies are formed at the $P_{O2}(hi)$ surface ledges by the reaction given in Eq. (11.3) and migrate to nearby GBs via surface diffusion. Annihilation of the aluminum vacancies at jogs on GB disconnections causes the formation of negatively charged jogs. The aluminum GB disconnections migrate toward the $P_{O2}(lo)$ surface. The aluminum vacancies are then reconstituted just beneath the $P_{O2}(lo)$ surface and undergo surface diffusion to the closest surface ledges, where they are annihilated by the reverse reaction of Eq. (11.3). Thus, the migration of aluminum GB disconnections means that aluminum diffuses from the $P_{O2}(lo)$ to $P_{O2}(hi)$ sides, which results in the formation of ridges near the GBs on the $P_{O2}(hi)$ surface.

The oxygen permeability constants for each P_{O2} region can be expressed in terms of Eqs. (11.8) and (11.9) [20–23].

For the low P_{O2} region (oxygen GB diffusion),

$$\frac{A_O}{S_{gb}}\left(P_{O_2}(hi)^{-1/6} - P_{O_2}(lo)^{-1/6}\right) = \frac{4PL}{S_{gb}}, \tag{11.8}$$

and for the high P_{O2} region (aluminum GB diffusion),

$$\frac{A_{Al}}{S_{gb}}\left(P_{O_2}(hi)^{3/16} - P_{O_2}(lo)^{3/16}\right) = \frac{4PL}{S_{gb}}. \tag{11.9}$$

Table 11.1 Frequency factors and activation energies for GB diffusion in alumina [21, 22]

Samples	$A_i^* \cdot S_{gb}^{-1}$ (mol s^{-1} Pa^{-n})		Q_i (kJ mol^{-1})	
	Oxygen ($\times 10^{-4}$)	Aluminum	Oxygen	Aluminum
Non-doped	15.49	23.14	467	604
Ln*-doped	4.547	23.14	467	604
Hf-doped	15.49	11.34	467	604

*Y or Lu

At temperatures above 1773 K, A_O and A_{Al} are normalized according to S_{gb} and are given by the following Arrhenius equation for non-doped, Ln-doped, and Hf-doped alumina, for which the concentration of each dopant was 0.2 cation% [21–23]:

$$\frac{|A_i|}{S_{gb}} = \frac{A_i^*}{S_{gb}} exp\left(\frac{-Q_i}{RT}\right),$$ (11.10)

where $A_i^* \cdot S_{gb}^{-1}$ and Q_i are the frequency factor and activation energy for oxygen and aluminum GB diffusion, respectively. Table 11.1 provides a summary of A_i^* S_{gb}^{-1} and Q_i [21–23].

Alumina scale formed on alloys is exposed to an extremely large $d\mu_O$, and scale growth proceeds by the interdiffusion of both oxygen and aluminum along the GBs. Accordingly, oxygen permeability constants were also measured at high temperatures under a $d\mu_O$ at which mutual GB diffusion proceeded in the samples. Figure 11.4 shows the oxygen permeability constants for the non-doped alumina as a function of P_{O2}(hi)/P_{O2}(lo) at 1923 K, in which P_{O2}(lo) was constant at 8×10^{-8} Pa. Lines a and b indicate the oxygen permeability constants related to the diffusion of aluminum and oxygen, respectively. Each line was calculated from Eqs. (11.8)–(11.10) with the values listed in Table 11.1. Line c is a sum of lines a and b, which is given by Eq. (11.11):

$$\frac{A_O}{S_{gb}}\left(P_{O_2}(hi)^{-1/6} - P_{O_2}(lo)^{-1/6}\right) + \frac{A_{Al}}{S_{gb}}\left(P_{O_2}(hi)^{3/16} - P_{O_2}(lo)^{3/16}\right) = \frac{4PL}{S_{gb}}.$$ (11.11)

The measured oxygen permeability constants were coincident with line c. Therefore, the experimental constants in Table 11.1 determined for either oxygen or Al diffusion are applicable to that with a large $d\mu_O$, where both oxygen and Al interdiffuse without any synergistic effect, which satisfies the Gibbs–Duhem equation. The contribution of aluminum GB diffusion to the oxygen permeation through non-doped alumina increases with the P_{O2}(hi)/P_{O2}(lo) ratio.

Figure 11.5 shows an SEM micrograph of the P_{O2}(hi) surface and cross section of non-doped alumina exposed to P_{O2}(hi)/P_{O2}(lo) = 10^5 Pa/8×10^{-8} Pa at 1923 K for 10 h, which corresponds to the condition shown by the arrow in Fig. 11.4 [20].

Fig. 11.4 Oxygen
permeability constants for
non-doped alumina as a
function of $P_{O2}(hi)/P_{O2}(lo)$ at
1923 K, where $P_{O2}(lo)$ is a
constant of 8×10^{-8} Pa at
1923 K. Lines a, b, and c
indicate the contributions
related to the GB diffusion of
oxygen, aluminum, and their
summation, respectively.
Each line was calculated from
Eqs. (11.8), (11.9), and
(11.11). The arrow
corresponds to the exposure
condition from Fig. 11.5 [20]

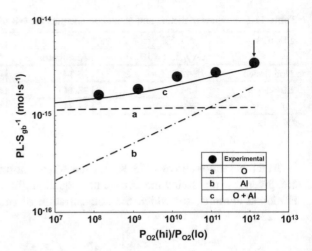

Fig. 11.5 SEM micrograph
of the $P_{O2}(hi)$ surface and
cross section of the
non-doped alumina exposed
to $P_{O2}(hi)/P_{O2}(lo) = 10^5$ Pa/
8×10^{-8} Pa at 1923 K for
10 h [20]

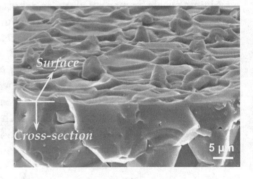

The $P_{O2}(hi)$ surface shown in Fig. 11.5 was exposed to the same $P_{O2}(hi)$ in
Fig. 11.3a; the amount of oxygen permeation related to the diffusion of aluminum is
predicted to be close to that in Fig. 11.3a, according to Eq. (11.11). This suggests
that the corresponding morphology of the $P_{O2}(hi)$ surfaces would be similar to each
other. However, the formation of GB ridges on the $P_{O2}(hi)$ surface is significantly
accelerated by the increase of the $d\mu_O$, especially at multi-junctions of the surface.
The large $d\mu_O$ may locally accelerate aluminum diffusivity near the GBs on the
$P_{O2}(hi)$ surface.

Figure 11.6 shows a SIMS-^{18}O map of a cross section in the vicinity of the
$P_{O2}(hi)$ surface of an alumina wafer exposed to $P_{18O2}(hi)/P_{16O2}(lo) = 10^4$ Pa/10^{-8}
Pa at 1873 K for 1 h. The triangular marks indicate the position of the $P_{O2}(hi)$

Fig. 11.6 SIMS-^{18}O map of the cross section in the vicinity of P_{O2}(hi) surface of alumina wafer exposed to P_{16O2}(hi)/P_{16O2}(lo) = 10^4 Pa/10^{-8} Pa at 1873 K for 9 h, and subsequent replacement of the $^{16}O_2$ at the P_{O2}(hi) side to the same partial pressure of the $^{18}O_2$ side for 1 h. The GBs used to determine the GB diffusion coefficients for oxide ions are surrounded by ellipses in the map. The arrowheads indicate the position of the P_{O2}(hi) surface [24]

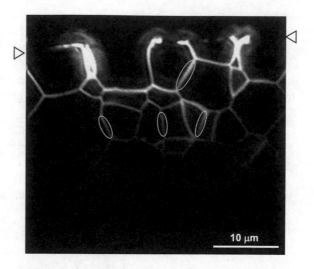

surface. ^{18}O was concentrated along the GBs from the P_{O2}(hi) surface to a depth of approximately 20 μm. A strongly concentrated region with a width of approximately 1 μm extended to a depth of approximately 5 μm in the vicinity of the P_{O2}(hi) surface. During oxygen permeation, ambient O_2 molecules were considered to dissociatively adsorb over the entire P_{O2}(hi) surface, and then immediately diffuse to the surface GBs. As a result, some reacted at the P_{O2}(hi) surface GBs with aluminum diffusing along the GBs from the P_{O2}(lo) side to the P_{O2}(hi) side to form GB ridges of new alumina, and the remaining oxygen diffused inward along the GBs [24]. The oxygen GB diffusion coefficients were measured from the ^{18}O line profiles along the GBs surrounded by ellipses in Fig. 11.6. The average value of the oxygen GB diffusion coefficient was determined to be 9.1 × 10^{-23} m/s.

11.3.2 GB Diffusion Under Oxygen Potential Gradients

The charged particle fluxes of oxygen and aluminum for oxygen permeation through the wafer, and from the spatial coordinate x = 0 to x = L, which correspond to the P_{O2}(lo) and P_{O2}(hi) surfaces, can be expressed in terms of the oxygen permeability constants [20–23]:

$$\int_0^L \frac{J_{TO}}{S_{gb}}dx = \int_0^L \frac{(J_O + J_{Al})}{S_{gb}}dx = \frac{4PL}{S_{gb}}, \tag{11.12}$$

where J_{TO} is the total flux of oxygen permeation through the wafer. J_O and J_{Al} correspond to the fluxes of oxygen and aluminum, respectively. The oxygen permeability constant at an arbitrary position x, along the depth direction of the wafer Px, is given by Eq. (11.13):

$$\int_0^x \frac{J_{TO}}{S_{gb}} dx = \int_0^x \frac{(J_O + J_{Al})}{S_{gb}} dx = \frac{4Px}{S_{gb}}, \tag{11.13}$$

where $P_{O2}(x)$ is the O_2 partial pressure in equilibrium with the chemical potential of oxygen at x. Combining Eqs. (11.12) and (11.13) gives Eq. (11.14):

$$\frac{x}{L} = \frac{\frac{A_{Al}}{S_{gb}}\left(P_{O_2}(x)^{3/16} - P_{O_2}(lo)^{3/16}\right) + \frac{A_O}{S_{gb}}\left(P_{O_2}(x)^{-1/6} - P_{O_2}(lo)^{-1/6}\right)}{\frac{A_{Al}}{S_{gb}}\left(P_{O_2}(hi)^{3/16} - P_{O_2}(lo)^{3/16}\right) + \frac{A_O}{S_{gb}}\left(P_{O_2}(hi)^{-1/6} - P_{O_2}(lo)^{-1/6}\right)} \tag{11.14}$$

The chemical potentials of oxygen (μ_O) and aluminum (μ_{Al}) are given by:

$$\mu_O = \frac{\mu_{O_2}^\circ + RT \ln P_{O_2}}{2}, \tag{11.15}$$

$$\mu_{Al} = \frac{2\mu_{Al_2O_3}^\circ - 3\left(\mu_{O_2}^\circ + RT \ln P_{O_2}\right)}{4}, \tag{11.16}$$

where $\mu_{O_2}^\circ$ and $\mu_{Al_2O_3}^\circ$ are the standard chemical potential energies per mole of molecular O_2 and pure alumina, respectively, R is the gas constant, and T is the absolute temperature. Thus, μ_O and μ_{Al} at x can be determined using Eqs. (11.15) and (11.16) with the $P_{O2}(x)$ values calculated from Eq. (11.14). The GB diffusion coefficients of oxygen and aluminum at x can be calculated using Eqs. (11.17) and (11.18) with the corresponding $P_{O2}(x)$.

$$D_O \delta = \frac{1}{6C_O \cdot t_{e'}} \frac{|A_O|}{S_{gb}} P_{O_2}^{-1/6}, \tag{11.17}$$

$$D_{Al} \delta = \frac{1}{12C_{Al} \cdot t_{e'}} \frac{A_{Al}}{S_{gb}} P_{O_2}^{3/16}, \tag{11.18}$$

where δ is the GB width. C_O and C_{Al}, the molar concentrations of the species per unit volume of alumina, are 1.168×10^5 and 7.787×10^4 mol/m^3, respectively. The experimental parameters $|A_O|$ and A_{Al} are related to the mobility of oxygen and aluminum, respectively. $t_{e'}$ is the electronic transference number, which was comparatively close to unity, as determined using Eq. (11.17) with the average value of the oxygen GB diffusion coefficients measured by the SIMS-^{18}O line profiling technique. That for alumina scale formed by the oxidation of β-NiAl alloy under high P_{O2} at 1373 K was reported to be approximately 0.9 [32]. Hence, in this study, the alumina subjected to dμ_O is assumed to be an electronic conductor, i.e., $t_{e'} = 1$.

Fig. 11.7 Distributions of
P_{O2}, chemical potentials, and
GB diffusion coefficients for
oxygen and aluminum in a
non-doped alumina wafer
exposed to P_{O2}(hi)/
P_{O2}(lo) = 10^5 Pa/10^{-8} Pa at
1873 K [23]

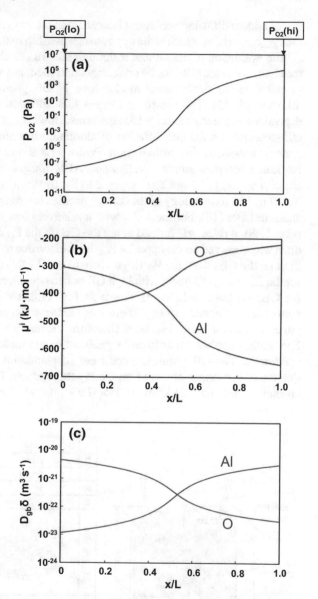

Figure 11.7 shows distributions of P_{O2}, chemical potentials, and GB diffusion coefficients for oxygen and aluminum in a non-doped alumina wafer exposed to P_{O2}(hi)/P_{O2}(lo) = 10^5 Pa/10^{-8} Pa at 1873 K. The P_{O2} plot is a sigmoid curve. μ_O increases with x/L in an inverse relationship to μ_{Al}, in accordance with the Gibbs–Duhem equation. The oxygen GB diffusion coefficient decreases with an increase in x/L, while the aluminum GB diffusion coefficient increases. As a result,

the aluminum diffusion coefficient is larger than the oxygen diffusion coefficient near the P_{O2}(hi) surface, which is the opposite relationship to that near the P_{O2}(lo) surface.

The evaluation of mass-transfer through the GBs in alumina bicrystals, in which the character of the GBs can be arbitrarily controlled, is a very powerful method used to elucidate the fundamental mechanisms of GB phenomena such as creep and diffusion [5, 15]. The measured oxygen GB diffusion coefficients were strongly dependent on the atomic-scale GB structures. However, the effect of the atomic-scale GB structures on the GB diffusion of aluminum has not yet been clarified for the reasons discussed in the Introduction. Figure 11.8 shows a schematic diagram of the fabricated bicrystal alumina wafers and AFM images of the surfaces of bicrystal alumina wafers ($\Sigma13$ and $\Sigma31$) exposed to P_{O2}(hi)/P_{O2}(lo) = 10^5 Pa/1 Pa at 1923 K for 10 h. The morphology of the surface profiles is strongly dependent upon the GB characteristics [18]. For the wafer with a relatively low GB coherence such as $\Sigma31$ (Fig. 11.8), a ridge was formed along the GB on the P_{O2}(hi) surface and a deep GB ditch was observed on the opposite P_{O2}(lo) surface due to the migration of aluminum through the GBs from the P_{O2}(lo) surface to the P_{O2}(hi) surface. On the other hand, for the $\Sigma13$ bicrystal wafer with high GB coherence, there is a shallow groove along the GBs on both surfaces, as shown in Fig. 11.8, similar to grooves formed by conventional thermal etching. There was neither a GB ridge on the P_{O2}(hi) surface nor a ditch on the P_{O2}(lo) surface. Therefore, the migration of aluminum through the $\Sigma13$ wafer does not occur to any significant extent under the present experimental conditions. The GB diffusion coefficient of aluminum was determined from the volume of GB ridges observed on the P_{O2}(hi) surface. The aluminum GB diffusion coefficient for the $\Sigma13$ GB (1.1×10^{-20} m^3/s) was similar to that for a

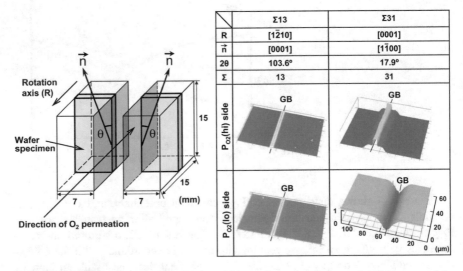

Fig. 11.8 Schematic diagram of the fabricated bicrystal alumina wafers and AFM images of the surfaces of bicrystal alumina wafers ($\Sigma13$ and $\Sigma31$) exposed to P_{O2}(hi)/P_{O2}(lo) = 10^5 Pa/1 Pa at 1923 K for 10 h

polycrystalline wafer (8.5×10^{-21} m^3/s). They had a tendency to be proportional to the GB energies and the mean bond lengths between oxygen and aluminum around the GB [18]. Mass-transfer during oxygen permeation is considered to progress preferentially along GBs with relatively low GB coherence.

Ogawa et al. investigated the switching behavior (P_{O2}-dependence) of the dominant diffusion species by quantum mechanical density functional theory (DFT) calculation of the formation energies for charged oxygen and aluminum vacancies [33]. The electronic structure of the $\Sigma 31$ bicrystal revealed significant narrowing of the band gap to approximately 60% of that for a single crystal ($E_g^B = 9.1$ eV). Figure 11.9 shows the effect of P_{O2} on the Fermi levels and formation energies of oxygen and aluminum vacancies at 1923 K for relative band gaps of 1.0 and 0.6 eV compared to that for the single crystal. Although the defect formation energies and the Fermi levels at the GB are not directly calculated, they exhibit different behavior for wide band gap and narrow band gap structures. For a wide band gap, the aluminum vacancies and holes are dominant, regardless of P_{O2}. However, a switchover in the formation energies of the two types of vacancies appears for a significantly narrow band gap. This suggests that GBs with low coherence in polycrystalline alumina, i.e., narrow band gap structures, is the origin of oxygen diffusion. In this case, the Fermi level at the P_{O2}(lo) side is only slightly higher than that at the P_{O2}(hi) side (+0.17 eV). This may support the assumption of the constant of $t_{e'}$ in alumina subjected to a dμ_O.

Fig. 11.9 Effect of P_{O2} on **a** Fermi levels and **b** formation energies of oxygen and aluminum vacancies at 1923 K for relative band gaps of 1.0 and 0.6 eV compared to that for a single crystal ($E_g^B = 9.1$ eV) [23]

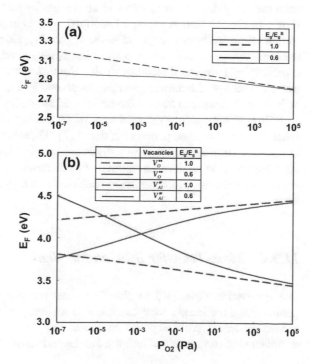

11.3.3 Design of Oxygen Shielding Capability and Structural Stability

The fluxes of oxygen and aluminum normalized according to L/S_{gb} at position x/L are given by:

$$\frac{J_O L}{S_{gb}} = 2\left(\frac{C_O \cdot t_{e'} \cdot D_O \delta}{RT}\right)\frac{\partial \mu_O}{\partial(x/L)}, \tag{11.19}$$

$$\frac{J_{Al} L}{S_{gb}} = -3\left(\frac{C_{Al} \cdot t_{e'} \cdot D_{Al}\delta}{RT}\right)\frac{\partial \mu_{Al}}{\partial(x/L)}. \tag{11.20}$$

Thus, each flux can be determined from Eqs. (11.19) and (11.20) with the calculated GB diffusion coefficients and differentials of the chemical potentials at x/L. In this study, $t_{e'}$ is assumed to be unity. Figure 11.10a shows that for non-doped alumina, the oxygen and aluminum fluxes at the outflow side are significantly larger than those at the inflow side. In this case, oxygen permeation from the diffusion of oxygen is comparable to that of aluminum. The dotted line in Fig. 11.10a represents the summation of both the fluxes and corresponds to the oxygen permeation in the steady state

As listed in Table 11.1, Lu-doping decreases only the frequency factor of oxygen to one-third of that for a non-doped alumina layer, while Hf-doping decreases only the frequency factor of aluminum by half. For the bilayer sample, as shown in Fig. 11.10b, in which a Ln-doped layer is exposed to the lower P_{O2} side and an Hf-doped layer is exposed to the higher P_{O2} side, and where each layer has the same thickness, the sum of both fluxes is decreased, i.e., the oxygen shielding capability and structural stability of the alumina bilayer are increased. However, when the bilayer structure is reversed, as shown in Fig. 11.10c, the summation of both fluxes is similar to that for the non-doped single layer. The integrated values of each flux with respect to the thickness of all the layers were consistent with four times the actual oxygen permeation data [22]. Therefore, these results suggest that to improve oxygen shielding and structural stability by the alumina bilayer, it is very important to achieve an optimal dopant arrangement that takes into consideration the behavior of the diffusion species and the role of the dopants within the layers.

11.3.4 Mass-Transfer in Alumina Scale

The approaches developed to elucidate mass-transfer in alumina during oxygen permeation experiments were extended to an analysis of the interdiffusion mechanisms in actual scale exposed to lower temperatures [23]. The GB diffusion coefficients of oxygen and aluminum are dependent on P_{O2}; therefore, a comparison

Fig. 11.10 Distributions of oxygen and aluminum fluxes in specimens exposed to $P_{O2}(hi)/P_{O2}(lo) = 10^5$ Pa/ 10^{-8} Pa at 1873 K: **a** non-doped sample, **b**, **c** double layered samples consisting of Ln-doped and Hf-doped layers. The dashed lines indicate the summation of both the oxygen and aluminum fluxes [23]

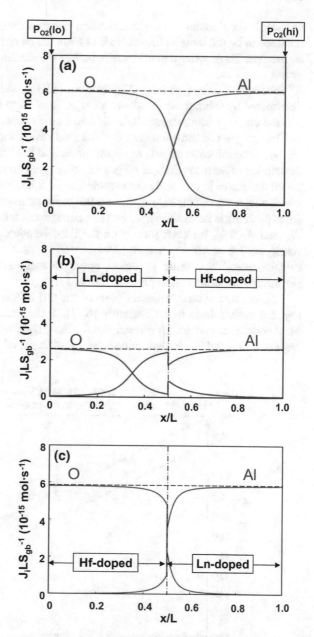

with the oxygen permeation data and those values obtained from ^{18}O depth profiling in the scale after two-stage oxidation experiments [6, 7] is required to estimate the P_{O2} value, in equilibrium with μ_O in the depth profiling zone. The activation energy for the oxygen GB diffusion coefficients from the oxygen permeation trials is close to that in scale. It is thus postulated that Eqs. (11.10) and (11.17) are applicable for alumina scale. The activation energies for oxygen in the scale are also assumed to be the same as those obtained from the oxygen permeation experiments, as listed in Table 11.1, regardless of whether the alumina scale was doped with Y or not.

Consequently, P_{O2} and A_O*/S_{gb} in Eqs. (11.10) and (11.17) for scale can be determined by solving the simultaneous equations using the profiling position (x/L) and the corresponding oxygen GB diffusion coefficients.

The oxygen GB diffusion data for Y-doped scale formed on ODS-MA956 alloy [6] was determined at an x/L of approximately 0.88–0.96; however, there was no description of the measurement ranges for other types of scales [7]. Thus, these ranges for all the scales in this study are assumed to be equal to that for Y-doped scale [6], which adopts the middle value (x/L = 0.92) of the measurement range because such a depth profiling is generally performed in a zone just near the scale surface. As a result, P_{O2} and A_O*/S_{gb} for scale formed on the RE-free alloy at 1373 K, i.e., non-doped scale, are 1.6×10^{-17} Pa and 13.94×10^{-4} mol s^{-1} Pa$^{-1/6}$, respectively. The calculated A_O*/S_{gb} value is almost equal to that determined from the oxygen permeation experiments, as given in Table 11.1.

Figure 11.11 shows Arrhenius plots of the GB diffusion coefficients for oxygen, together with data from the literature [6, 7]. Table 11.2 summarizes the measurement conditions and activation energies for the GB diffusion data in Fig. 11.11. The dashed line a, which is determined by substitution of $P_{O2} = 1.6 \times 10^{-17}$ Pa in

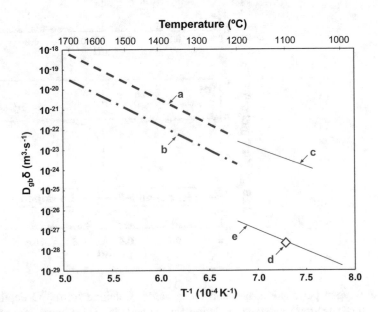

Fig. 11.11 Arrhenius plots of the GB diffusion coefficients for oxygen, together with data from the literature [6, 7]

Table 11.2 Summary of measurement conditions and activation energies for the oxygen GB diffusion data in Fig. 11.11

	Lines	Samples	Methods	P_{O2} (Pa)	Q_i (kJ mol^{-1})
This work	a	Non-doped	Oxygen permeation	2×10^{-17}	467
	b	Ln*-doped		7×10^{-13}	
Chevalier et al. [7]	c	"Non-doped" scale formed on Fe–20Cr–5Al	Isotopic tracer (SIMS)	–	323
	d	"Y-doped" scale formed on Fe–20Cr–5Al–0.1Y		–	–
Messaoudi et al. [6]	e	"Y-doped" scale formed on ODS-MA956	Isotopic tracer (SIMS)	–	391

*Y or Lu

Eq. (11.17), when extrapolated to lower temperature is consistent with that reported for scale (line c). Thus, the oxygen GB diffusion mechanism for non-doped alumina is considered to be independent of temperature. The oxygen GB diffusion coefficients for Y-doped scale (point d and line e) is approximately $1/10^4$ of that for the non-doped scale (line c) shown in Fig. 11.11.

P_{O2} and A_O*/S_{gb} for the Y-doped scale were also calculated at 1373 K using a similar method to that for the non-doped scale, and were determined as 6.7×10^{-13} Pa and 2.175×10^{-6} mol s^{-1} Pa$^{-1/6}$, respectively. Therefore, the significant retardation of oxygen GB diffusivity due to Y-doping is probably related to a decrease of A_O*/S_{gb} and an increase of μ_O in the vicinity of the scale surface, which results in a decrease of the driving forces for both oxygen and aluminum diffusion according to the Gibbs–Duhem relationship. Line b in Fig. 11.11 at $P_{O2} = 6.7 \times 10^{-13}$ Pa, when extrapolated to a lower temperature, is significantly deviated from the data for the Y-doped scale (d and e in Fig. 11.11), despite the almost identical activation energies. The magnitude of the reduction in oxygen diffusivity due to the presence of Y suggests a discontinuous decrease with an increase in temperature. A similar phenomenon was reported for the evolution of a bimodal Y-doped alumina structure by characterization of the grain growth of both normal and unimpinged abnormal grains as a function of time [34]. The discontinuous change of the GB mobility at approximately 1773 K is considered to be caused by transition of the GB structures, i.e., so-called complexion to produce an equilibrium interfacial state. However, the corresponding activation energies were constant during the complexion transition, so that there may be other possible causes for this phenomenon. This requires further examination of the discontinuity with respect to the temperature dependence of the GB diffusivity.

11.4 Conclusions

The oxygen permeability of polycrystalline alumina wafers, with and without RE dopants such as Ln (Y, Lu) and Hf, served as model alumina scale for evaluation under a $d\mu_O$ at temperatures up to 1923 K. Oxygen permeation occurred by the GB diffusion of oxygen from the P_{O2}(hi) surface side to the P_{O2}(lo) surface side, while simultaneous GB diffusion of aluminum proceeded in the opposite direction. A bilayer wafer with a Ln-doped layer on the P_{O2}(lo) side and a Hf-doped layer on the P_{O2}(hi) side decreased the oxygen permeability. When the sign of $d\mu_O$ was reversed, the wafer did not exhibit a decrease in oxygen permeability and instead exhibited behavior similar to that of a non-doped wafer. Furthermore, the approaches developed to elucidate the mass-transfer in alumina during oxygen permeation experiments were extended to analysis of the interdiffusion mechanisms in actual scale exposed to lower temperatures. Y segregated at the GBs in the scale was considered to decrease the oxygen frequency factor and the driving forces for both oxygen and aluminum diffusion in the vicinity of the P_{O2}(hi) surface.

Acknowledgements This work was partially supported by a Grant-in-Aid for Scientific Research on Priority Area "Nano Materials Science for Atomic Scale Modification 474" and Innovative Areas "Nano Informatics" (No. JP25106008) from the Japan Society for the Promotion of Science (JSPS) and by the Advanced Low Carbon Technology Research and Development Program of the Japan Science and Technology Agency (JST).

References

1. A.H. Heuer, T. Nakagawa, M.Z. Azar, D.B. Hovis, J.L. Smialek, B. Gleeson, N.D.M. Hine, H. Guhl, H.-S. Lee, P. Tangney, W.M.C. Foulkes, M.W. Finnis, Acta Mater. **61**, 6670 (2013)
2. A.H. Heuer, M.Z. Azar, Scripta Mater. **102**, 15 (2015)
3. J.A. Nychka, D.R. Clarke, Oxid. Metals **63**, 325 (2005)
4. D. Plot, M. Le Gall, B. Lesage, A.M. Huntz, C. Monty, Philos. Mag. A **73**, 935 (1996)
5. T. Nakagawa, I. Sakaguchi, N. Shibata, K. Matsunaga, T. Mizoguchi, T. Yamamoto, H. Haneda, Y. Ikuhra, Acta Mater. **55**, 6627 (2007)
6. K. Messaoudi, A.M. Huntz, B. Lesage, Mater. Sci. Eng. A **247**, 248 (1998)
7. S. Chevalier, B. Lesage, C. Legnos, G. Borchardt, G. Stnehl, M. Kilo, Defect Diffusion Forum **237–40**, 899 (2005)
8. A.H. Heuer, J. Eur. Ceram. Soc. **28**, 1495 (2008)
9. B.A. Pint, A.J. Garratt-Reed, L.W. Hobbs, J. Am. Ceram. Soc. **81**, 305 (1998)
10. K.P.R. Reddy, J.L. Smialek, A.R. Cooper, Oxid. Met. **17**, 429 (1982)
11. W.J. Quadakkers, A. Elschner, W. Speier, H. Nickel, Applied. Surf. Sci. **52**, 271 (1991)
12. B.A. Pint, J.R. Martin, L.W. Hobbs, Oxid. Met. **39**, 167 (1993)
13. D. Naumenko, B. Gleeson, E. Wessel, L. Singheiser, W.J. Quadakkers, Metal. Mater. Trans. **38A**, 2974 (2007)
14. B.A. Pint, J. Am. Ceram. Soc. **86**, 686 (2003)
15. J.P. Buban, K. Matsunaga, J. Chen, N. Shibata, W.Y. Ching, T. Yamamoto, Y. Ikuhara, Science **311**, 212 (2006)
16. S. Kitaoka, T. Matsudaira, M. Wada, Mater. Trans. **50**, 1023 (2009)
17. T. Matsudaira, M. Wada, T. Saitoh, S. Kitaoka, Acta Mater. **58**, 1544 (2010)

18. T. Matsudaira, S. Kitaoka, N. Shibata, T. Nakagawa, Y. Ikuhara, J. Mater. Sci. **46**, 4407 (2011)
19. T. Matsudaira, M. Wada, T. Saitoh, S. Kitaoka, Acta Mater. **59**, 5440 (2011)
20. M. Wada, T. Matsudaira, S. Kitaoka, J. Ceram. Soc. Jpn. **119**, 832 (2011)
21. T. Matsudaira, M. Wada, S. Kitaoka, J. Am. Ceram. Soc. **96**, 3243 (2013)
22. S. Kitaoka, T. Matsudaira, M. Wada, T. Saito, M. Tanaka, Y. Kagawa, J. Am. Ceram. Soc. **97**, 2314 (2014)
23. S. Kitaoka, J. Ceram. Soc. Jpn. **124**, 1100 (2016)
24. S. Kitaoka, T. Matsudaira, T. Nakagawa, N. Shibata, Y. Ikuhara, Mater. Sci. Forum **879**, 966 (2017)
25. K. Matsunaga, T. Tanaka, T. Yamamoto, Y. Ikuhara, Phys. Rev. B **68**, 085110 (2003)
26. Y. Ikuhara, H. Yoshida, T. Sakuma, Mater. Sci. Eng. **A319–321**, 24 (2001)
27. H. Yoshida, Y. Ikuhara, T. Sakuma, Acta Mater. **50**, 2955 (2002)
28. S. Kitaoka, T. Matsudaira, D. Yokoe, T. Kato, M. Takata, J. Am. Ceram. Soc. **100**, 3217 (2017)
29. M. Wada, T. Matsudaira, N. Kawashima, S. Kitaoka, M. Takata, Acta Mater. **135**, 372–381 (2017)
30. A.D. Le Claire, Brit. J. Appl. Phys. **62**, 351 (1963)
31. A.H. Heuer, M.Z. Azar, H. Guhl, W.M.C. Foulkes, B. Gleeson, T. Nakagawa, Y. Ikuhara, M.W. Finnis, J. Am. Ceram. Soc. **99**, 1 (2016)
32. J. Balmain, A.M. Huntz, Oxid. Metals **45**, 183 (1996)
33. T. Ogawa, A. Kuwabara, C.A.J. Fisher, H. Moriwake, K. Matsunaga, K. Tsuruta, S. Kitaoka, Acta Mater. **69**, 365 (2014)
34. P.R. Cantwell, M. Tang, S.J. Dillon, J. Luo, G.S. Rohrer, M.P. Harmer, Acta Mater. **62**, 1 (2014)

Chapter 12
Structural Relaxation of Oxide Compounds from the High-Pressure Phase

Hitoshi Yusa

Abstract In this chapter, several types of structural relaxation of oxide compounds from the high-pressure phase are systematically introduced in terms of high-pressure comparative crystallography. Structural relaxation of various ABO_3 compounds from the perovskite phase to the lithium niobate phase is explained in detail from rotation of the BO_6 octahedral frameworks. Depressurized amorphization of $ASiO_3$ perovskites containing large divalent cations (A = Ba^{2+}, Sr^{2+}, and Ca^{2+}) is elucidated by the characteristics of the hexagonal and cubic perovskite structures. The unquenchable $Rh_2O_3(II)$ phases of group-13 sesquioxides, such as Ga_2O_3 and In_2O_3, are confirmed by both experimental and computational studies. Ab initio calculations of Y_2O_3 show that the unquenchable pressure-induced phase (A-type structure) is not the stable phase under high pressure. Knowledge about the unquenchable and/or metastable phases in recovered high-pressure products is beneficial for advanced computational materials design.

Keywords High-pressure experiments · Structural relaxation · Quenchability Amorphization · Ab initio calculation

12.1 General

Under high pressure, typical ABO_3 oxide compounds undergo a phase transition with the coordination of the B atoms changing from tetrahedral to octahedral. For the most popular $MgSiO_3$ compounds, which are believed to be one of the most abundant constituent minerals in the Earth's mantle, the crystal structure changes from pyroxene to spinel (ringwoodite) plus stishovite, ilmenite (akimotoite), garnet (majorite), perovskite (bridgmanite), and postperovskite [1, 2]. All of the structures are quenchable, except for the postperovskite structure ($CaIrO_3$ structure) which appears under ultrahigh pressure above 140 GPa [2]. Therefore, the physical

H. Yusa (✉)
National Institute for Materials Science, Namiki 1-1, Tsukuba, Ibaraki, Japan
e-mail: YUSA.Hitoshi@nims.go.jp

© The Author(s) 2018
I. Tanaka (ed.), *Nanoinformatics*, https://doi.org/10.1007/978-981-10-7617-6_12

259

properties of most of the recovered structures can be investigated under ambient pressure. In this case, even the equilibrium phase boundary can be thermodynamically determined by measuring the enthalpy and heat capacity at ambient pressure [1, 3]. However, the high-pressure phase is not always quenchable. Because high-pressure phases tend to undergo structural relaxation during decompression, the high-pressure structures cannot be characterized from the recovered products. The structure can be elucidated by in situ X-ray observation under pressure. In particular, a synchrotron radiation X-ray source combined with a diamond anvil cell (DAC) can shed light on the real structure of the unquenchable phase under pressure.

Some high-pressure perovskites in ABO_3 compounds exhibit unquenchable phenomena during decompression to atmospheric pressure. There are two types of structure instability: conversion to perovskite-related structures and amorphization. Structural relaxation in the former case accompanies a symmetry change to a non-centrosymmetric structure, retaining the ferroelectricity. The representative example is structural relaxation from the orthorhombic perovskite structure to the lithium niobate structure. Many compounds with the lithium niobate structure have been found by high-pressure synthesis.

In other simple oxides, there are peculiar high-pressure phases in sesquioxides that revert to a lower pressure phase under room temperature. In some cases, there are definite crystallographic relationships between their lower pressure phases.

Ab initio computational studies are indispensable to confirm whether the phase appearing by structural relaxation is metastable. Recent computational studies have predicted novel materials with high-performance functionalities. In particular, a data-driven material design approach has identified many candidates for high-pressure synthetic materials. However, the predicted materials are not always realized in the recovered products because of structural relaxation during decompression. To enhance the capability of material design by computational approaches, systematic information about structure relaxation would be highly beneficial.

In this chapter, we focus on relaxation structure and quenchability from the high-pressure phase. By classifying the relaxation process, we discuss the recovery compounds from high-pressure synthesis.

12.2 Phase Transition from the Perovskite Structure to the Lithium Niobate Structure

12.2.1 Crystal Structure Relationship Among Lithium Niobate, Perovskite, and Ilmenite Phases

The typical lithium niobate phase of Li-bearing compounds, which is represented by $LiNbO_3$ and $LiTaO_3$, is only found in similar lithium-bearing compounds, such as $LiUO_3$ [4] and $LiReO_3$ [5], and all of these lithium niobate phases show stability

Fig. 12.1 Lithium niobate
structure

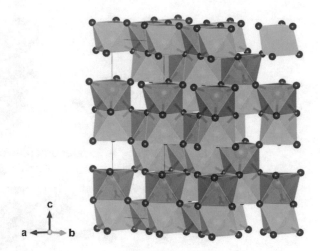

under ambient conditions. In contrast, high-pressure synthesis makes it possible to
crystallize lithium niobate phases of various Li-free compounds, such as
$A^{2+}B^{4+}O_3$-type [6–13] and $A^{4+}B^{2+}O_3$-type [14] oxides. One of the lithium niobate
structures is shown in Fig. 12.1. It is widely known that lithium niobate phases
appear with retrogressive transition from high-pressure perovskite phases. Such a
hidden perovskite phase is difficult to confirm with only the recovered
high-pressure products, but it has been directly elucidated by in situ experiments
under high pressure [6–9, 12–14].

It should be noted that these lithium niobate phases convert from the perovskite
structure with structural relaxation during decompression, which is closely related
to the rotation of BO_6 octahedra. This is a first-order transformation accompanied
by a 2–3% volume change. The typical structural relationship among the ilmenite,
perovskite, and lithium niobate phases is shown in Fig. 12.2. As shown in
Fig. 12.2, where a specific crystallographic orientation is chosen, the transformation
from lithium niobate to perovskite appears to be much easier than that from the
ilmenite structure to the perovskite structure. In other words, there must be large
displacement of the BO_6 octahedra to trigger the ilmenite–perovskite transition,
where atomic rearrangement should be controlled by diffusion under high temperature. In fact, for many ABO_3 compounds, the perovskite to ilmenite transition is
not observed at room temperature throughout the pressure range even though the
density of ilmenite is smaller than that of lithium niobate.

12.2.2   Perovskite Tolerance Factor

It is believed that such instability is closely correlated with the ionic radii of the A-
and B-site cations forming the perovskite structure. The Goldschmidt tolerance

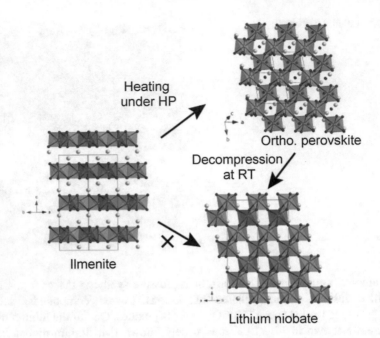

Fig. 12.2 Structural relationship among ilmenite, lithium niobate, and orthorhombic perovskite

factor [15] indicates the distortion from ideal cubic perovskite and it is also applicable to such instabilities during decompression: $t = (r_A + r_o)/\sqrt{2}(r_B + r_o)$, where r is the effective ionic radius of each element [16]. The tolerance factor is determined from the geometrical relationship of the ionic radii, as shown in Fig. 12.3. The right-hand side figures show the polyhedral types of the A-site cations. Ideal cubic perovskite ($t = 1$) is composed of cubo-octahedral coordinated A cations. Orthorhombic distortion ($t < 1$) incorporates A-site cations, forming square-antiprism-type polyhedra.

The Goldschmidt diagram is useful for understanding the degree of distortion from the ideal perovskite structure. The cation radius ratios of various ABO_3 compounds are plotted in the Goldschmidt diagram in Fig. 12.4. In Fig. 12.4, the white arrow indicates compounds in the lower right region that tend to convert to the lithium niobate phase, whereas the black arrow indicates compounds in the upper left region that tend to retain the perovskite structure. This trend means that orthorhombic distortion induces conversion to the lithium niobate phase. Orthorhombic distortion is derived from rotation of BO_6 octahedra. Therefore, rotation of BO_6 octahedra can be used to understand the degree of rotation for conversion to the lithium niobate structure. O'Keeffe et al. [17] suggested that a single rotation Φ about the triad [111] axis of a pseudocubic perovskite lattice (the direction is indicated in Fig. 12.3) can be represented as rotation of the BO_6

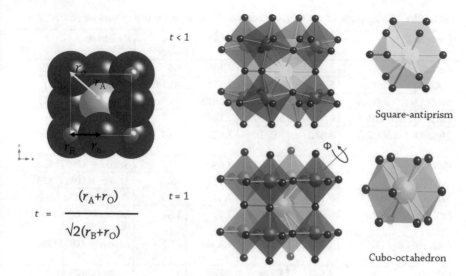

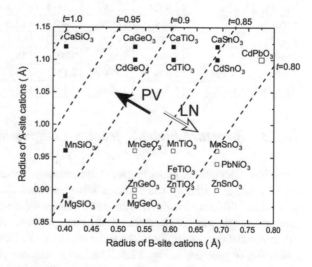

Fig. 12.3 Left: geometrical explanation of the perovskite tolerance factor. Right: orthorhombic ($t < 1$) and cubic perovskite ($t = 1$) polyhedra

Fig. 12.4 Goldschmidt diagram with tolerance factor (t) of ABO_3 compounds. The tolerance factors (dashed lines) were calculated from the ionic radii of the six-fold coordinated B cations (x axis) and eight-fold coordinated A cations (y axis). Open squares are compounds that convert to the lithium niobate structure under decompression. Solid squares are compounds that quench as the perovskite structure at ambient pressure

octahedra. The angle can be calculated from the atomic coordinates [18] or estimated from the cell dimensions: $\Phi = \cos^{-1}(\sqrt{2}c^2/ab)$ [17, 19]. According to the calculated Φ values of the various perovskite compounds listed in Table 12.1, the critical angle for conversion is estimated to be $15°–16°$, except for $MgSiO_3$ perovskites. This value is useful for exploring compositions that may adopt the lithium niobate structure.

Table 12.1 Tilting angle of the BO_6 octahedra and lattice parameters of various perovskites

Compound	a (Å)	b (Å)	c (Å)	Φ (deg)	Reference
$CaGeO_3$	5.2607	5.2688	7.4452	3.9	Susaki + (1983) [20]
$CdGeO_3$	5.209	5.253	7.434	10.7	Susaki (1989) [21]
$MnGeO_3$ (21.8 GPa)	4.9298	5.0586	7.0873	16.5	Yusa + (2014) [22]
$ZnGeO_3$ (15.6 GPa)	4.8796	5.0129	7.1233	19.4	Yusa + (2006) [9]
$MgGeO_3$ (17.9 GPa)	4.832	5.031	7.022	20.8	Leinenweber + (1994) [7]
$CaTiO_3$	5.3796	5.4423	7.6401	10.2	Sasaki + (1987) [23]
$CdTiO_3$	5.3053	5.4215	7.6176	15.5	Sasaki + (1987) [23]
$MnTiO_3$ (4.45 GPa)	5.1048	5.3046	7.4180	20.5	Ross + (1989) [24]
$FeTiO_3$ (16.2 GPa)	5.04	5.17	7.27	17.1	Leinenweber + (1991) [6]
$(Mg, Fe)TiO_3$ (21 GPa)	4.9852	5.2104	7.2305	21.1	Linton + (1991) [8]
$CdSnO_3$	5.547	5.577	7.867	7.3	Smith (1960) [25]
$MnSnO_3$ (7.35 GPa)	5.301	5.445	7.690	18.4	Leinenweber +(1991) [6]
$MnSiO_3$ (40 GPa)	4.7559	4.8073	6.7723	10.7	Fujino + (2009) [26]
$ZnTiO_3$ (21 GPa)	4.9324	5.1148	7.2514	21.9	Akaogi + (2015) [27]
$CaSnO_3$	5.5142	5.6634	7.8816	15.6	Zhao + (2004) [28]
$MgSiO_3$	4.7754	4.9292	6.8969	18.4	Horiuchi + (1987) [29]
$PbNiO_3$ (c.a. 1 atm)	5.358	5.463	7.705	15.3	Inaguma + (2010) [30]
$CdPbO_3$ (1.4 GPa)	5.639	5.724	8.135	15.0	Yusa +, in preparation
$ZnSnO_3$ (20.2 GPa)	5.1640	5.2546	7.4455	15.4	Yusa +, in preparation

12.2.3 Structure Stability from a Computational Viewpoint

Ab initio calculations provide useful information about the phase stability under high pressure. Enthalpy calculations have revealed the structural stability of the perovskite, lithium niobate, and ilmenite phases of several compounds. All of the lithium niobate phases are metastable under pressure. As an example, the relative differences of the enthalpies of the three phases of $ZnGeO_3$ perovskite are plotted as a function of pressure in Fig. 12.5. The lower pressure phase (the imenite structure) directly change to the perovskite structure. Therefore, we can conclude that the lithium niobate phase is a metastable structure of $ZnGeO_3$ [22]. Similar trends have been found for $MnTiO_3$ [31], $MgGeO_3$ [32], and $ZnTiO_3$ [10] by enthalpy calculations. A further transformation from the perovskite structure to the postperovskite structure has been confirmed for $ZnGeO_3$ [22] and $MgGeO_3$ [32].

Fig. 12.5 Enthalpy differences of the ilmenite (blue triangles) and lithium niobate (red squares) phases of ZnGeO₃ relative to the perovskite phase (black diamonds)

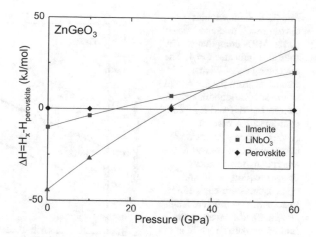

12.3 Amorphization from Cubic and Hexagonal Silicate Perovskites

12.3.1 Phase Transition Sequence of Silicate Perovskites

For a tolerance factor less than one, as represented by $MgSiO_3$ perovskite, the BO_6 octahedra in the perovskite structure tilts to make an allowance for the small divalent cations in the BO_6 octahedral corner-sharing framework. The tilting in perovskite has been discussed in detail by many researchers (e.g., Glazer [33]), where rotation does not disrupt the corner-sharing connectivity. As mentioned in Sect. 12.2, if rotation of the BO_6 octahedra reaches a limit, conversion to the lithium niobate phase occurs with a displacive-type phase transition. In contrast, perovskites bearing large divalent cations, which is formally expressed as a tolerance factor of greater than one (as shown in Fig. 12.6), cannot make enough space for such large cations in tilting of BO_6 octahedra. Therefore, silicate perovskites containing Ca^{2+}, Sr^{2+}, and Ba^{2+} cations are stabilized as hexagonal and/or cubic forms under high pressure [34–38]. These transformations have been confirmed by high-pressure experiments. The phase transition sequence is summarized in Table 12.2.

12.3.2 Crystal Structures of Hexagonal Perovskite and Structural Relation with Cubic Perovskite

Perovskites containing large divalent cations tend to expand and form a BO_6 face-sharing octahedral framework to accommodate the large cations, where the B^{4+} ions in the face-sharing octahedra cause oxygen anions to move to closer to

Fig. 12.6 Goldschmidt diagram with tolerance factor (t) for ABO$_3$ compounds. The solid line indicates $t = 1$. The tolerance factors were calculated from the ionic radii of the six-fold coordinated B cations (x axis) and eight-fold coordinated A cations (y axis). Open diamonds are compounds that convert to the amorphous form under decompression. Solid and open squares are compounds that quench as the perovskite structure and convert to the lithium niobate structure at ambient pressure, respectively

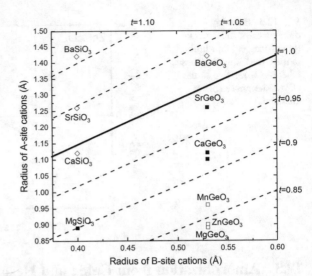

Table 12.2 High-pressure phase transition sequences of ASiO$_3$ (A = Ca, Sr, and Ba) and transition pressures

CaSiO$_3$	SrSiO$_3$	BaSiO$_3$
wollastonite [39]	pseudowollastonite [40]	High-T BaGeO$_3$ type [41]
~3 GPa	~4 GPa	~4 GPa
walstromite [39, 42]	walstromite [43]	BaSiO$_3$(II) [44]
~9 GPa	~9 GPa	~12 GPa
Ca$_2$SiO$_4$ + CaSi$_2$O$_5$ [42]	Sr$_2$SiO$_4$ + SrSi$_2$O$_5$ [36, 45]	Ba$_2$SiO$_4$ + BaSi$_2$O$_5$ [37]
~13 GPa	~20 GPa	~18 GPa
cubic perovskite [34]	6H perovskite [35]	9R perovskite [37]
	~38 GPa	~40 GPa
	cubic perovskite [38]	6H perovskite [37]
		~140 GPa
		cubic perovskite [46]

reduce their repulsion. Stacking of the face-sharing framework in the c-axis direction results in a hexagonal unit cell. Examples of the hexagonal crystal structures of BaSiO$_3$ [37] are shown in Fig. 12.7.

In Fig. 12.7, both the SiO$_6$ octahedra and barium atoms are shown along the c-axis direction to clarify the relationships of the stacking sequences. The 9R phase (space group $R\bar{3}m$) resembles the 6H phase (space group $P6_3/mmc$) in that the SiO$_6$ octahedra are periodically connected by face sharing. The difference is the periodicity of the face- and corner-sharing of SiO$_6$ octahedra. In the c-axis direction, 9R perovskite exhibits a $(chh)_3$ sequence whereas 6H perovskite exhibits a $(cch)_2$ sequence, where c and h correspond to corner- and face-sharing octahedra, respectively. For perovskites, It is known that such hexagonal polytypes lie in a sequence from 9R to 3C (space group $R\bar{3}m$) cubic perovskites. In this hexagonal sequence, pressure increases the frequency of corner-sharing octahedra. This relation can be extended to cubic perovskite (3C), which only consist of corner-sharing octahedral, as shown in Fig. 12.8. For $BaSiO_3$, the density increases for the transitions from 9R to 6H and 6H to 3C are 3.5% and 1.4%, respectively.

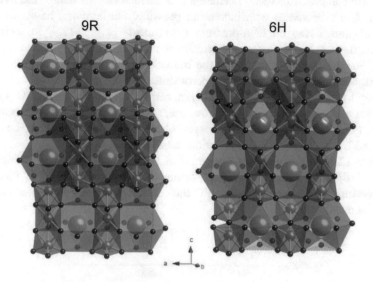

Fig. 12.7 Crystal structures of 9R and 6H $BaSiO_3$

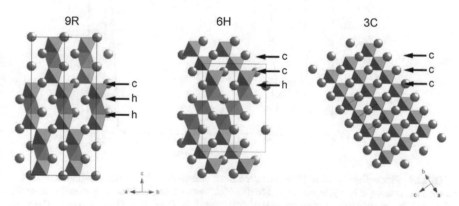

Fig. 12.8 Structural relationship among 9R, 6H, and 3C perovskites in terms of the BO_6 stacking sequence

12.3.3 Phase Diagrams: Experiments and Ab Initio Calculations

The ionic radius can be controlled under high pressure. In particular, larger A-site cations in perovskites, such as Sr^{2+} and Ba^{2+}, are sensitive to pressure. The A-site cations are compressed to SiO_6 octahedra and the face-sharing octahedral frequency then gradually decreases with increasing pressure. Furthermore, as shown in the phase diagram based on high-pressure experiments in Fig. 12.9, for cubic perovskites, there is a systematic relation between the transition pressure and the A^{2+} radius. For the $BaSiO_3$ compound, the transition occurs above 130 GPa [46]. In contrast, the transitions of the cubic perovskites $CaSiO_3$ and $SrSiO_3$ occur at significantly lower pressures of 15 and 38 GPa, respectively. Note that $SrSiO_3$ does not transform to a 9R-type hexagonal perovskite, such as that of $BaSiO_3$. Furthermore, no hexagonal perovskites are found for $CaSiO_3$. These results can be simply explained by the difference of the cation radii in the A sites.

Figure 12.10 shows the phase diagram of $BaSiO_3$ at 0 K from ab initio calculations [46]. The phase transition sequence is consistent with that from high-pressure experiments, although the calculated transition pressures are underestimated.

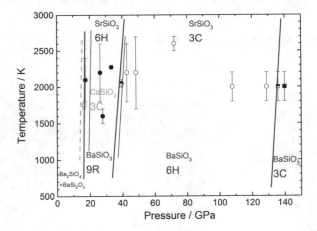

Fig. 12.9 Phase diagram of $BaSiO_3$ estimated from data plots of high pressure–high temperature experiments using a laser-heated DAC. Solid circles, open circles, and solid squares represent 9R, 6H, and 3C perovskites, respectively. Half-filled symbols indicate a phase mixture. The open square symbol at low pressure represents phase disproportionation of $Ba_2SiO_4 + BaSi_2O_5$. The estimated phase boundaries of $BaSiO_3$ (red solid lines), $SrSiO_3$ (blue thin lines), and $CaSiO_3$ (green broken line) are indicated for comparison

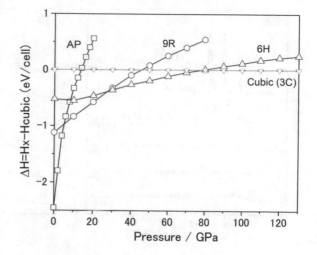

Fig. 12.10 Relative enthalpy differences of the BaSiO₃ phases [46] as a function of pressure. AP represents an ambient phase

12.3.4   Amorphization Under Decompression at Room Temperature

In the cubic and hexagonal perovskites stabilized under high pressure, the A-site cations are compressed to retain the BO_6 framework structure. In other words, the cations expand under decompression. Among the high-pressure phases of silicate perovskites, the first reported example was amorphization of $CaSiO_3$ perovskite, which was confirmed at a pressure very close to 1 atm. Because the ambient wollastonite phase is composed of a SiO_4 tetrahedral chain structure, the cubic perovskite structure cannot revert to the ambient structure at room temperature. The corner-sharing BO_6 framework can be adjusted for smaller cations, as suggested by conversion to the lithium niobate structure. However, the framework is not as flexible for larger cations. Therefore, expansion of the A-site cations disrupts the framework and makes the structure amorphous. Amorphization of the cubic perovskite structure has also been observed for $SrSiO_3$ [38] and $BaSiO_3$ [46]. Considering the structural similarity, the hexagonal perovskite structures could become amorphous during decompression. The pressure for amorphization is believed to be related to the A-site cation size in the hexagonal structure because the BO_6 face-sharing frequency of hexagonal perovskites is correlated with the cation size. The experimental results for $BaSiO_3$ are shown in Fig. 12.11. The 6H phase begins to decompose at 21.9 GPa. In contrast, the 9R phase persists at 8.9 GPa and suddenly changes to amorphous at 4.8 GPa. At 1.8 GPa, both of the phases completely change to amorphous. As a result, we can conclude that the stability of 9R is higher than that of 6H. However, this type of amorphization has not been elucidated by computational approaches. If the ionic radii are determined under pressure, this type of structural instability related to amorphization could be clarified.

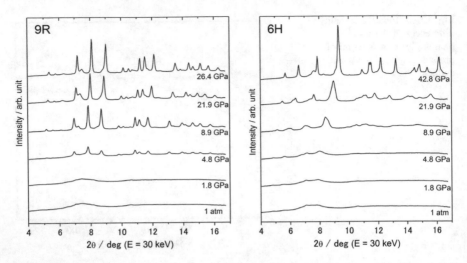

Fig. 12.11 In situ X-ray diffraction profiles of the 9R and 6H perovskites of BaSiO₃ during decompression

12.4 Relaxation Structures from the High-Pressure Phases of Sesquioxides

12.4.1 Rh₂O₃(II) Structure Reverting to the Corundum Structure in Group 13 Sesquioxides

Group-13 sesquioxides, such as aluminum oxide, gallium oxide, and indium oxide, have been widely investigated as attractive electroceramics. Their most stable phases under ambient conditions, corundum (Al_2O_3), monoclinic β-Ga_2O_3, and cubic In_2O_3 (bixbyite-type structure, C-type rare earth sesquioxide structure, hereafter denoted as C-RES), are used for many application, such as lasers and transparent electronic devices [47, 48]. It is believed that their dense phase is the corundum structure [49]. However, in situ X-ray diffraction experiments have revealed that the Rh_2O_3(II) structure that appears as a post-corundum phase under pressure reverts to the corundum structure under decompression. In Al_2O_3, the corundum structure that transforms to the Rh_2O_3(II) phase under very high pressure above 95 GPa reverts to the corundum structure at ambient pressure after decompression [50, 51]. In other instances, the Rh_2O_3(II) phase in Ga_2O_3 identified under pressure transforms to the corundum phase after decompression rather than changing to β-Ga_2O_3 [52], as shown in Fig. 12.12.

Figure 12.13 shows the crystal structures of the Rh_2O_3(II)-type and corundum structures of Ga_2O_3 with a specific direction for comparison. A twin-like relation between the Rh_2O_3(II) and corundum phases can be seen in the vertical direction. Considering the structural resemblance between Rh_2O_3(II) and corundum, we

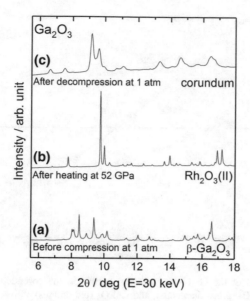

Fig. 12.12 X-ray diffraction profiles of Ga$_2$O$_3$ samples. **a** Starting β-Ga$_2$O$_3$ structure at ambient pressure, **b** Rh$_2$O$_3$(II) structure after laser heating at 52 GPa, and (c) corundum structure after decompression at ambient pressure

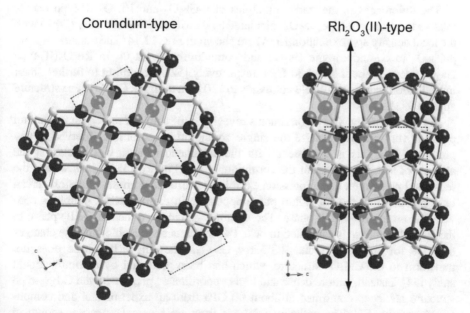

Fig. 12.13 Projections of the corundum structure of Ga$_2$O$_3$ along the hexagonal *a* axis (left) and the Rh$_2$O$_3$(II) structure of Ga$_2$O$_3$ along the *a* axis (right)

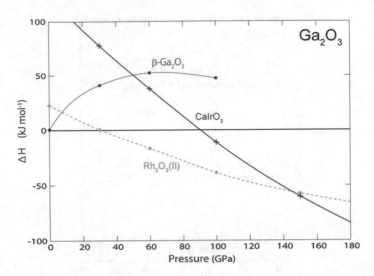

Fig. 12.14 Enthalpies of the Ga_2O_3 polymorphs relative to the corundum structure: β-Ga_2O_3 (blue circles), $Rh_2O_3(II)$ (green diamonds), and $CaIrO_3$ (red crosses)

conclude that the $Rh_2O_3(II)$ structure is appropriate for the post-corundum phase of Ga_2O_3.

The differences in the static enthalpies of β-Ga_2O_3 and $Rh_2O_3(II)$-type Ga_2O_3 relative to corundum-type Ga_2O_3 calculated by density functional theory (DFT) with the local density approximation(LDA) are shown in Fig. 12.14. The transitions from β-Ga_2O_3 to corundum-type Ga_2O_3 and corundum-type Ga_2O_3 to $Rh_2O_3(II)$-type Ga_2O_3 occur at about 0 and 30 GPa, respectively [52]. According to further phase investigation, the stability field continues to 130 GPa until the $CaIrO_3$-type structure appears [53].

For In_2O_3, in situ X-ray experiments reveal that the stability region for corundum phase is very narrow because the single corundum phase is not observed at any pressure [52]. This is consistent with the calculated results, which suggest the absence of a stability area for the corundum phase (Fig. 12.15) [52]. However, the recovered phase after decompression exhibits the corundum phase. Therefore, it can be concluded that the corundum phase appearing in the recovered sample is con-verted from the $Rh_2O_3(II)$ phase. The volume change from the $Rh_2O_3(II)$ phase to the corundum phase is estimated to be 2.1%, which is comparable with the changes of 3.1% for Al_2O_3 [51] and 2.3% for Ga_2O_3. The $Rh_2O_3(II)$ phase does not transform to the $CaIrO_3$ structure, which had been predicted by a computational study [54]. Instead, a more dense and higher coordinated phase with the Gd_2S_3-type structure has been confirmed at about 40 GPa from an experimental and compu-tational study [55]. The enthalpy relations from DFT calculations are shown in Fig. 12.15.

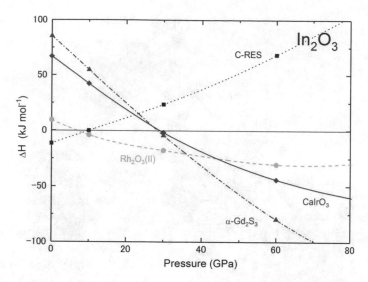

Fig. 12.15 Enthalpies of the In_2O_3 polymorphs relative to the corundum structure: C-RES (black squares), $Rh_2O_3(II)$ (green circles), Gd_2S_3 (blue triangles), and $CaIrO_3$ (red diamonds)

12.4.2 A-RES Structure of Y_2O_3 Reverting to the B-RES Structure

Yttrium has a similar ionic radius to the ionic radii of lanthanides, so lanthanide ions can be incorporated into yttria to make optical ceramics, such as $Eu^{3+}:Y_2O_3$ phosphor [56] and $Yb^{3+}:Y_2O_3$ laser [57]. Yttria crystallizes in the bixbyite structure (C-RES) under ambient conditions, similar to lanthanide sesquioxides. B-RES has been confirmed as the high-pressure phase in the recovery sample from high-pressure experiments. The A-RES phase was not found, which is expected to be part of the phase transformation sequence of lanthanide sesquioxides [58]. In situ X-ray diffraction experiments performed at room temperature using a DAC revealed the existence of the A-RES phase [59]. Back transformation to the B-RES structure was also confirmed. The reversible transformation mechanism from B-RES to A-RES can be explained from a crystallographic viewpoint, as shown in Fig. 12.16.

The B-RES structure of yttria consists of three different yttrium sites. Among these sites, only the Y3 site can be considered to possess six-fold oxygen coordination because the Y3–O2 distance is too long to be classified as seven-fold coordination, as shown in Fig. 12.16b. With increasing pressure, O2 moves closer to Y3, which results in the formation of seven-fold polyhedra. Upon further compression to 15–20 GPa, the Y3–O2 distance becomes shorter than the average Y3–O distance. The B-RES structure finally changes to the structure shown in Fig. 12.16c, which is equivalent to the A-RES structure. This means that the A-RES structure can be directly derived from the B-RES structure. The volume

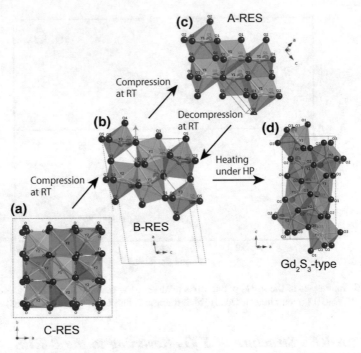

Fig. 12.16 Structural relationship among the high-pressure polymorphs of Y_2O_3

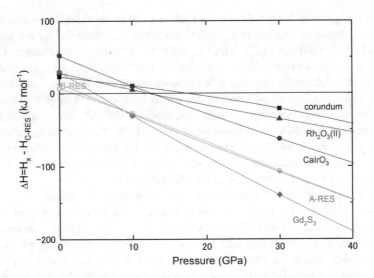

Fig. 12.17 Enthalpies of the Y_2O_3 polymorphs relative to the C-RES structure: B-RES (open green circles), A-RES (green crosses), Gd_2S_3 (green diamonds), $CaIrO_3$ (red circles), $Rh_2O_3(II)$ (blue triangles), and corundum (black squares)

change from the B-RES structure to the A-RES structure (2.5%) is characteristic of a first-order phase transition.

Contrary to confirmation of the A-RES structure by compression experiments at room temperature, enthalpy calculations performed by DFT with the LDA indicate no stability region of the A-RES structure (Fig. 12.17) [59]. The transition to the other high-coordination structure (Gd_2S_3-type structure, Fig. 12.16d) occurs before the appearance of the A-RES phase. In fact, laser heating experiments under high pressure result in Y_2O_3 crystallizing in the Gd_2S_3 structure at about 10 GPa. Therefore, it can be concluded that the A-RES structure appearing under room temperature compression is a metastable phase.

12.5 Concluding Remarks

Large volume high-pressure apparatus (e.g., cubic, belt, and KAWAI-type presses) is a fundamental tool for materials scientists, because high-pressure methods enable the synthesis of novel materials under ambient conditions. High-pressure synthesis provides the opportunity to obtain high density and/or highly coordinated compounds. However, the recovered product does not always reflect the structure under pressure. If a new structure is found, the stability relation with the lower pressure phase(s) should be evaluated using computational approaches, such as ab initio calculations. If the structure is a metastable phase, the structure should be examined for crystallographic similarity with an objective structure. Conversion to the metastable phase would be clarified by structural relaxation. A trace amount of a high-pressure phase is sometimes found in the recovered products as a defect origination from twin structures. This is also an indication to identify the unquenchable high-pressure phase.

In situ X-ray diffraction is the most powerful approach to determine structures under pressure. In some cases, recompression of the metastable phase gives the high-pressure structure. During structural relaxation, symmetry change likely occurs, as exemplified by the transition from the perovskite to the lithium niobate phase as described in Sect. 12.2. Relaxation from a centrosymmetric to a non-centrosymmetric structure is important to determine the functionality, such as ferroelectricity.

As mentioned in Sect. 12.3, amorphization is a usual phenomenon for high-pressure products under decompression. Therefore, if there is a complete or part of an amorphous-like pattern in the X-ray diffraction profile of the recovered product, the amorphous structure is an indication of an unquenchable high-pressure phase. In situ X-ray experiments using a laser-heated DAC reveal the structure of the unquenchable phase. Amorphization can be triggered by the expansion of specific cations during decompression. In particular, elucidation of the compression behavior for relatively large cations, such as K^+, Ca^{2+}, Sr^{2+}, and Ba^{2+}, would aid in understanding the quenchability of high-pressure structures containing such cations. Therefore, an approach to determine the ionic radii under pressure is required for prediction of the quenchability.

Acknowledgements I am deeply grateful to Prof. I. Tanaka and Dr. T. Taniguchi for their advice on the topics discussed in this chapter. I thank Profs. T. Tsuchiya and H. Hiramatsu for discussion on the computational studies. The X-ray diffraction experiments were performed under SPring-8 and KEK proposals. This work was supported in part by Innovative Areas "Nano Informatics" (Grant No. 25106006) and JSPS KAKENHI (Grant No. 16H04078).

References

1. H. Yusa, M. Akaogi, E. Ito, J. Geophys. Res.-Solid Earth **98**, 6453 (1993)
2. M. Murakami, K. Hirose, K. Kawamura, N. Sata, Y. Ohishi, Science **304**, 855 (2004)
3. M. Akaogi, E. Ito, Geophys. Res. Lett. **20**, 1839 (1993)
4. Y. Hinatsu, T. Fujino, N. Edelstein, J. Solid State Chem. **99**, 182 (1992)
5. R.J. Cava, A. Santoro, D.W. Murphy, S. Zahurak, R.S. Roth, J. Solid State Chem. **42**, 251 (1982)
6. K. Leinenweber, W. Utsumi, Y. Tsuchida, T. Yagi, K. Kurita, Phys. Chem. Miner. **18**, 244 (1991)
7. K. Leinenweber, Y.B. Wang, T. Yagi, H. Yusa, Am. Mineral. **79**, 197 (1994)
8. J.A. Linton, Y.W. Fei, A. Navrotsky, Am. Mineral. **84**, 1595 (1999)
9. H. Yusa, M. Akaogi, N. Sata, H. Kojitani, R. Yamamoto, Y. Ohishi, Phys. Chem. Miner. **33**, 217 (2006)
10. Y. Inaguma, A. Aimi, Y. Shirako, D. Sakurai, D. Mori, H. Kojitani, M. Akaogi, M. Nakayama, J. Am. Chem. Soc. **136**, 2748 (2014)
11. Y. Inaguma, M. Yoshida, T. Tsuchiya, A. Aimi, K. Tanaka, T. Katsumata, D. Mori, J. Phys: Conf. Ser. **215**, 012131 (2010)
12. T. Hattori, T. Matsuda, T. Tsuchiya, T. Nagai, T. Yamanaka, Phys. Chem. Miner. **26**, 212 (1999)
13. W. Wang, S. Wang, D. He, J.A. Xu, Solid State Commun. **196**, 8 (2014)
14. Y. Inaguma, K. Tanaka, T. Tsuchiya, D. Mori, T. Katsumata, T. Ohba, K.-I. Hiraki, T. Takahashi, H. Saitoh, J. Am. Chem. Soc. **133**, 16920 (2011)
15. V.M. Goldschmidt, Naturwissenschaften **14**, 477 (1926)
16. R.D. Shannon, Acta Crystallographica Section A **32**, 751 (1976)
17. M. O'keeffe, B.G. Hyde, J.O. Bovin, Phys. Chem. Miner. **4**, 299 (1979)
18. Y. Zhao, D.J. Weidner, J.B. Parise, D.E. Cox, Phys. Earth Planet. Inter. **76**, 17 (1993)
19. Y. Zhao, D.J. Weidner, J.B. Parise, D.E. Cox, Phys. Earth Planet. Inter. **76**, 1 (1993)
20. S. Sasaki, C.T. Prewitt, R.C. Liebermann, Am. Mineral. **68**, 1189 (1983)
21. J. Susaki, Phys. Chem. Miner. **16**, 634 (1989)
22. H. Yusa, T. Tsuchiya, M. Akaogi, H. Kojitani, D. Yamazaki, N. Hirao, Y. Ohishi, T. Kikegawa, Inorg. Chem. **53**, 11732 (2014)
23. S. Sasaki, C.T. Prewitt, J.D. Bass, W.A. Schulze, Acta Crystallogr. Sect. C-Struct. Chem. **43**, 1668 (1987)
24. N.L. Ross, J.D. Ko, C.T. Prewitt, Phys. Chem. Miner. **16**, 621 (1989)
25. A. Smith, Acta Crystallogr. **13**, 749 (1960)
26. K. Fujino, D. Nishio-Hamane, K. Suzuki, H. Izumi, Y. Seto, T. Nagai, Phys. Earth Planet. Inter. **177**, 147 (2009)
27. M. Akaogi, K. Abe, H. Yusa, H. Kojitani, D. Mori, Y. Inaguma, Phys. Chem. Miner. **42**, 421 (2015)
28. J. Zhao, N.L. Ross, R.J. Angel, Phys. Chem. Miner. **31**, 299 (2004)
29. H. Horiuchi, E. Ito, D.J. Weidner, Am. Mineral. **72**, 357 (1987)
30. Y. Inaguma, M. Yoshida, T. Tsuchiya, A. Aimi, K. Tanaka, T. Katsumata, D. Mori, J. Phys: Conf. Ser. **215**, 012131 (2010)
31. F. Zhu, X. Wu, S. Qin, Solid State Commun. **152**, 984 (2012)

32. T. Tsuchiya, J. Tsuchiya, Phys. Rev. B **76**, 092105 (2007)
33. A.M. Glazer, M. Ahtee, H.D. Megaw, Acta Crystallogr. Sect. A **28**, S179 (1972)
34. H.K. Mao, L.C. Chen, R.J. Hemley, A.P. Jephcoat, Y. Wu, W.A. Bassett, J. Geophys. Res.-Solid Earth Planets **94**, 17889 (1989)
35. H. Yusa, M. Akaogi, N. Sata, H. Kojitani, Y. Kato, Y. Ohishi, Am. Mineral. **90**, 1017 (2005)
36. M. Akaogi, H. Kojitani, H. Yusa, R. Yamamoto, M. Kido, K. Koyama, Phys. Chem. Miner. **32**, 603 (2005)
37. H. Yusa, N. Sata, Y. Ohishi, Am. Mineral. **92**, 648 (2007)
38. W.S. Xiao, D.Y. Tan, W. Zhou, J. Liu, J. Xu, Am. Mineral. **98**, 2096 (2013)
39. T. Gasparik, K. Wolf, C.M. Smith, Am. Mineral. **79**, 1219 (1994)
40. F. Nishi, Acta Crystallogr. Sect. C-Cryst. Struct. Commun. **53**, 534 (1997)
41. W. Hilmer, Acta Crystallogr **15**, 1101 (1962)
42. M. Akaogi, M. Yano, Y. Tejima, M. Iijima, H. Kojitani, Phys. Earth Planet. Inter. **143**, 145 (2004)
43. K. Machida, G. Adachi, J. Shiokawa, M. Shimada, M. Koizumi, K. Suito, A. Onodera, Inorg. Chem. **21**, 1512 (1982)
44. Y. Shimizu, Y. Syono, S. Akimoto, High Temperatures-High Pressures **2**, 113 (1970)
45. H. Kojitani, M. Kido, M. Akaogi, Phys. Chem. Miner. **32**, 290 (2005)
46. H. Hiramatsu, Y. Hitoshi, I. Ryo, O. Yasuo, T. Kamiya, H. Hosono, Inorg. Chem. **56** (2017)
47. K. Matsuzaki, H. Yanagi, T. Kamiya, H. Hiramatsu, K. Nomura, M. Hirano, H. Hosono, Appl. Phys. Lett. **88**, 092106 (2006)
48. H. Ohta, K. Nomura, H. Hiramatsu, K. Ueda, T. Kamiya, M. Hirano, H. Hosono, Solid-State Electron. **47**, 2261 (2003)
49. C.T. Prewitt, R.D. Shannon, D.B. Rogers, A.W. Sleight, Inorg. Chem. **8**, 1985 (1969)
50. N. Funamori, R. Jeanloz, Science **278**, 1109 (1997)
51. J.F. Lin, O. Degtyareva, C.T. Prewitt, P. Dera, N. Sata, E. Gregoryanz, H.K. Mao, R. J. Hemley, Nat. Mater. **3**, 389 (2004)
52. H. Yusa, T. Tsuchiya, N. Sata, Y. Ohishi, Phys. Rev. B **77** (2008)
53. T. Tsuchiya, H. Yusa, J. Tsuchiya, Phys. Rev. B **76** (2007)
54. R. Caracas, R.E. Cohen, Phys. Rev. B **76**, 184101 (2007)
55. H. Yusa, T. Tsuchiya, J. Tsuchiya, N. Sata, Y. Ohishi, Phys. Rev. B **78** (2008)
56. S.L. Jones, D. Kumar, R.K. Singh, P.H. Holloway, Appl. Phys. Lett. **71**, 404 (1997)
57. J. Kong et al., Appl. Phys. Lett. **82**, 2556 (2003)
58. V.M.U. Goldschmidt, F. Barth, T.S.k.R, Nor, V. Akad. Kl. 1: Mat. Naturvindensk. **5** (1925)
59. H. Yusa, T. Tsuchiya, N. Sata, Y. Ohishi, Inorg. Chem. **49**, 4478 (2010)

Chapter 13
Synthesis and Structures of Novel Solid-State Electrolytes

Ryoji Kanno, Genki Kobayashi, Kota Suzuki, Masaaki Hirayama, Daisuke Mori and Kazuhisa Tamura

Abstract Two classes of new materials possessing ion conductivity have been developed: a lithium ion conductor and a hydride ion conductor. Conventional perovskite and ordered rock-salt structures were adopted as frameworks for lithium migration, and electrochemically stable elements such as Al, Ga, Ta, and Sc were used in the materials to facilitate their use as low-potential negative electrodes. New compositions of $(Li_{0.25}Sr_{0.625}V_{(Li,Sr)0.125})(Ga_{0.25}Ta_{0.75})O_3$, and $Li_{0.9}Sc_{0.9}Zr_{0.1}O_2$ were found to be novel oxide-based lithium ion conductors. Oxyhydrides with K_2NiF_4-type structures were synthesized via a high-pressure synthesis method and their use in pure hydride ion conduction was demonstrated. The $La_{2-x-y}Sr_{x+y}LiH_{1-x+y}O_{3-y}$ oxyhydrides showed wide composition ranges of solid solution formation and the conductivity increased with anion vacancies or the introduction of interstitial hydride ions. The performance of an all-solid-state TiH_2/o-La_2LiHO_3 ($x = y = 0$, o: orthorhombic)/Ti cell provided conclusive evidence of pure H^- conduction.

Keywords Solid electrolyte · Hydride ion conductor · Lithium ion conductor · Material search

R. Kanno (✉) · K. Suzuki · M. Hirayama
Department of Chemical Science and Engineering, School of Materials and Chemical Technology, Tokyo Institute of Technology, Yokohama 226-8502, Japan
e-mail: kanno@echem.titech.ac.jp

G. Kobayashi
Institute for Molecular Science, Research Center of Integrative Molecular Systems (CIMoS), 38 Nishigonaka, Myodaiji, Okazaki, Aichi 444-8585, Japan

D. Mori
Department of Chemistry for Materials, Graduate School of Engineering, Mie University, Tsu 514-8507, Japan

K. Tamura
Synchrotron Radiation Research Center, Kansai Research Establishment, Japan Atomic Energy Agency, Sayo-Gun, Hyogo 679-5148, Japan

I. Tanaka (ed.), *Nanoinformatics*, https://doi.org/10.1007/978-981-10-7617-6_13

13.1   Novel Solid-State Electrolytes

Solid materials exhibiting purely ionic conduction are used as solid-state electrolytes in a wide variety of electrochemical devices and chemical sensors, with the corresponding charge carriers being specific ions such as H^+, Cu^+, Ag^+, Na^+, Li^+, F^-, and O^{2-}. The resulting charged ion flow in electrolytes creates an electric current that drives the device, the characteristics and performance of which are thus influenced by the nature of the charge carriers. Generally, in view of their small ionic radii, cations migrate easily in solid electrolytes, showing facile diffusion. For example, silver and copper ion solid electrolytes, such as $RbAg_4I_5$ and $Rb_4Cu_{16}I_7Cl_{13}$, show extremely high ionic conductivities of > 100 mS cm^{-1} at room temperature [1–3]. Moreover, the recently developed lithium ion conductors ($Li_{10}GeP_2S_{12}$, LGPS) have achieved room temperature conductivities of >10 mS cm^{-1} [4, 5], with Li-based all-solid-state batteries reported to exhibit exceptionally good power characteristics. On the other hand, newly developed materials such as hydride ion conductors have expanded the research field and the scope of available energy devices [6, 7]. In this section, we focus on Li^+ and H^- as charge carriers and describe the structural characteristics of the corresponding newly developed materials.

13.2   Lithium Ion Conductors

Lithium ion conductors continue to attract much attention owing to their practical applications in all-solid-state lithium batteries [5, 8]. A wide variety of such conductors exists (e.g., LISICON, perovskite, garnet, glass, glass ceramics, thio-LISICON, and LGPS), some of which were developed in the 1970s [4, 9–14]. For instance, LGPS-based materials ($\sigma > 10$ mS cm^{-1} at 25 °C) enable high-power operation of solid-state lithium batteries; this is an intrinsic merit of solid-state systems, in addition to their safety and reliability. However, sulfide-based solids are sensitive to atmospheric moisture. As a result, most current research focuses on oxide-based materials, in order to satisfy the requirements of practical applications and engineering processes.

Novel ion conductors are typically developed using three methods: (i) element substitution-based, (ii) structure-based, and (iii) composition-based material searches. Approach (i) relies on existing materials with ionic conductivity of the target charge carrier [15], which are amenable to tuning of their physical and electrochemical properties [16]. Therefore, although it is relatively easy to find new materials using this method, remarkable performance improvements are difficult to achieve. Approach (ii) is initiated by selecting a suitable crystal structure candidate for ion diffusion [6, 11], which can be complicated by the fact that the diffusion of the target ion in the selected structure has usually not been demonstrated.

Finally, approach (iii) is the most challenging, but also has the greatest potential to afford new materials with unique structures and properties. This approach starts with the selection of a suitable phase diagram [17]. Subsequently, materials corresponding to the chosen region in this diagram are synthesized and characterized; in certain cases, they exhibit unique structures and properties [4, 18]. In this chapter, some examples of material searches are introduced.

13.2.1 Novel Lithium Ion-Conducting Perovskite Oxides [15]

Lithium ion-conducting solids are key materials for all-solid-state lithium batteries, which, compared with conventional liquid electrolyte-based lithium batteries, exhibit improved energy density, stability, safety, and reliability. Among the solid electrolytes that have been developed, the oxide-based ones are among the most promising candidates, owing to their high ionic conductivities and good chemical stabilities over a wide range of operating temperatures [11, 19]. Lithium ion-conducting perovskites such as $La_{(2/3)-x}Li_{3x}TiO_3$ (which exhibits an ionic conductivity above 10^{-3} S cm^{-1} at room temperature) are considered to be particularly attractive [19]. However, the interfacial reduction of Ti^{+4} to Ti^{+3} during the electrochemical process or upon contact with lithium metal gives rise to undesirable electronic conduction. On this basis, novel perovskite-structured materials were examined, in typical example of an element substitution-based material search. As a result, the Li-Sr-Ta-M-O system (M = Al, Ga) was postulated to be ideal for achieving high ionic conductivity, with $(Li_xSr_{1-x})(M_{(1-x)/2}Ta_{(1+x)/2})O_3$ mixed oxides expected to exhibit superior characteristics owing to (i) the presence of largely non-reducible metals (Ta and Al/Ga) in their structures, (ii) the presence of a large cation (Sr) at the A-site, (iii) the limited distortion of BO_6 octahedra exhibited by Ta at the B-site, (iv) the presence of a small B-site cation (Al, Ga), and (v) the availability of controlled vacancies introduced by adjusting the concentration of the B-site cation. Thus, $(Li_xSr_{1-x})(M_{(1-x)/2}Ta_{(1+x)/2})O_3$ (M = Al, Ga) and $(Li_xSr_{1-x-y}V_{(Li,Sr)y})(Ga_{[(1-x)/2]-y}Ta_{[(1+x)/2]+y})O_3$ systems were synthesized by solid-state reactions involving Li-rich starting materials to obtain a single phase. These were subsequently subjected to electrochemical examination and crystallographic analysis by X-ray and neutron diffraction Rietveld analysis.

The temperature-dependent conductivities of $(Li_{0.2}Sr_{0.65}V_{(Li,Sr)0.15})(Ga_{0.25}Ta_{0.75})O_3$ (x = 0.2, y = 0.15) and $(Li_{0.25}Sr_{0.625}V_{(Li,Sr)0.125})(Ga_{0.25}Ta_{0.75})O_3$ (x = 0.25, y = 0.125) are shown in Fig. 13.1, with the corresponding activation energies (E_a) calculated as 35.04 and 34.64 kJ mol^{-1}, respectively. The comparable E_a values of these systems indicate that they both feature the same Li$^+$ conduction mechanism. The highest conductivity exhibited by the $(Li_{0.25}Sr_{0.625}V_{(Li,Sr)0.125})(Ga_{0.25}Ta_{0.75})O_3$ (x = 0.25, y = 0.125) sample equaled 1.85×10^{-3} S cm^{-1} at 250 °C, which was the highest value measured for Li-Sr-Ga-Ta-O perovskite materials. The structure of

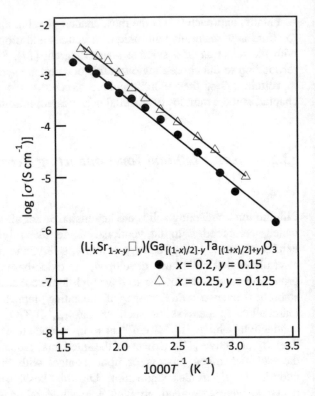

Fig. 13.1 Arrhenius plot showing the temperature-dependent conductivities of $(Li_xSr_{1-x-y}V_{(Li,Sr)y})$ $(Ga_{[(1-x)/2]-y}Ta_{[(1+x)/2]+y})O_3$ with $x = 0.2$, 0.25 and $y = 0.125$, 0.15

this sample was determined by powder neutron diffraction and the obtained data were refined using a structural model of a cubic perovskite-type material with Pm-$3m$ symmetry.

Figure 13.2 shows the thermal ellipsoid structure model obtained by neutron Rietveld analysis and Table 13.1 summarizes the corresponding interatomic distances, revealing that the O–O distance of 2.79411 Å in $(Li_{0.25}Sr_{0.625}V_{(Li,Sr)0.125})$ $(Ga_{0.25}Ta_{0.75})O_3$ was slightly larger than that determined for cubic perovskite-structured $Li_{0.5}La_{0.5}TiO_3$ (2.7358 Å). This increase was most likely due to substitution by the larger Sr^{2+} cation at the A-site, which is responsible for widening the bottleneck for lithium ion diffusion in the structure. The A–O distance of 2.79411 Å was larger than the calculated sum of the Li and O ionic radii of 2.32 Å (Li^+ (CN 8): 0.92 Å and O^{2-} (CN 6): 1.4 Å), making the Li cation more ionic in nature and, therefore, more mobile. The tolerance factor t, calculated based on the ionic radius of Sr^{2+} at the A-site, equaled 0.9855, which was close to unity and indicated an ideal cubic perovskite-type structure.

The increased ionic conduction was confirmed to result from the introduction of vacancies at A-sites. Average bond valence sum (BVS) values were calculated for each site of the perovskite structure using refined structural data, with the average BVS for A-sites equaling 1.98. The larger average BVS of $(Li_{0.25}Sr_{0.625}V_{(Li,Sr)0.125})$ $(Ga_{0.25}Ta_{0.75})O_3$ compared with that of $La_{(2/3)-x}Li_{3x}TiO_3$ (0.95–1.57) [20]

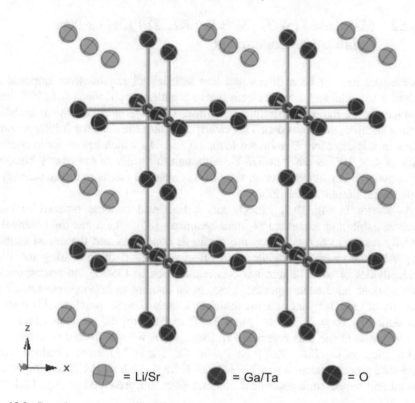

Fig. 13.2 Crystal structure model of $(Li_{0.25}Sr_{0.625}V_{(Li,Sr)0.125})(Ga_{0.25}Ta_{0.75})O_3$ based on neutron Rietveld analysis

Table 13.1 Selected bond distances for $(Li_{0.25}Sr_{0.625}V_{(Li,Sr)0.125})(Ga_{0.25}Ta_{0.75})O_3$

Distance/Å
A–O 2.79411(0)
B–O 1.97573(0)
A–B 3.42207(0)
O–O 2.79411(0)

A = Li and Sr, B = Ga and Ta

corresponded to the greater activation energy required for ion conduction in the former. Figure 13.2 shows the perovskite structure with thermal ellipsoids for each site, revealing their approximately isotropic thermal nature. The presence of vacancies in conjunction with Li ions at the A-sites suggests that Li diffusion proceeds according to the vacancy mechanism.

13.2.2 M-Doped LiScO₂ (M = Zr, Nb, Ta) [21] as New Lithium Ion Conductors

No material has yet been discovered that satisfies all requirements imposed on lithium ion conductors as solid electrolytes for battery applications (i.e., high ionic conductivity at room temperature, chemical stability, electrochemical stability, thermal stability, and low cost). This clearly indicates the need for further research efforts in this direction. Herein, we focus on $LiScO_2$, which has an ionic conductivity of 4×10^{-9} S cm^{-1} at 573 K. Although this value is not overly high, the above material is still attractive in view of its enhanced thermodynamic stability in contact with lithium metal [22].

As shown in Fig. 13.3, $LiScO_2$ has a fractional cationic ordered rock-salt structure exhibiting tetragonal $I4_1/amd$ symmetry [23], which has the potential to partially rearrange depending on the synthesis conditions and the doped element [24]. Although element doping is an effective method of increasing the ionic conductivities of solid lithium ion conductors such as $LiScO_2$, no corresponding investigations have been reported. Thus, in an attempt to improve the ionic conductivity of $LiScO_2$ by introducing lithium vacancies into its structure, this material was doped by $M = Zr^{4+}$, Nb^{5+}, and Ta^{5+}, and the crystal structures and ionic conductivities of the thus prepared $Li_{1-y}Sc_{1-x}M_xO_2$ were evaluated in detail.

$Li_{1-y}Sc_{1-x}M_xO_2$ ($M = Zr^{4+}$, Nb^{5+}, or Ta^{5+}; $x = 0.1$) were obtained via a solid-state reaction (sintering at 1073–1623 K for 1–12 h in air). Their impedance spectra and temperature-dependent conductivities are presented in Fig. 13.4. The

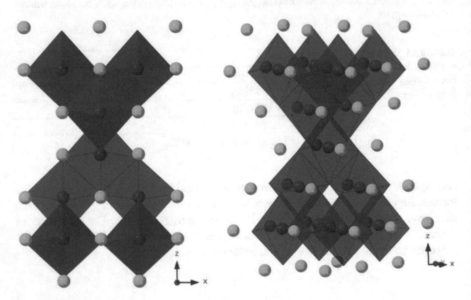

Fig. 13.3 Crystal structure of $LiScO_2$ [23], with blue octahedra and green spheres indicating ScO_6 and Li, respectively

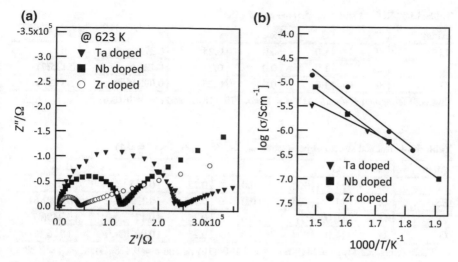

Fig. 13.4 a Representative impedance plots at 623 K and **b** Arrhenius plots showing the temperature-dependent conductivities of doped $Li_{1-y}Sc_{1-x}M_xO_2$ ($M = Zr^{4+}$, Nb^{5+}, or Ta^{5+})

conductivities were calculated from the corresponding impedance spectra, which comprised semicircles and spikes corresponding to contributions of the bulk and grain boundary and the electrode, respectively. The bulk and grain boundary contributions could not be separated and were therefore calculated together.

Resistances were calculated from the diameter of the aforementioned semicircles and used to determine conductivities. The diameters of these semicircles decreased upon doping, indicating the suitability of impedance spectroscopy to survey and evaluate ionic conductivities, with capacitance values corresponding to the observed semicircles being in the range of $10^{-10}-10^{-12}$ F. Table 13.2 summarizes the ionic conductivities and activation energies of $Li_{1-y}Sc_{1-x}M_xO_2$ ($x = 0.1$) at 573 K, along with the values previously reported for $LiScO_2$. All doped samples showed higher conductivities than the parent compound, owing to the formation of solid solutions upon aliovalent cation doping. Furthermore, this doping decreased the activation energies by more than 10%, indicating that the formation of lithium vacancies in the $LiScO_2$ lattice reduced the energy barrier of lithium diffusion.

Table 13.2 Ionic conductivities (573 K) and activation energies of $Li_{1-y}Sc_{1-x}M_xO_2$ ($M = Zr^{4+}$, Nb^{5+}, or Ta^{5+})

Doped element	Composition	Conductivity σ, S cm^{-1} @573 K	Activation energy E_a, kJ mol^{-1}
–	$LiScO_2$ [22]	4.2×10^{-9}	101
Ta	$Li_{0.8}Sc_{0.9}Ta_{0.1}O_2$	5.79×10^{-7}	57 ± 11
Nb	$Li_{0.8}Sc_{0.9}Nb_{0.1}O_2$	6.01×10^{-7}	85 ± 3
Zr	$Li_{0.9}Sc_{0.9}Zr_{0.1}O_2$	9.73×10^{-7}	88 ± 5

Table 13.3 Refined structural parameters of LiScO$_2$

Atom	Site	G	x	y	z	B, Å^2
Li	4b	1.0	0.0	0.25	0.375	2.0
Sc	4a	1.0	0.0	0.75	0.125	0.135(7)
O	8e	1.0	0.0	0.25	0.09895(3)	1.352(6)

Unit cell: tetragonal $I4_1/amd$(141); $a = b = 4.1791(18)$ Å and $c = 9.3610(4)$ Å; $R_{wp} = 11.55$

Table 13.4 Refined structural parameters of Li$_{1-x}$(Sc$_{1-x}$Zr$_x$)O$_2$ ($x = 0.1$)

Atom	Site	G	x	y	z	B, Å^2
Li	4b	1 − [g(Zr)]	0.0	0.25	0.375	2.0
Sc	4a	1 − [g(Zr)]	0.0	0.75	0.125	0. 240(6)
Zr	4a	0.0968(7)	0.0	0.75	0.125	0.240(6)
O	8e	1.0	0.0	0.25	0.09793(3)	1.116 (6)

Unit cell: tetragonal $I4_1/amd$(141); $a = b = 4.1804(16)$ Å, and $c = 9.4186(3)$ Å; $R_{wp} = 7.28$

Zr^{4+}-doped samples showed the highest ionic conductivities, with a maximum value of 9.73×10^{-7} S cm^{-1} observed at 573 K. In order to verify the changes in ionic conductivities caused by Zr^{4+} doping, the corresponding crystal structures were evaluated in detail.

Tables 13.3 and 13.4 summarize the refinement-determined structural parameters for $x = 0.0$ and $x = 0.1$, respectively. All diffraction peaks were indexed to the $I4_1/amd$(141) space group with tetragonal symmetry, with the exception of reflections ascribed to impurities. The lattice parameters of LiScO$_2$ were determined as $a = b = 4.1791(18)$ Å and $c = 9.3610(4)$ Å, making them nearly identical to the reported values of $a = b = 4.182$ Å and $c = 9.318$ Å [23]. The lattice parameters calculated for $x = 0.1$ ($a = b = 4.1804(16)$ Å and $c = 9.4186(3)$ Å) were increased by doping with Zr^{4+}, with refinement results showing that 10% Zr^{4+} was doped at Sc^{3+} sites in the above structure, in agreement with the ratio of utilized reactants. Concomitantly, lithium vacancies were probably formed to maintain the charge balance in LiScO$_2$, since the doped Zr^{4+} ion has a higher charge than Sc^{3+}. These results demonstrate that the ionic conductivity of LiScO$_2$ was markedly improved by substitution with certain aliovalent cations, owing to the resulting lattice expansion and formation of lithium vacancies.

13.3 Development of Hydride Ion Conductors

Hydride ion conduction is particularly attractive, as H$^-$ is similar in size to fast ionic conduction-suitable oxide and fluoride ions, while exhibiting strong reducing properties (standard H$^-$/H$_2$ redox potential $= -2.3$ V), comparable to those of

Fig. 13.5 Standard redox potentials of typical charge carriers, showing that the highly negative potential of H_2/H^- is similar to that of Mg^{2+}/Mg

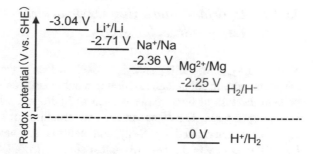

Mg/Mg^{2+} (-2.4 V) (Fig. 13.5). Thus, hydride ion conductors may be applied in energy storage/conversion devices with high energy densities. To indicate a new direction for next-generation battery systems beyond lithium ion batteries and fuel cells, we herein focus on hydride ion conduction in solids.

Hydride ion conduction in CaH_2 was first described by Andresen et al. in 1977 [25], with similar reports on other materials following in later years [26–31]. However, experimental evidence of H^- conduction was not obtained until Irvine et al. determined the transport number of BaH_2 by electromotive force measurements in 2015 [7]. Although alkaline earth metal hydrides such as BaH_2 act as pure H^- conductors, they are also strong reducing agents. This complicates their use as solid electrolytes of energy devices, in which electrochemical stability to both oxidation and reduction is required. Indeed, these metal hydrides have not yet been applied to battery reactions. From the viewpoint of material design, the structural inflexibility of metal hydrides complicates the control of their lattice structure (which is required to create smooth transport pathways) and their conducting hydride ion content. Thus, little progress has been achieved in the development of H^- conductors. We have considered oxyhydrides, in which hydride and oxide ions share anion sublattices, as prospective hydride conductors with flexible anion sublattices. Known oxyhydrides include $A_2BH_xO_{4-x}$ (K_2NiF_4 structure; A = La, Ce, Nd, Pr, Sr; B = Co, V, Li; $0 < x \leq 1$), $Sr_3Co_2O_{4.33}H_{0.84}$ (Ruddlesden-Popper structure), $ATiO_{3-x}H_x$ (perovskite structure; A = Ba, Sr, Ca) [32–37], and $[Ca_{24}Al_{28}O_{64}]^{4+} \cdot 4H^-$ (mayenite structure) [38–40]. However, none of these materials display pure H^- conductivity, since hydride ions have been reported to act as electron donors in oxide-based materials [38–42], donating electrons to their lattice and thus causing electron conduction accompanied by a characteristic change in hydrogen charge from H^- to H^+. Indeed, perovskite- and mayenite-type oxyhydrides predominantly exhibit electron conduction caused by the dissociation of hydride ions into electrons and protons [33, 38–40, 43]. Taking this into consideration, preventing the above electron donation may be important for achieving pure H^- conduction in the oxide framework structure. Herein, we attempted to synthesize a series of K_2NiF_4-type oxyhydrides, $La_{2-x-y}Sr_{x+y}LiH_{1-x+y}O_{3-y}$ ($0 \leq x \leq 1$, $0 \leq y \leq 2$, $0 \leq x + y \leq 2$) featuring cation sublattices that contain cations that are more electron-donating than H^- and anion sublattices that allow flexible storage of H^-, O^{2-}, and vacancies.

13.3.1 Hydride-Conducting Oxyhydrides
$La_{2-X-Y}Sr_{x+Y}H_{1-X+Y}O_{3-Y}$

Novel $La_{2-x-y}Sr_{x+y}LiH_{1-x+y}O_{3-y}$ oxyhydrides were synthesized by a high-temperature solid-state reaction in a cubic anvil cell [6] under high pressure to prevent the loss of light elements such as hydrogen, which can easily vaporize at high temperatures. The compositions and structures of $La_{2-y}Sr_yLiH_{1+y}O_{3-y}$ ($y = 0$, 1, 2) were determined by X-ray and neutron Rietveld analyses (Fig. 13.6). In La_2LiHO_3 ($x = y = 0$). The two apical sites of LiX_6 ($X = H^-$, O) octahedra were occupied only by O^{2-}, with the four in-plane apexes occupied by O^{2-} and H^-. These results indicate that highly charged cations, i.e., La^{3+} and Sr^{2+}, need to be surrounded by highly charged anions. $LaSrLiH_2O_2$ ($x = 0$, $y = 1$) was composed of tetragonal $(LiH_2)^-$ and $(LaSrO_2)^+$ layers alternately stacked along the c-axis, with the further increased hydride content of Sr_2LiH_3O resulting in the formation of $(Sr_2HO)^+$ layers. Considering the above series of compositions, it should be noted that there exists a K_2NiF_4-type H^-–free oxide, $La_2LiO_{3.5}$, in which the anion vacancies are randomly distributed in basal $(LiO_{0.75})^{0.5-}$ layers [44].

Remarkably, t-La_2LiHO_3 contains anion vacancies ($V_{(H,O)}$), which are best represented as $La_2Li(H_{0.53}O_{1.21}V_{(H,O)0.26})O_2$ and exhibit H^-, O^{2-}, and $V_{(H,O)}$ disorder at the axial sites of LiX_6 octahedra. By contrast, the orthorhombic phase, o-La_2LiHO_3, is stoichiometric, with H^- and O^{2-}located in axial anion sites. This symmetry change can be attributed to the order–disorder transition of H^- and O^{2-} in axial sites, both with and without vacancies. The crystal structures of the anion-deficient series, $La_{2-x}Sr_xLiH_{1-x}O_3$ and $La_{1-x}Sr_{1+x}LiH_{2-x}O_2$, were also determined by Rietveld analysis. Representative results obtained for

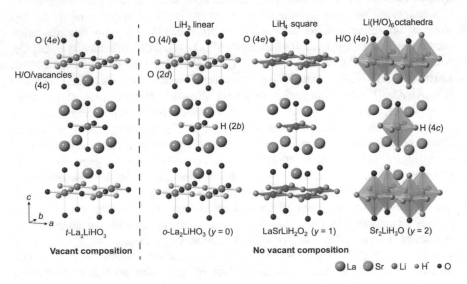

Fig. 13.6 Crystal structures of t-La_2LiHO_3 and $La_{2-y}Sr_yLiH_{1+y}O_{3-y}$ ($y = 0$, 1, 2)

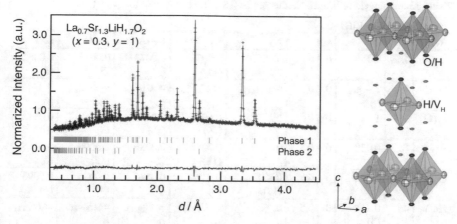

Fig. 13.7 Neutron Rietveld analysis of anion-deficient $La_{0.7}Sr_{1.3}LiH_{1.7}O_2$ ($x = 0.3$, $y = 1$)

$La_{0.7}Sr_{1.3}LiH_{1.7}O_2$ ($x = 0.3$, $y = 1$) are shown in Fig. 13.7 and Table 13.5. The H and O occupancy parameters were calculated as $g_{H1} = 0.938(2)$, $g_{H2} = 0.118(3)$, and $g_{O1} = 0.882(3)$, leading to a composition of $La_{0.7}Sr_{1.3}Li(H_{1.88}V_{(H)0.12})$ $H_{0.24}O_{1.76}$. Thus, doping resulted in the generation of vacancies in the LiH_4 plane and caused H^-/O^{2-} anion mixing at apical sites.

The valence states of all constituent atoms in La_2LiHO_3, $LaSrLiH_2O_2$, and Sr_2LiH_3O were determined by valence charge integration over the corresponding Voronoi cells (Table 13.6). The valences of hydrogen and oxygen in all materials were estimated at approximately -0.8 to -1.0 and -1.3 to -1.6, respectively, indicating that these elements were present as H^- and O^{2-}. Electronic density of states calculations corroborated the presence of hydride ions (Fig. 13.8), with their localized electrons located between approximately 0 and -5 eV below the Fermi level, confirming the ionic nature of the Li–H^- bond.

13.3.2 Hydride Ion Conductivity of $La_{2-X-Y}Sr_{X+Y}H_{1-X+Y}O_{3-Y}$

The ionic conductivities of $La_{2-x-y}Sr_{x+y}LiH_{1-x+y}O_{3-y}$ were examined by impedance measurements. The Arrhenius plots of conductivities are shown in Fig. 13.9. In the case of $La_{2-y}Sr_yLiH_{1+y}O_{3-y}$ ($x = 0$), conductivity increased with increasing H^- content, with the highest value of 3.2×10^{-5} S cm^{-1} at 573 K observed for Sr_2LiH_3O ($y = 2$) (Fig. 13.9a). Thus, introduction of hydride ions into the anion sites of the K_2NiF_4 structure improved ionic conductivity, confirming that these ions were primary charge carriers. Conduction was further facilitated by the introduction of vacancies, indicating that structural defects can affect ionic diffusion, as can be seen for $La_{2-x}Sr_xLiH_{1-x}O_3$ ($y = 0$) and $La_{1-x}Sr_{1+x}LiH_{2-x}O_2$ ($y = 1$),

Table 13.5 Rietveld refinement results for $La_{0.7}Sr_{1.3}LiH_{1.7}O_2$ ($x = 0.3$, $y = 1$)

Phase 1: $La_{0.7}Sr_{1.3}LiH_{1.7}O_2$

Atom	Site	g	x	Y	z	U_{11} =/Å²	U_{22}/Å²	U_{33}/Å²	U/Å³
La(1)	4e	0.354	0	0	0.35620 (9)	0.0072 (4)	0.0072 (4)	0.0045 (5)	0.0062 (3)
Sr(1)	4e	$1 - g_{La}$ (1)	0	0	$= z_{La(1)}$	$= U_{11\,La}$ (1)	$= U_{22\,La}$ (1)	$= U_{33\,La}$ (1)	$= U_{La}$ (1)
Li(1)	2a	1	0	0	0	0.0037 (19)	0.0037 (19)	0.0352 (5)	0.0142 (6)
H(1)	4c	0.938 (2)	0	0.5	0	0.0258 (18)	0.0086 (14)	0.0090 (15)	0.0145 (6)
O(1)	4e	0.882 (3)	0	0	0.17138 (13)	0.0126 (8)	0.0126 (8)	0.0035 (9)	0.0096 (4)
H(2)	4e	0.118 (3)	0	0.5	0	$= U_{11\,O}$ (1)	$= U_{22\,O}$ (1)	$= U_{33\,O}$ (1)	$= U_{O(1)}$ (1)

Unit cell: tetragonal $I4/mmm$, $a = 3.65672(4)$, $c = 13.3066(2)$ Å

Phase 2: Li_2O

Atom	Site	g	x	y	z	U/Å²
Li(1)	8c	1	0.25	0.25	0.25	0.0028(3)
O(1)	4a	1	0	0	0	0.0049(5)

Unit cell: cubic Fd-3 m, $a = 4.61290(11)$ Å

$R_{wp} = 1.92$, $R_p = 1.47$, $R_e = 3.12$, $S = 1.63$, $R_B = 5.63$, $R_F = 5.21$

Table 13.6 Valence states of elements in $La_{2-y}Sr_yLiH_{1+y}O_{3-y}$ ($y = 0, 1, 2$) derived from valence charge integrations over Voronoi cells, with (ax) and (ap) representing axial and apical anion sites, respectively

La_2LiHO_3 ($y = 0$)		$LaSrLiH_2O_2$ ($y = 1$)		Sr_2LiH_3O ($y = 2$)	
Atom	Valence	Atom	Atom	Atom	Valence
La	2.57	La	2.57	Sr (1)	1.8
Li	0.27	Sr	1.68	Sr (2)	1.69
H (ax)	−1.00	Li	0.43	Li	0.44
O (ax)	−1.26	H (ax)	−0.87	H (ax)	−0.80
O (ap)	−1.57	O (ap)	−1.47	H (ap)	−0.97
				O (ap)	−1.36

with conductivities of up to 2.1×10^{-4} S cm^{-1} observed for $La_{0.6}Sr_{1.4}LiH_{1.6}O_2$ at 590 K (activation energy ~ 68.4 kJ mol^{-1}) (Fig. 13.9b, c).

To further identify the nature of the charge carriers, the electrical conductivity of $La_{0.6}Sr_{1.4}LiH_{1.6}O_2$ ($x = 0.4$, $y = 1.0$) was evaluated by the Hebb-Wagner polarization method [45] at 480 and 590 K using an asymmetric (−) Pd/ $La_{0.6}Sr_{1.4}LiH_{1.6}O_2$/Mo (+) cell, with the total electrical conductivities (electrons + holes) at the irreversible Mo-electrolyte interface (2.9×10^{-8} and 4.1×10^{-7} S cm^{-1}, respectively) showing that $La_{0.6}Sr_{1.4}LiH_{1.6}O_2$ is a purely ionic conductor (Fig. 13.10 and Table 13.7).

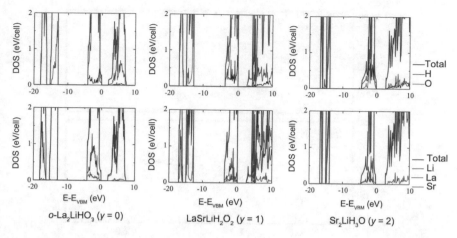

Fig. 13.8 Electronic densities of states for $La_{2-y}Sr_yLiH_{1+y}O_{3-y}$ ($y = 0, 1, 2$) determined by first principles calculations

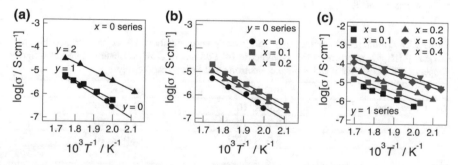

Fig. 13.9 Temperature-dependent ionic conductivities of $La_{2-x-y}Sr_{x+y}H_{1-x+y}O_{3-y}$. **a** $La_{2-y}Sr_yLiH_{1+y}O_{3-y}$ ($x = 0$, $y = 0$, 1, and 2) with a fixed cation/anion ratio of $(A_2B)/X_4$, where A, B, and X are La(Sr), Li, and O(H), respectively. Anion-deficient series: **b** $La_{2-x}Sr_xLiH_{1-x}O_3$ ($y = 0$, $0 \leq x \leq 0.2$) and **c** $La_{1-x}Sr_{1+x}LiH_{2-x}O_2$ ($y = 1$, $0 \leq x \leq 0.4$)

13.3.3 Development of Electrochemical Devices Based on Hydride Ion Conduction

To verify the occurrence of H^- conduction in $La_{2-x-y}Sr_{x+y}LiH_{1-x+y}O_{3-y}$, we constructed a Ti/o-La_2LiHO_3/TiH_2 all-solid-state cell and subjected it to galvanostatic discharge, with an electrode configuration (powdered mixture of electrode and electrolyte materials) similar to that previously used in an all-solid-state lithium battery [46]. Figure 13.11a shows the discharge curve of the cell, revealing a constant discharge current of 0.5 μA at 300 °C. Moreover, the cell showed an initial open circuit voltage of 0.28 V, which was consistent with the theoretical value calculated from the standard Gibbs energy of formation of TiH_2 [47]. During

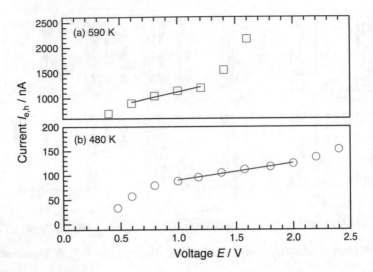

Fig. 13.10 Hebb-Wagner polarization curves of the (−) Pd/La$_{0.6}$Sr$_{1.4}$LiH$_{1.6}$O$_2$/Mo (+) cell at **a** 590 K and **b** 480 K

Table 13.7 Partial conductivities (σ) and transference numbers (t) at the irreversible Mo/La$_{0.6}$Sr$_{1.4}$LiH$_{1.6}$O$_2$ interface of the asymmetric (−) Pd/La$_{0.6}$Sr$_{1.4}$LiH$_{1.6}$O$_2$/Mo (+) cell at different temperatures. Subscripts e and h denote electrons and holes, respectively

	480 K	590 K
$(\sigma_e + \sigma_h)$/S cm^{-1}	2.9×10^{-8}	4.1×10^{-7}
σ_{H_-}/S cm^{-1}	9.7×10^{-6}	2.1×10^{-4}
t_{e+h}	3.0×10^{-3}	2.0×10^{-3}
t_{H_-}	1.0	1.0

the electrochemical reaction, the cell voltage rapidly dropped from 0.28 to 0.06 V and then gradually decreased to 0.0 V. The initial steep drop corresponded to an increase in hydride ion content at the anode, owing to the following constant current discharge reaction:

$$Ti + xH^- \rightarrow TiH_x + xe^-$$

with the cathode reaction represented as:

$$TiH_2 + xe^- \rightarrow TiH_{2-x} + xH^-$$

The occurrence of these discharge reactions was confirmed by analysis of the produced phases. Figure 13.11b shows the synchrotron X-ray diffraction patterns of the cathode, electrolyte, and anode materials before and after the reaction. The absence of any variation in the diffraction patterns of the electrolyte indicates that the La$_2$LiHO$_3$ electrolyte was stable in contact with the Ti and TiH$_2$ electrodes

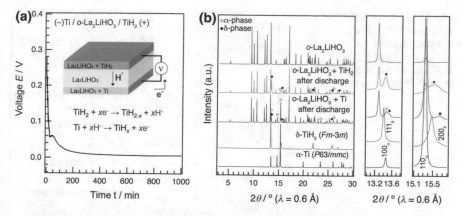

Fig. 13.11 All-solid-state cell fabricated for verification of H$^-$ conduction in La$_{2-x-y}$Sr$_{x+y}$LiH$_{1-x+y}$O$_{3-y}$. **a** Discharge curve for a Ti/o-La$_2$LiHO$_3$/TiH$_2$ solid-state battery, with the inset showing a schematic illustration of the cell and the proposed electrochemical reaction. **b** X-ray diffraction patterns of electrolyte (o-La$_2$LiHO$_3$), cathode (TiH$_2$ + o-La$_2$LiHO$_3$), and anode (Ti + o-La$_2$LiHO$_3$) materials after the reaction; the two right panels show expanded ranges of 13–13.8° and 15.1–15.8°

during the reaction. Conversely, phase changes were observed for the cathode and anode materials, as expected from the Ti–H phase diagram [47], where the δ-TiH$_2$ ($Fm\overline{3}m$) phase releases hydrogen and is transformed into α-Ti ($P6_3/mmc$), passing through a two-phase (α-TiH$_b$ + δ-TiH$_{2-a}$) coexistence region found below ∼ 573 K. In the case of the cathode, additional diffraction peaks corresponding to species with $P6_3/mmc$ symmetry were detected, and the signals of TiH$_2$ shifted to a higher angle, indicating that the release of hydrogen from TiH$_2$ induced lattice shrinkage. In the case of the anode, peaks corresponding to species with $Fm\overline{3}m$ symmetry were detected. Thus, the results indicate that during the electrochemical reaction, hydride ions were released from the TiH$_2$ cathode and diffused into the Ti anode through o-La$_2$LiHO$_3$. The present success in the construction of an all-solid-state electrochemical cell exhibiting H$^-$ diffusion confirms not only the ability of oxyhydrides to act as H$^-$ solid electrolytes, but also the possibility of developing electrochemical solid devices based on H$^-$ conduction.

13.3.4 Ambient-Pressure Synthesis of H$^-$-Conductive Oxyhydrides

The abovementioned high-pressure method is efficient for synthesizing oxyhydrides, owing to its ability to inhibit hydrogen desorption from the starting materials during sintering. However, in order to apply H$^-$ conductors to electrochemical devices, a simple synthetic protocol needs to be established for oxyhydrides, in parallel with the development of highly H$^-$-conductive novel materials. Here, we

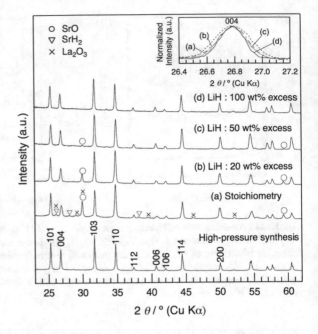

Fig. 13.12 X-ray diffraction profiles of LaSrLiH$_2$O$_2$ synthesized using various amounts of LiH: **a** stoichiometric, **b** 20, **c** 50, and **d** 100 wt% excess

describe the synthesis of LaSrLiH$_2$O$_2$ by a conventional solid-state reaction under ambient pressure and characterize its electrochemical properties.

The starting materials (which were identical to those used in the high-pressure method) were pelletized and placed in a sealed sample container made of stainless steel, with subsequent sintering performed at 650 °C for 6 h under H$_2$.

Figure 13.12 shows the X-ray diffraction patterns of LaSrLiH$_2$O$_2$ synthesized using different amounts of LiH (stoichiometric, 20, 50, and 100 wt% excess), with the main diffraction peaks corresponding to the space group of LaSrLiH$_2$O$_2$, i.e., $I4/mmm$. However, small diffraction peaks indexed to SrO, SrH$_2$, and/or La$_2$O$_3$, which were present in the raw starting materials, were observed for samples synthesized using a small excess or no excess of LiH (stoichiometric, 20 wt%, and 50 wt%). The amount of residual starting materials decreased as the amount of LiH increased, with LaSrLiH$_2$O$_2$ obtained as a single phase only at a 100 wt% excess. In addition, excess LiH improved the crystallinity of LaSrLiH$_2$O$_2$, i.e., the magnification of the normalized 004 peaks (Fig. 13.12) showed that their full width at half maximum decreased as the amount of LiH increased. Therefore, excess LiH not only prevented the loss of lithium and hydrogen during the synthesis of LaSrLiH$_2$O$_2$, but also acted as a flux for reducing the synthesis temperature.

Crystal structure analysis revealed that the sample prepared under ambient pressure had almost the same structure as the high-pressure one, with the refined site occupancies of each atom indicating that the former exhibited a nearly stoichiometric composition without vacancies. However, mixing of H$^-$ and O in the 4c axial anion site (g(H1) = 0.9361(5) and g(O1) = 1 − g(H1)) in Li octahedra

Fig. 13.13 Arrhenius and Cole-Cole plots of LaSrLiH$_2$O$_2$. The inset shows an equivalent circuit used to model impedance data

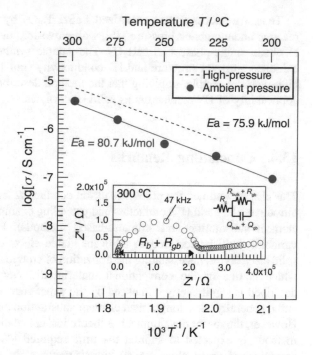

was detected for the ambient-pressure sample, whereas this site was exclusively occupied by H$^-$ in the high-pressure sample.

The ionic conductivity of LaSrLiH$_2$O$_2$ synthesized at ambient pressure was evaluated by AC impedance measurements. The corresponding impedance and Arrhenius plots are shown in Fig. 13.13, with the conductivity of the high-pressure LaSrLiH$_2$O$_2$ synthesized in our previous study also plotted for comparison [6]. The impedance plot exhibited a typical form, comprising a semicircle in the high-frequency range and a spike in the low-frequency range, which corresponded to contributions of the bulk and grain boundary and the electrode, respectively. The former contribution was estimated by fitting impedance spectra using an equivalent circuit, as shown in Fig. 13.13. For the ambient-pressure sample, the activation energy of ionic conduction was calculated as 80.7 kJ mol^{-1}, which is nearly equal to that observed for the high-pressure sample [6]. The total conductivity (bulk + grain boundary) of the ambient-pressure sample was determined as 3.2 × 10^{-6} S cm^{-1} at 300 °C, slightly less than that of the high-pressure sample. Given the crystal structure of LaSrLiH$_2$O$_2$, in which tetragonal (LiH$_2$)$^-$ and (LaSrO$_2$)$^+$ layers are alternately stacked along the c-axis, the hydride ions were expected to exhibit two-dimensional diffusion in the LiH$_4$ plane. Hence, the movement of H$^-$ in the crystal lattice of ambient-pressure LaSrLiH$_2$O$_2$ may have been inhibited by the presence of oxide ions in the (LiH$_2$)$^-$ layer.

Thus, we successfully synthesized $LaSrLiH_2O_2$ by a conventional solid-state reaction under ambient pressure [48], with a two-fold molar excess of LiH required to obtain single-phase $LaSrLiH_2O_2$. The sample synthesized at ambient pressure exhibited a crystal structure and H^- conductivity similar to those observed for the high-pressure sample, implying that the method described here should increase the applicability of H^- conductors as solid electrolytes.

13.4 Concluding Remarks

This chapter outlined the properties of ion conductors and material search methods, introducing Li^+ and H^- conductors and providing examples of material search (e.g., element substitution and structure-based methods). However, broader material variability will be required to fabricate viable electrochemical devices based on solid electrolytes, necessitating the utilization of composition-based material search, which is one of the conventional material discovery methods. The approach described is significantly influenced by the experience and intuition of researchers, and it generally takes longer than element substitution and structure-based methods. However, the recent development of theoretical calculation and material informatics methods is expected to shorten the time required [49–53], allowing high-speed screening of prospective compositions/structures. In some cases, this approach might be misleading, since not all theoretically predicted compositions or structures can be obtained by the present synthetic techniques, as exemplified by the failure of the composition/structure-based search in the case of Li^+ and H^- conductors. Thus, the area of materials informatics for composition-based material search is still in its infancy, but it holds promise for the future.

Acknowledgements This research was supported by JST, PRESTO, and Grant-in-Aid for Young Scientists (A) no. 15H05497 and (B) no. 24750209; Grant-in-Aid for Challenging Exploratory Research no. 15K13803, 23655191, and 25620180; and Grant-in-Aid for Scientific Research on Innovative Areas no. 25106005 and 25106009, from the Japan Society for the Promotion of Science. Synchrotron and neutron radiation experiments were carried out as four projects approved by the Japan Synchrotron Radiation Research Institute (JASRI) (proposals no. 2013A1704, 2015A1778, and 2015B1768), the Japan Proton Accelerator Research Complex (J-PARC) (proposal no. 2010A0058), the Spallation Neutron Source (SNS) in the Oakridge National Laboratory (proposal no. IPTS5808 and 10030), and the Neutron Scattering Program Advisory Committee of IMSS, KEK (proposal no. 2014S10). Part of the neutron experiments (proposal no. 2014S10) was performed at the BL09 Special Environment Neutron Powder Diffractometer (SPICA) developed by the Research and Development Initiative for Scientific Innovation of New Generation Batteries (RISING) project of the New Energy and Industrial Technology Development Organization (NEDO). Supercomputing time at the Academic Center for Computing and Media Studies (ACCMS) at Kyoto University is gratefully acknowledged. Further information regarding the materials and methods is included in the supplementary materials.

References

1. B.B. Owens, G.R. Argue, Science **157**, 308 (1967)
2. T. Takahashi, O. Yamamoto, S. Yamada, S. Hayashi, J. Electrochem. Soc. **126**, 1654 (1979)
3. T. Takahashi, R. Kanno, Y. Takeda, O. Yamamoto, Solid State Ion. **3**, 283 (1981)
4. N. Kamaya, K. Homma, Y. Yamakawa, M. Hirayama, R. Kanno, M. Yonemura, T. Kamiyama, Y. Kato, S. Hama, K. Kawamoto, A. Mitsui, Nat. Mater. **10**, 682 (2011)
5. Y. Kato, S. Hori, T. Saito, K. Suzuki, M. Hirayama, A. Mitsui, M. Yonemura, H. Iba, R. Kanno, Nat. Energy **1**, 16030 (2016)
6. G. Kobayashi, Y. Hinuma, S. Matsuoka, A. Watanabe, M. Iqbal, M. Hirayama, M. Yonemura, T. Kamiyama, I. Tanaka, R. Kanno, Science **351**, 1314 (2016)
7. M.C. Verbraeken, C. Cheung, E. Suard, J.T.S. Irvine, Nat. Mater. **14**, 95 (2015)
8. V. Thangadurai, S. Narayanan, D. Pinzaru, Chem. Soc. Rev. **43**, 4714 (2014)
9. U.V. Alpen, M.F. Bell, W. Wichelhaus, K.Y. Cheung, G.J. Dudley, Electrochim. Acta **23**, 1395 (1978)
10. Y. Inaguma, L. Chen, M. Itoh, T. Nakamura, Solid State Ion. **70–71** Part 1, 196 (1994)
11. V. Thangadurai, H. Kaack, W.J.F. Weppner, J. Am. Ceram. Soc. **86**, 437 (2003)
12. A. Hayashi, S. Hama, H. Morimoto, M. Tatsumisago, T. Minami, Chem. Lett. **30**, 872 (2001)
13. R. Kanno, M. Murayama, J. Electrochem. Soc. **148**, A742 (2001)
14. K. Minami, A. Hayashi, M. Tatsumisago, J. Am. Ceram. Soc. **94**, 1779 (2011)
15. T. Phraewphiphat, M. Iqbal, K. Suzuki, Y. Matsuda, M. Yonemura, M. Hirayama, R. Kanno, J. Solid State Chem. **225**, 431 (2015)
16. Y. Sun, K. Suzuki, S. Hori, M. Hirayama, R. Kanno, Chem. Mater. **29**, 5858 (2017)
17. R. Kanno, T. Hata, Y. Kawamoto, M. Irie, Solid State Ion. **130**, 97 (2000)
18. Y. Inoue, K. Suzuki, N. Matsui, M. Hirayama, R. Kanno, J. Solid State Chem. **246**, 334 (2017)
19. Y. Inaguma, C. Liquan, M. Itoh, T. Nakamura, T. Uchida, H. Ikuta, M. Wakihara, Solid State Commun. **86**, 689 (1993)
20. Y. Inaguma, T. Katsumata, M. Itoh, Y. Morii, T. Tsurui, Solid State Ion. **177**, 3037 (2006)
21. G. Zhao, I. Muhammad, K. Suzuki, M. Hirayama, R. Kanno, Mater. Trans. **57**, 1370 (2016)
22. E.E. Hellstrom, W. Van Gool, Solid State Ion. **2**, 59 (1981)
23. C.J.M. Rooymans, Z. Anorg, Allg. Chem. **313**, 234 (1961)
24. M. Tabuchi, S. Tsutsui, C. Masquelier, R. Kanno, K. Ado, I. Matsubara, S. Nasu, H. Kageyama, J. Solid State Chem. **140**, 159 (1998)
25. A.F. Andresen, A.J. Maeland, D. Slotfeldt-Ellingsen, J. Solid State Chem. **20**, 93 (1977)
26. B. Wegner, R. Essmann, J. Bock, H. Jacobs, P. Fischer, Eur. J. Solid State Inorg. Chem. **29**, 1217 (1992)
27. F. Altorfer, W. Buhrer, B. Winkler, G. Coddens, R. Essmann, H. Jacobs, Solid State Ion. **70–71**, 272 (1994)
28. B. Zhu, X. Yang, Electrochem. Commun. **1**, 411 (1999)
29. B. Zhu, X. Liu, Electrochem. Commun. **2**, 10 (2000)
30. F.W. Poulsen, Solid State Ion. **145**, 387 (2001)
31. M.C. Verbraeken, E. Suard, J.T.S. Irvine, J. Mater. Chem. **19**, 2766 (2009)
32. R.M. Helps, N.H. Rees, M.A. Hayward, Inorg. Chem. **49**, 11062 (2010)
33. Y. Kobayashi, O.J. Hernandez, T. Sakaguchi, T. Yajima, T. Roisnel, Y. Tsujimoto, M. Morita, Y. Noda, Y. Mogami, A. Kitada, M. Ohkura, S. Hosokawa, Z. Li, K. Hayashi, Y. Kusano, J.E. Kim, N. Tsuji, A. Fujiwara, Y. Matsushita, K. Yoshimura, K. Takegoshi, M. Inoue, M. Takano, H. Kageyama, Nat. Mater. **11**, 507 (2012)
34. T. Sakaguchi, Y. Kobayashi, T. Yajima, M. Ohkura, C. Tassel, F. Takeiri, S. Mitsuoka, H. Ohkubo, T. Yamamoto, J.E. Kim, N. Tsuji, A. Fujihara, Y. Matsushita, J. Hester, M. Avdeev, K. Ohoyama, H. Kageyama, Inorg. Chem. **51**, 11371 (2012)
35. H. Schwarz, Ph.D. Thesis, University of Karlsruhe, 1991

36. J. Bang, S. Matsuishi, H. Hiraka, F. Fujisaki, T. Otomo, S. Maki, J.-I. Yamaura, R. Kumai, Y. Murakami, H. Hosono, J. Am. Chem. Soc. **136**, 7221 (2014)
37. M.A. Hayward, E.J. Cussen, J.B. Claridge, M. Bieringer, M.J. Rosseinsky, C.J. Kiely, S.J. Blundell, I.M. Marshall, F.L. Pratt, Science **295**, 1882 (2002)
38. K. Hayashi, S. Matsuishi, T. Kamiya, M. Hirano, H. Hosono, Nature **419**, 462 (2002)
39. K. Hayashi, P.V. Sushko, A.L. Shluger, M. Hirano, H. Hosono, J. Phys. Chem. B **109**, 23836 (2005)
40. S. Matsuishi, K. Hayashi, M. Hirano, H. Hosono, J. Am. Chem. Soc. **127**, 12454 (2005)
41. C.G. Van de Walle, Phys. Rev. Lett. **85**, 1012 (2000)
42. C.G. Van de Walle, J. Neugebauer, Nature **423**, 626 (2003)
43. J. Zhang, G. Gou, B. Pan, J. Phys. Chem. C **118**, 17254 (2014)
44. J.P. Attfield, G. Férey, J. Solid State Chem. **80**, 112 (1989)
45. B.J. Neudecker, W. Weppner, J. Electrochem. Soc. **143**, 2198 (1996)
46. N. Kamaya, K. Homma, Y. Yamakawa, M. Hirayama, R. Kanno, M. Yonemura, T. Kamiyama, Y. Kato, S. Hama, K. Kawamoto, A. Mitsui, Nat. Mater. **10**, 682 (2011)
47. A. San-Martin, F.D. Manchester, Bull. Alloy Phase Diagrams **8**, 30 (1987)
48. A. Watanabe, G. Kobayashi, N. Matsui, M. Yonemura, A. Kubota, K. Suzuki, M. Hirayama, R. Kanno, Electrochemistry **85**, 88 (2017)
49. M. Nishijima, T. Ootani, Y. Kamimura, T. Sueki, S. Esaki, S. Murai, K. Fujita, K. Tanaka, K. Ohira, Y. Koyama, I. Tanaka, Nat. Commun. **5**, 4553 (2014)
50. Y. Hinuma, T. Hatakeyama, Y. Kumagai, L.A. Burton, H. Sato, Y. Muraba, S. Iimura, H. Hiramatsu, I. Tanaka, H. Hosono, F. Oba, Nat. Commun. **7**, 11962 (2016)
51. A.D. Sendek, Q. Yang, E.D. Cubuk, K.-A.N. Duerloo, Y. Cui, E.J. Reed, Energy Environ. Sci. **10**, 306 (2017)
52. W.D. Richards, Y. Wang, L.J. Miara, J.C. Kim, G. Ceder, Energy Environ. Sci. **9**, 3272 (2016)
53. A. Seko, H. Hayashi, K. Nakayama, A. Takahashi, I. Tanaka, Phys. Rev. B **95**, 144110 (2017)